Thermodynamic Processes 1

Series Editor
Jean-Claude Charpentier

Thermodynamic Processes 1

Systems without Physical State Change

Salah Belaadi

iSTE

WILEY

First published 2020 in Great Britain and the United States by ISTE Ltd and John Wiley & Sons, Inc.

ISTE Ltd
27-37 St George's Road
London SW19 4EU
UK

www.iste.co.uk

John Wiley & Sons, Inc.
111 River Street
Hoboken, NJ 07030
USA

www.wiley.com

Library of Congress Control Number: 2019953627

British Library Cataloguing-in-Publication Data
A CIP record for this book is available from the British Library
ISBN 978-1-78630-513-8

MIX
Paper from responsible sources
FSC® C013604

Contents

Foreword 1

Circular Economy and Engineering

Circular economy and engineering: process thermodynamics as an essential chemical engineering tool for the design and control of the processes encountered in the factory of the future within the framework of Industry 4.0

Process engineering involves the sciences and technologies that optimally transform matter and energies into products required by a consumer and into non-polluting wastes. Today, it takes part in the framework of circular economy and engineering (monitoring of products and processes from cradle to grave), and the optimal transformations of matter and energies must be carried out to design the factory of the future, taking into account the emergence of Industry 4.0 and the voluminous amount of data (*Big Data movement*).

Modern (green) process engineering is deliberately oriented toward process intensification (i.e., producing much more and better, with use of much less resources). This involves a physical-chemistry multidisciplinary and multiscale approach to modelling and computer simulation, in terms of time and space, from the atomic and molecular scales, from the equipment and the reactor scales, up to the scales of the overall factory (i.e., the design of a refinery, a chemical, a textile or a cement complex plant from Schrödinger equations).

To meet this multidisciplinary and multiscale approach, the preponderant and irreplaceable concept and background of chemical thermodynamics appears in all its splendour, and more generally, this concerns the thermodynamics of processes for the multiscale control of these processes.

It is clear that studies that discuss thermodynamics of processes must cover chemical thermodynamics (open or closed systems with or without chemical reaction, phase equilibrium) and the energetics of processes (thermal cycles, heat pump, degraded energy, exergy). However, these studies must also be illustrated with examples of real multiscale physicochemical applications. This will prepare or help or contribute to the design, the development and the control of the processes that will be encountered in the factory of the future, by means of methodologies and techniques to obtain reliable thermodynamic data that will contribute to the abundance of data (*Big Data/Industry 4.0*).

A big thank you to Professor Salah Belaadi, leading expert in education and research in the field of thermodynamics of processes, for offering such an instructional and didactic book, whose chapters mainly present exercises oriented towards industrial applications.

This book on thermodynamics and energetics of processes is a guide (a *vademecum*), which I am personally convinced will be of great benefit to a large number of university teachers and researchers, and engineers and technicians active in today's economy sector and in the very near future.

Jean-Claude CHARPENTIER
Former director of ENSIC Nancy and ESCPE Lyon, France
Former president of the European Federation of Chemical Engineering
Laboratoire Réactions et Génie des Procédés
CNRS/ENSIC/University of Lorraine

Foreword 2

Thermodynamics is a universal science that is of great interest in all its applications. The premises of thermodynamics are not always easy to understand, in the eyes of students nor in those of seasoned researchers, nor are the numerous developments that result from it.

Nature is complex, the scientist must be humble with regard to what he/she sees. He/she must scrupulously observe, carefully reflect, attempt to interpret and undertake modelling, a theory which will be deemed valid only until a new observation, or a new experiment comes to question them. From this perspective, it is necessary for the educator to show conviction, insight and passion in order to best convey the thirst for effort and for accomplishing work, to promote vocations and to discover talents.

Professor Salah Belaadi has spent many years teaching thermodynamics. He has enriched his courses with many exercises entirely dedicated to understanding this science and potential applications for the industrial world. Over the years, he has been able to select the most interesting and relevant exercises. The collection he proposes today is therefore an assortment of carefully selected topics for thermodynamic reflection and culture.

I wish him the success he deserves, and I hope that a very large number of readers will enjoy this content.

Dominique RICHON
Emeritus Professor at Mines ParisTech
Former director of Thermodynamics
and Phases Equilibrium Laboratory.

Preface

The aim of this book is to reduce apprehension toward thermodynamics and to make it more familiar to those who have to use it, both on completion of apprenticeships, training and retraining as well as to those conducting research and reflection on the evolution of processes at the time of transformation of matter and/or energy.

The need to write this book was apparent to me, after so many years of teaching at various university levels, after the unequivocal statement: the difficulty encountered by students – or engineers working in companies or research groups – to solve concrete problems in thermodynamics comes from the fact that the manuals, which cover applications of the concepts of this discipline, are too didactic.

Hence why I propose an original approach for this book – to use thermodynamics as a resolution tool – indispensable for mastering a process of energy transformation and/or matter using one or more thermodynamic concepts. Thus, this book is not structured according to the progression of the teaching of the concepts of thermodynamics, but rather according to the evolution of the scientific difficulty compared to the state of the thermodynamic system studied from closed systems to energy processes; thermodynamics is above all "the science of the evolution of the states of a system, whatever it is".

The book derives its interest from the very definition of this science, accepted by the scientific community for a long time as the "mother of sciences". Indeed, everyone agrees that: "*If thermodynamics does not solve everything, without it, we will not solve anything*"; this is all the more true for the physicochemical processes of matter and energy transformation.

Salah BELAADI
November 2019

1

Basic Concepts of Thermodynamics

1.1. Exercises

EXERCISE 1.1.– Number of molecules

An electronic tube with a volume of 100 cm^3 is placed in a vacuum at 27 °C and under a pressure of 1.2×10^{-5} torr.

What is the number of molecules remaining in the tube?

Answer: $n \# 4 \times 10^{13}$.

EXERCISE 1.2.– Total volume

A certain mass of gas occupies a volume of 200 cm^3 STP.

What volume will be occupied at 273 °C at a pressure of 4 atm?

Answer: $V = 100$ cm^3.

EXERCISE 1.3.– Density

What is the density of methane (CH_4) at 27 °C and a pressure of 5 atm?

Answer: $\rho = 3.25$ g/L.

EXERCISE 1.4.– Density of a body

What is the density, with regard to air, of carbon monoxide (CO) at –20 °C and at a pressure of 2.5 atm? What conclusion can be drawn here?

Answer: $d_{/air} = 0.965$; see correction.

EXERCISE 1.5.– Mass and flow

A steel bottle with a capacity of 24 L contains argon at 17 °C, deemed an ideal gas, with a molar mass of 40 g, compressed to 50 atm.

1) What mass of gas does it contain?

The bottle is placed in contact with an apparatus under ambient conditions at 17 °C and 1 atm, so that the flow rate of the gas, measured under these same conditions, is constant and equal to 1 L/min (it will be assumed that the gas temperature remains constant and equal to the ambient T).

2) How long does it take for the bottle pressure to half?

Answer: (1) $m = 2{,}017.2$ g; (2) $t = 10$ h.

EXERCISE 1.6.– Atomic mass and density

The ratio of density ρ, in g/L, phosphorus hydride (PH_3) at pressure P (atm) is given as a function of P, at $T = 0$ °C in the following table:

P	1.000	0.750	0.500	0.250
ρ/P	1.5307	1.5272	1.5238	1.5205

Calculate the atomic mass of phosphorus, using 1.008 g for the atomic mass of hydrogen.

Answer: $M_P = 30.96$ g.

EXERCISE 1.7.– Volumic percentage and partial pressure

A gas mixture, whose volume is 65% nitrogen, 15% oxygen and 20% carbon dioxide, is under a total pressure of 760 torr.

What are the molar fraction and partial pressures of the three constituents?

Answer: $x_1 = 0.65$ and $P_1 = 494$ torr; $x_2 = 0.15$ and $P_2 = 114$ torr; $x_3 = 0.20$ and $P_3 = 152$ torr.

EXERCISE 1.8.– Average molar mass

In a 22.4 L balloon, initially filled with air at a pressure of 1 atm, 4 g of hydrogen is introduced, keeping the temperature constant and equal to 0 °C.

Knowing that air contains approximately 80% of nitrogen and 20% of oxygen in moles, what is the average molar mass of the air/hydrogen mixture?

Answer: M_{moy} # 11 g/mol.

EXERCISE 1.9.– Molar fraction and partial pressure

Two balloons (A) and (B), with volumes of 250 cm^3 and 1,500 cm^3, respectively, contain nitrogen at a pressure of 500 torr, and hydrogen at a pressure of 250 torr, respectively. We spontaneously mix them.

1) Calculate the pressure of the mixture at equilibrium.

2) Deduce the molar fraction of the mixture at equilibrium, and the partial pressure of each gas.

Answer: (1) $P_t = 286$ torr; (2) $x_1 = 0.25$ and $x_2 = 0.75$; $P_1 = 71.50$ torr and $P_2 = 214.50$ torr.

EXERCISE 1.10.– Density by the Dumas method

Analyzing an organic substance gives the following results, in mass percentages: C = 54.5%, H = 9.1% and O = 36.4%.

On the other hand, we measure the density of steam using the Dumas method: we have a non-expandable balloon with a volume of 272 cm^3, a mass (when filled with air at 18 °C) equal to 70.00 g. We introduce a small quantity of substance to study. The flask is heated to a temperature of 100 °C and then sealed. It then returns to a temperature of 18 °C and the new mass is measured: 70.45 g. Knowing that the experiment is done at normal atmospheric pressure and that 1 L of air weighs 1.29 g in STP:

1) calculate the molar mass of the studied compound;

2) deduce its formula.

Answer: (1) M # 88 g/mol; (2) $C_4H_8O_2$.

EXERCISE 1.11.– Density and altitude

A balloon-probe, with a diameter of 60 cm, destined for stratospheric explorations, consists of a perfectly spherical waterproof covering made of very thin elastic rubber. It is filled with H_2 under the following conditions: $P = 765$ torr, $T = 17$ °C and $h = 0$ km.

1) What mass of gas does it contain?

In the stratosphere, approximately 15 km < h < 50 km, T remains constant and equal to –50 °C, whereas P evolves according to the law: $P = P_0e^{-ah}$ with $P_0 = 1$ if P is in atm and a = 0.13 if h is in km. The balloon bursts as soon as its diameter reaches 2 m.

2) What is the maximum altitude it can reach?

Assumption: Assume that P inside the balloon remains equal at all times to the P outside.

Data: H_2 ideal gas with atomic mass of 1 g.

Answer: (1) m = 9.57 g; (2) h # 30 km.

EXERCISE 1.12.– Pressure at equilibrium

Two balloons, non-expandable, of equal volumes, are connected by a tube of negligible volume and containing 0.70 mol of hydrogen at a pressure of 0.5 atm and uniform temperature of 27 °C. One of the balloons is immersed in an oil bath at 127 °C, and the other is maintained at 27 °C.

1) Calculate the number of moles of H_2 in each balloon.

2) Deduce the pressure at equilibrium.

Answer: (1) $n_{27} = 0.4$ mol and $n_{127} = 0.3$ mol; (2) $P' = 0.57$ atm.

EXERCISE 1.13.– Kinetic energy

1) A vehicle is rolling at a speed of 110 km/h and suddenly hits an obstacle so that its speed is cancelled out; the resulting energy then launches it in the air. How high h would it be launched?

2) A metallic projectile of C_p = 0.13 cal/g.K is propelled at a speed of 700 m/s when it is stopped by a protective shield. Calculate the rise in temperature?

Answer: (1) $h = 47.6$ m; (2) $\Delta T = 451$ °C.

EXERCISE 1.14.– Potential energy

For a hydraulic turbine to work, a flow of water is required, with a flow rate of 100 L/s and a drop of 30 m.

What is the maximum power that can be obtained if all the energy produced is recovered?

Answer: $P = 29.4$ kW.

EXERCISE 1.15.– Electrical energy

An oil bath maintained at 50 °C emits heat to the exterior at 1,000 calories per minute; the temperature is maintained owing to an electrical resistance of 50 Ω, heated to 110 V and connected to a control thermostat.

What is the percentage of time taken for the electric current to flow?

Answer: $\tau = 28.8\%$.

EXERCISE 1.16.– Molar fraction and temperature

A 2.24 L container was filled with a mass of 7.10 g of chlorine gas, at atmospheric pressure and at temperature T. Then it is placed in a thermostat whose temperature is 30 °C greater than T; a certain quantity of chlorine is allowed to escape, so as to restore P to its initial value. The container then contains only 6.39 g of Cl_2.

1) Calculate the initial temperature.

2) Calculate the final pressure.

Data: Atomic mass of chlorine = 35.5 g.

Answer: (1) $T = 270$ K; (2) $P = 0.99$ atm.

1.2. Problems

PROBLEM 1.1.– Chemical reaction

A mixture of potassium chlorate and sodium chlorate weighing a total of 65.50 g is made to react cold on the concentrated hydrochloric acid; a volume of 40.32 L chlorine is measured at STP.

1) Write the reaction equations involved.

2) What are the respective masses of potassium chlorate and sodium chlorate contained in the initial mixture?

Answer: (1) see corrections; (2) $m_{KClO_3} = 12.26$ g and $m_{NaClO_3} = 53.24$ g.

PROBLEM 1.2.– Bunsen calorimeter

The Bunsen ice calorimeter is an isothermal device, which allows us to determine a quantity of heat by simply measuring the volume change during the melting of the ice, at constant temperature (0 °C), by heat input. It can be schematized as shown in Figure 1.1.

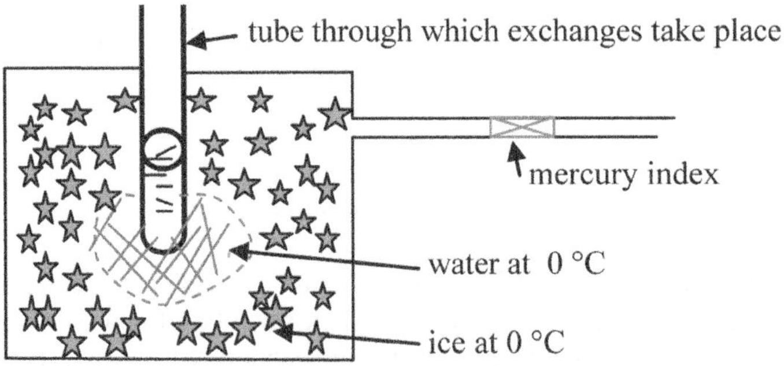

Figure 1.1. *Bunsen ice calorimeter*

1) Describe what happens when you put 100 g of water in the tube at 20 °C.

2) In what direction and how much will the index move?

3) What conclusions can be made?

Data: The capillary has a cross-section of 4×10^{-2} cm^2; water: $C_p = 1$ cal·g^{-1}·K^{-1} and $\rho = 1$ g·cm^{-3}; ice: $\rho = 0.9$ g·cm^{-3} and $\Delta H_f = 80$ cal·g^{-1}.

Answer: (1) see corrections; (2) left by 68.7 cm; (3) see corrections.

PROBLEM 1.3.– McLeod gauge

The McLeod gauge is a device for measuring low pressures of around 10^{-2} mm Hg; it is shown schematically in Figure 1.2.

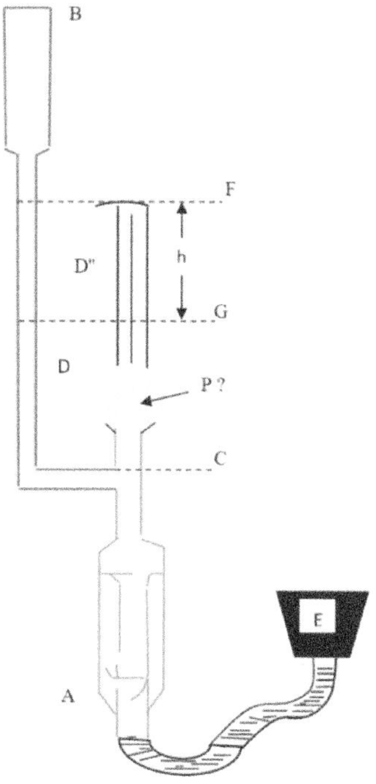

Figure 1.2. *The McLeod gauge*

COMMENTS ON FIGURE 1.2.– (i) Part (A) is connected by a rubber tube to a closed tank (E) filled with mercury, whose vertical displacement makes it possible to raise the mercury in the gauge; (ii) part (B) is in relation to the container D of gas whose pressure P is to be measured, the whole apparatus is thermostatically controlled; (iii) as the tank (E) is raised, the mercury is raised in the gauge; when it reaches point C, it fills part (D), a volume V_0, of gas at pressure P, the volume V_0 is a function of the dimensions of part (D) that consists of a ball topped with a cylindrical tube with cross-section S_0; (iv) the mercury is continued up to point F; in part (D), the mercury compresses the gas and consequently stops at level G, lower than F: the difference in level h is measured using a magnifier.

1) Express the pressure P as a function of the characteristics of the apparatus and the measured difference in levels h.

2) What level of accuracy can be achieved with such a device?

Answer: (1) $P = \dfrac{h^2 S_0}{V_0}$; (2) see corrections.

PROBLEM 1.4.– U-shaped tube

At atmospheric pressure ($P_0 = 760$ torr) and at ambient temperature ($T_0 = 300$ K), a U-shaped tube is partially filled with mercury, the two vertical branches 1 and 2 have the same cross-section S. Then both sections are sealed, enclosing two columns of air of the same height h. Branch 1 is warmed slowly, keeping branch 2 at room temperature, until the mercury levels in both branches differ by a certain amount h′.

1) Schematically represent the experiment.

2) Express the temperature of gas in branch 1 as a function of T_0, P_0, h and h′.

3) Calculate this temperature for h = 1 m and h′= 20 cm.

Assumption: Air is deemed an ideal gas.

Answer: (1) see corrections; (2) $T = T_0 \dfrac{h + \frac{h'}{2}}{h - \frac{h'}{2}} \left(1 + \dfrac{h'}{P} \dfrac{h - \frac{h'}{2}}{h} \right)$; (3) $T = 454$ K.

PROBLEM 1.5.– Density by the Meyer method

The composition of a liquid homogeneous mixture of methanol (1) and ethanol (2) can be determined from the relative density of the vapor of the mixture relative to air. The experimental device comprises an non-expandable container (A), placed in a thermostat maintained at 100 °C, connected by a capillary of negligible volume to a test piece that returns to a water tank.

When 421 mg of the liquid mixture in (A) is evaporated, the container is filled with 242 cm^3 of air, in the following conditions: water level in the container: 30 mm above the water level in the tank; gas temperature: 18 °C and outside pressure: 765 torr; saturation vapor pressure of the water at 18 °C: 15.8 torr.

To measure the density, a vessel (r) containing the liquid to be vaporized is placed in the container (A) and it is sealed. The air it contains is at 18 °C and 765 torr. Then the thermostat is heated until the container and its contents are brought to the uniform temperature of 100 °C, with the gas pressure still at 765 torr. The volume of air leaving the container during this heating, measured in STP, is equal to 392 cm^3. At this moment the vessel (r) breaks, which makes it possible to measure the density of the vapor: a gas mixture of oxygen, nitrogen, methanol and ethanol is then obtained in (A).

1) Make a clear diagram of the experimental setup.

2) Calculate the density of the vapor with respect to the air.

3) Calculate the composition of the liquid mixture (1)/(2) in molar fractions and in mass percentages.

4) What is the volume of container (A)?

5) Calculate the molar fraction, the partial pressure and the molarity of each of the constituents of the mixture at the end of the experiment.

Data: Density of mercury: 13.6 g/cm^3; density of air in STP: 1.293 g/L; composition of air: 20% O_2 and 80% N_2; M1 = 32 g/mol and M2 = 46 g/mol.

Answer: (1) see corrections; (2) d = 1.46; (3) $x_1 = 0.27$; $x_2 = 0.73$; $\%X_1 = 20.7$; $\%X_2 = 79.3$; (4) $V_A = 1.89$ L; (5) $x_{air} = 0.052$; $x_{O_2} = 0.168$; $x_{N_2} = 0.671$; $x_1 = 0.044$; $x_2 = 0.117$; $P_{O_2} = 128.5$ torr; $P_{N_2} = 513.3$ torr; $P_1 = 33.7$ torr; $P_2 = 89.5$ torr; $m_{O_2} = 5.5 \times 10^{-3}$; $m_{N_2} = 22 \times 10^{-3}$; $m_1 = 1.43 \times 10^{-3}$; $m_2 = 3.86 \times 10^{-3}$.

PROBLEM 1.6.– Yield, composition and partial pressure

To determine the yield of sulfuric anhydride obtained in a catalytic furnace, operating at 600 °C and under a constant pressure of 1 atm, two operations are performed.

Operation I: A certain volume of gas is taken from a graduated cylindrical tube maintained at 25 °C, then this gas is bubbled into a flask containing 10 cm^3 of a deci-normal solution of iodine, water and drops of starch: 640 cm^3 of gas is required at 25 °C to change color of the solution.

Operation II: A liter of gas measured at 25 °C is taken, an excess of hydrogen peroxide is added and, after boiling to destroy the excess hydrogen peroxide, an excess of barium chloride solution is poured into the precipitate obtained and dried; it weighs 0.685 g.

If we assume that the samples do not change the composition of the gas mixture and that the partial pressure of the sulfur dioxide at 25 °C is lower than the vapor pressure of the condensed phase at the same temperature:

1) Write the chemical reaction equations involved in each case.

2) Calculate the yield of sulfuric anhydride at 600 °C.

3) Determine the percentage composition of the gases in the equilibrium mixture, if before the reaction they had the following volumes: 83% N_2, 10% O_2 and 7% of anhydride.

4) Calculate the partial pressures of the various components coming out of the oven.

5) Calculate the equilibrium constant at 600 °C, defined by $K_P = \frac{P_{SO_3}^2}{P_{SO_2}^2 . P_{O_2}}$.

What would be the value if we replaced the partial pressures by molar concentrations?

6) What would the yield be at equilibrium at the same temperature, beginning with the stoichiometric mixture without nitrogen, for a constant pressure of 1 atm and 100 atm?

Data: M_{BaSO_4} = 233 g/mol; 0 °C = 273 K.

Answer: (1) see corrections; (2) $\rho = 73.5\%$; (3) 85.19% of N_2; 7.62% of O_2; 1.89% of SO_2 and 5.30% of SO_3; (4) $P_{N_2} = 0.8519$ atm; $P_{O_2} = 0.0762$ atm; $P_{SO_2} = 0.0189$ atm and $P_{SO_3} = 0.0530$ atm; (5) $K_P = 100.29$ atm and $K_c = 7{,}148.4$ L; (6) $\rho_1 = 76.5\%$ and $\rho_{100} = 94\%$.

1.3. Tests

TEST 1.1.– Density with respect to air

A 2 L balloon contains water vapor (humid air) at 27 °C under a total pressure of 760 torr and a vapor pressure of water of 16 torr. Dry air is composed of 20% oxygen and 80% nitrogen in moles; it is considered to be an ideal gas.

1) Calculate the density with respect to air in the water vapor.

2) The balloon is subjected to a temperature of 0.1 °C at which the saturated vapor pressure of water is 5 torr. Calculate the mass of condensed water and the new pressure in the balloon.

TEST 1.2.– Partial pressure

The pressure of a gaseous mixture of nitrogen (1), oxygen (2) and carbon dioxide (3), deemed ideal, in composition: $x_1 = 0.60$; $x_2 = 0.20$ and $x_3 = 0.20$, is contained in a non-dilatable balloon with a volume of 237 L, under a pressure of 760 torr at a temperature of 107 °C.

1) What is the partial pressure of each constituent in the balloon at $T = 427$ °C?

2) 76.8 g of O_2 is added to the balloon. What is the partial pressure of O_2 at 107 °C?

TEST 1.3.– Pressure

Two balloons (A) and (B) with a volume $V_A = 2$ L and $V_B = 1$ L are joined by a tube of negligible volume and have an isolation valve. Initially, they are isolated and the temperature is 27 °C in both; (A) contains nitrogen under a pressure of 760 torr and (B) hydrogen under a pressure of 280 torr, both deemed ideal gases.

1) The valve is opened. Calculate the total pressure in each balloon.

2) What are the partial pressures of N_2 and H_2?

3) Balloon (B) is then heated to 100 °C, while still connected to (A) which remains at 27 °C. What is the final pressure in each balloon?

TEST 1.4.– Total volume

A balloon with a valve is filled with air, considered an ideal gas, at a temperature of 20 °C and under atmospheric pressure of 754 torr: the mass of the balloon is 627.001 g.

The valve is opened and the balloon is heated to $T = 100$ °C. The valve is closed and the balloon cools to the initial temperature: the mass of the balloon is 626.350 g.

1) Determine the volume of the non-dilatable balloon.

2) What is the final air pressure in the balloon?

1.4. Detailed corrections

EXERCISE 1.1.– Number of molecules

From the ideal gas equation $PV = nRT$, we take $n = \frac{PV}{RT}$ where $R = \frac{P_0V_0}{T_0}$ therefore $n = \frac{PVT_0}{P_0V_0T}$, $P_0 = 1$ atm, $V_0 = 22.4$ L, $T_0 = 273.15$ K, $V = 0.1$ L, $T = 300.15$ K and $P = \frac{1.2\times10^{-5}}{760}$ atm. This gives $n = \frac{1.2\times10^{-5}\times0.1\times273.15}{760\times22.4\times300.15} = 0.63\times10^{-10}$ mol. The number of molecules (n) is obtained by multiplying it by Avogadro's number $\mathcal{N} = 6.023 \times 10^{23}$, therefore n = 3.86×10^{13} molecules.

EXERCISE 1.2.– Total volume

Under STP, $P_0V_0 = nRT_0$ and under other conditions, $PV = nRT$. We can therefore write $\frac{P_0V_0}{PV} = \frac{T_0}{T}$, i.e. $V = \frac{P_0V_0}{T_0}\cdot\frac{T}{P}$ with $V_0 = 200$ cm^3, $P_0 = 1$ atm, $T_0 = 273.15$ K, $P = 4$ atm and $T = 546.15$ K, and thus:

$$V = \frac{200\times1\times546.15}{4\times273.15} \# 100 \text{ cm}^3$$

EXERCISE 1.3.– Density

By definition, the density is: $\rho = \frac{M}{V}$ and under STD $\rho_0 = \frac{M}{V_0}$ therefore, from $PV = nRT$, we can write $\frac{\rho_0 T_0}{P_0} = \frac{\rho T}{P}$, i.e. $\rho = \rho_0 \cdot \frac{T_0}{P_0} \cdot \frac{P}{T}$.

Methane has the formula CH_4 and $M = 16$ g/mol with $V_0 = 22.4$ L and so $\rho_0 = \frac{16}{22.4}$ g/L and $\rho = \frac{16}{22.4}\cdot\frac{273.15}{300.15}\cdot\frac{5}{1} = 3.25$g/L.

EXERCISE 1.4.– Density of a body

By definition, the density (d) of a gas (1) with respect to another (2) is: $\mathrm{d} = \frac{\rho_1}{\rho_2}$ and under standard conditions of T and P, we saw in the previous exercise that $\rho = \rho_0 \cdot \frac{T_0}{T} \cdot \frac{P}{P_0}$, which gives: for air $\rho_2 = 1.29 \frac{273.15xP}{T}$, for CO $\rho_1 = \frac{M}{22.4} \frac{273.15.P}{T}$. Hence, $\mathrm{d} = \frac{\rho_1}{\rho_2} = \frac{M}{22.4 \times 1.29}$, therefore $\mathrm{d} = \frac{M}{29}$ (which shows that it neither depends on T nor P), hence $d_{CO} = \frac{28}{29} = 0.966$.

EXERCISE 1.5.– Mass and flow

1) The equation of state of the ideal gas allows us to write:

$$n = \frac{PV}{RT} = \frac{50 \times 24 \times 273.15}{22.4 \times 290.15} = 50.458 \text{ mol}$$

therefore $m = 40n = 40 \times 50.458 = 2{,}018.35$ g.

2) If V and T are constant, when P halves n will be divisible by 2. At 17 °C and 1 atm, 1 mole of argon occupies $v = \frac{22.4 \times 290.15}{273.15} = 23.79$ L, and the volume of gas that escaped $V = \frac{50.458 \times 23.79}{2} \# 600$ L. Therefore, for a flow of 1 L/min, we get a time of t = 600 min = 10 h.

EXERCISE 1.6. – Atomic mass and density

By definition, the molar volume is $v = \frac{M}{\rho}$ and, for 1 mole of ideal gas $Pv = RT$, i.e. $P\frac{M}{\rho} = RT$ and therefore $M = \frac{\rho}{P} RT$. However, this relationship cannot be applied to PH_3, which is a real gas; it will only be valid when P tends toward 0 (this is a property of real gases when $P \to 0$); the curve $\left(\frac{\rho}{P}\right) = f(P)$, shown in

Figure 1.3, is plotted and it is extrapolated to $P = 0$ to calculate M from $\left(\frac{\rho}{P}\right)_{P=0}$.

Reading the Y-axis $\left(\frac{\rho}{P}\right)_{P=0} = 1.517$ we obtain the molar masses: $M_{PH_3} = 1.517 \times 1.987 \times 273.15 \# 34$ g/mol and $M_P = M_{PH_3} - 3\ M_H = 34 - 3 \# 31$ g/mol.

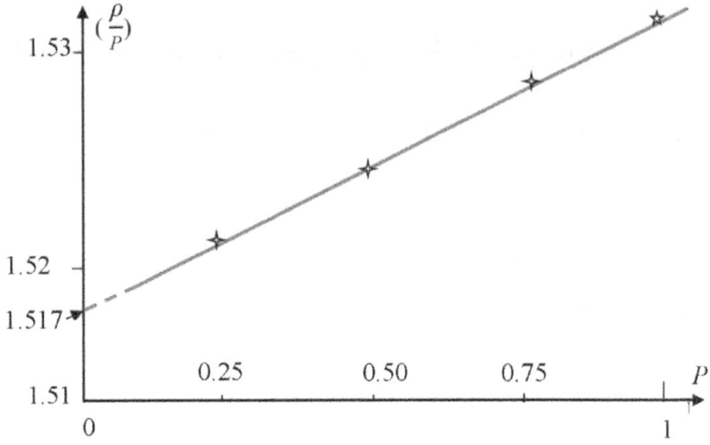

Figure 1.3. *Curve* $\left(\frac{\rho}{P}\right) = f(P)$

EXERCISE 1.7.– Volumic percentage and partial pressure

Remember that the molar fraction is identical to the volumic percentage (*), such that the following is always written: $P(V_1 + V_2 + V_3) = (n_1 + n_2 + n_3)RT$ therefore $n_1 = 0.65$; $n_2 = 0.15$; $n_3 = 0.20$ and $n = 1$ therefore $x_1 = n_1$, $x_2 = n_2$ and $x_3 = n_3$.

By definition, the partial pressure $P_i = x_i P$, i.e. $P_1 = 0.65 \times 760 = 494$ torr, $P_2 = 0.15 \times 760 = 114$ torr and $P_3 = 0.20 \times 760 = 152$ torr.

(*) V_1% and V_2% are the percentages by volume of each component in the mixture. $V_1 + V_2 = 100$ and $V = \frac{V}{100} V_1 + \frac{V}{100} V_2$ with $n_1 + n_2 = n$ and $PV = nRT$. This gives: $P\frac{V}{100}(V_1 + V_2) = (n_1 + n_2)RT \Rightarrow PV\left(\frac{V_1}{100} + \frac{V_2}{100}\right) = (n_1 + n_2)RT$. By dividing term by term in both expressions, we obtain: $\frac{V_1}{100} + \frac{V_2}{100} = \frac{n_1}{n} + \frac{n_2}{n} = x_1 + x_2$ $\Rightarrow x_i = \frac{V_i}{100} = Vi\%$. The molar fraction is therefore equal to % in volume.

EXERCISE 1.8.– Average molar mass

The average molar mass of air is: $M_{\text{air}} = 29^{(*)}$, and 4 g of H_2 corresponds to 2 mol; the balloon contains 22.4 L of air under STP conditions, i.e. 1 mole of air. In total, the balloon contains at present 3 mol, and hence the molar mass of the mixture: $M_{moy} = 29\frac{1}{3} + 2\frac{2}{3} = \frac{33}{3} = 11\,\text{g}^{(**)}$.

(*) We can verify that: $M = 0.8 \times 28 + 0.2 \times 32 = 28.80$ g/mol.

(**) Using 28.8 g instead of 29 g for air, we obtain 10.93 g.

EXERCISE 1.9.– Molar fraction and partial pressure

1) Before mixing: for N_2: $P_1V_1 = n_1RT$ and for H_2: $P_2V_2 = n_2RT$.

After mixing, in $V = V_1 + V_2$, the total number of moles is $n = n_1 + n_2$ such that: $PV = nRT = (n_1 + n_2)RT$, i.e. $n_1 + n_2 = \frac{(P_1V_1 + P_2V_2)}{RT}$ therefore:

$$P = \frac{P_1V_1 + P_2V_2}{V_1 + V_2} = \frac{500 \times 250 + 250 \times 1.500}{1.750} = \frac{5 \times 10^4}{175} = 285.74 \text{ torr}$$

i.e. $x_1 = \frac{n_1}{n_1 + n_2} = \frac{P_1V_1}{P_1V_1 + P_2V_2} = \frac{500 \times 250}{5 \times 10^5} = 0.25$ and $x_2 = 1 - x_1 = 0.75$.

2) Then $P_1 = x_1P = 0.25 \times 286 = 71.5$ torr and $P_2 = 0.75 \times 286 = 214.5$ torr.

EXERCISE 1.10.– Density by the Dumas method

1) At 18 °C, under 1 atm, $\rho_{air} = 1.29\frac{273.15}{291.15} = 1.213$ g/L, therefore the mass of air contained in the balloon is: $m_{air} = \rho V = 1.213 \times 0.272 = 0.33$ g. The mass of the empty balloon is $m = 70 - 0.33 = 69.67$ g. The mass of the full (of vapor) balloon is 70.45 g, i.e. $m_{vap} = 70.45 - 69.67 = 0.78$ g. The ideal gas equation is written as $PV = \frac{m}{M}RT$ and therefore: $M = \frac{0.78 \times 22.4 \times 373.15}{273.15 \times 0.272} = 88$ g/mol and $d = \frac{M}{29} = \frac{88}{29} = 3.03$.

2) The compound formula is: $C_xH_yO_z$, i.e. $\frac{12x}{54.5}=\frac{y}{9.1}=\frac{z}{36.4}=\frac{88}{100}$ whose solution is $x = 4$, $y = 8$ and $z = 2$, which gives a substance with the formula $C_4H_8O_2$.

EXERCISE 1.11.– Density and altitude

1) The volume of the balloon on the ground is $V=\frac{4}{3}\pi r^3=\frac{4\times 3.14\times 30^3}{3}=113{,}040\ \text{cm}^3$, and from the ideal gas equation:

$$n=\frac{PV}{RT}=\frac{765\times 113.040\times 273.15}{760\times 22.4\times 290.15}=4.78\ \text{mol, i.e. } m_{H_2}=9.57\ \text{g}$$

2) In the stratosphere, the balloon bursts when its volume reaches the following value: $V=\frac{4}{3}\pi r^3=\frac{4\times 3.14}{3}1^3\ \text{m}^3$, with n remaining unchanged, i.e. $P=\frac{nRT}{V}$ with $T = 223.15$ K and $R=\frac{22.4\times 10^{-3}}{273.15}$. We obtain:

$$P=\frac{4.78\times 22.4\times 10^{-8}\times 223.15\times 3}{273.15\times 4\pi}=0.021\ \text{atm}$$

This is the same pressure as outside the balloon; so, the maximum altitude that the balloon can reach is: $\text{h}=-\frac{\ln P}{a}=\frac{\ln(0.021)}{0.13}\#30$ km.

EXERCISE 1.12.– Pressure at equilibrium

1) We can represent the problem with two states given in Figure 1.4.

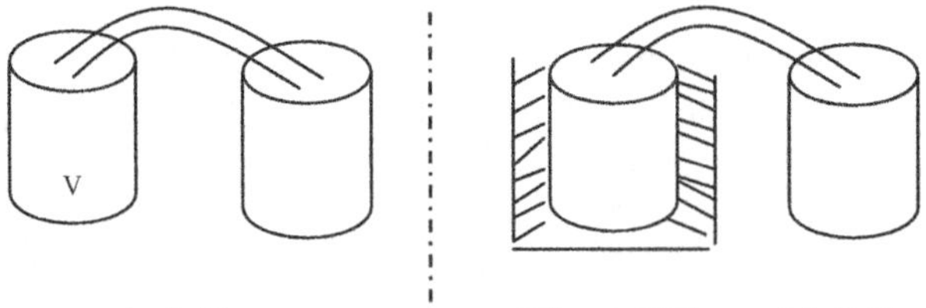

(n, P, T)
Initial state

$(0.7 - n', P', T')$ (n', P', T)
Final state

Figure 1.4. *Schematic showing two equilibrium states*

The two respective relations are as follows:

– for the initial state: $P \times 2VnRT = 0.7 \times 300.15 \times R$;

– for the final state, P it is equilibrated to $P' : P'V = 400.15 \times R\ (0.7 - n') = 300.15 \times R \times n'$.

We therefore take the values of $n' : n' = \begin{cases} \dfrac{0.7 \times 400.15}{300.15 + 400.15} = 0.4 \text{ mol at } 300.15\text{ K} \\ (0.7 - 0.4) = 0.3 \text{ mol at } 400.15\text{ K} \end{cases}$.

2) So: $P'V = 300.15R\,n' = 0.3 \times 400.15\,R = 120.45\,R$. Therefore $P' = \dfrac{120.45R}{V}$ with $P \times 2V = 0.7 \times 300.15\,R$, i.e. $V = \dfrac{300R \times 0.7}{2 \times 0.5}$ hence $P' = \dfrac{120 \times 2 \times 0.5}{300.15 \times 0.7} = 0.57$ atm.

EXERCISE 1.13.– Kinetic energy

1) The kinetic energy of a vehicle $E_c = \frac{1}{2}\text{mv}^2$ is instantly transformed to overcome the W forces of gravity W_g = mgh, i.e. $\frac{1}{2}\text{mv}^2 = \text{mgh}$ and so:

$$\text{h} = \frac{1}{2}\frac{\text{v}^2}{\text{g}} = \frac{1}{2}\left(\frac{110 \times 10^3}{3{,}600}\right)^2 \frac{1}{9.81} = 47.6 \text{ m}$$

2) The kinetic energy of the projectile $E_c = \frac{1}{2}\text{mv}^2$ is transformed into heat $Q = \text{m}C_P\Delta T$, i.e. $\frac{1}{2}\text{mv}^2 = \text{m}C_P\Delta T$ and so:

$$\Delta T = \frac{1}{2}\frac{\text{v}^2}{C_P} = \frac{700^2}{2 \times 0.13 \times 4.184 \times 10^3} = 451\ °\text{C}.$$

EXERCISE 1.14.– Potential energy

The mass of water when falling generates a potential energy E_P = mgh, then the maximum energy will be: $W_g = E_P = 100 \times 9.81 \times 30 = 29.4$ kJ/s = 29.4 kW.

EXERCISE 1.15.– Electrical energy

At 50 °C, the quantity of heat dissipated toward the exterior is $Q = 10^3$ cal/min, i.e. $Q = \frac{4,184}{60}$ J/s = 69.73 W.

So, to maintain T = 50 °C, the heating resistance $\mathcal{R}$ compensates for these losses, and must provide a minimum power of: $\mathcal{P} = \frac{\mathcal{U}^2}{\mathcal{R}} = \frac{110^2}{50} = 242$ W for a particular amount of time $\tau = \frac{69.73}{242} = 28.8$ %.

EXERCISE 1.16.– Molar fraction and temperature

The experiment can be shown as we can see in Figure 1.5.

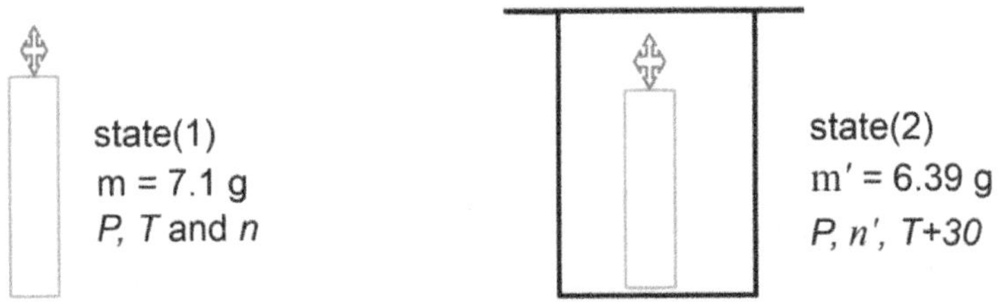

Figure 1.5. *Experiment diagram*

1) In the first case: $n = \frac{m}{M} = \frac{7.1}{2 \times 35.5} = 0.1$ mol and in the second case: $n' = \frac{m'}{M} = \frac{6.39}{2 \times 35.5} = 0.09$ mol giving the relations for each case: $PV = nRT$ and $PV = n'R(T + 30)$, i.e. $0.1T = 0.09\,(T + 30)$ and therefore $0.01T = 2.7$, i.e. $T = 270$ K.

2) The final pressure is:

$$P = \frac{nRT}{V} = \frac{0.1 \times 22.4 \times 270}{2.24 \times 273.15} = \frac{270}{273.15} = 0.989 \# 0.99 \text{ atm}$$

PROBLEM 1.1.– Chemical reaction

1) The reactions occurring are as follows:

$$6HCl + KClO_3 \rightarrow KCl + 3Cl_2 + 3H_2O$$

$$6HCl + NaClO_3 \rightarrow NaCl + 3Cl_2 + 3H_2O$$

So, in ions: $6H^+ + 6Cl^- + ClO_3^- \rightarrow Cl^- + 3Cl_2 + 3H_2O$.

The balanced equation is: $6H^+ + 5Cl^- + ClO_3^- \rightarrow 3Cl_2 + 3H_2O$, which shows that 1 mole of $KClO_3$ is required for 3 moles of Cl_2.

2) If we have 40.32 L of Cl_2, this gives $\frac{40.32}{22.4} = 1.8 \text{ mol}$ so, to begin with we have: $\frac{1.8}{3} = 0.6 \text{ mol}$ of a hypothetical compound with the formula: $K_xNa_{(1-x)}ClO_3$ and initial mass of 65.5 g whose molar mass is: $M = \frac{65.5}{0.6} = 109.17 \text{ g/mol}$.

We can calculate x since we know the atomic masses of each element ($K = 39$, Na = 23, Cl = 35.5 and O = 16), i.e.

$$39x + 23(1-x) + 35.5 + 48 = 16x + 106.5 = 109.17$$

then $x = 0.167$, which gives $KClO_3$ a molar mass of 122.5 g/mol, and $1 - x = 0.833$ gives $NaClO_3$ a molar mass of 106.5 g/mol.

So in 1 mole of mixed compound, we obtain:

$0.167 \times 122.5 = 20.44$ g of $KClO_3$

$0.833 \times 106.5 = 88.73$ g of $NaClO_3$

So for a mass of 65.5 g of product we obtain:

$$\frac{m_{KClO_3}}{20.44} = \frac{m_{NaClO_3}}{88.73} = \frac{65.5}{109.17}, \text{ i.e. } \begin{cases} m_{KClO_3} = 12.26 \text{ g} \\ m_{NaClO_3} = 53.24 \text{ g} \end{cases}$$

PROBLEM 1.2.– Bunsen calorimeter

1) When hot water is added to the tube, a certain amount of ice melts by absorbing heat; as the liquid water has a different volume to that of the ice (dilation during freezing), the index will move in the direction of variation of the total volume.

2) The heat transferred by the water, from 20 °C to 0 °C, is received by the ice and we obtain the following: $mC_P\Delta T = 100 \times 1(20-0) = 2 \times 10^3$ cal, i.e. $m'\Delta H_f = 2 \times 10^3$ cal, therefore, the amount of ice that melted is: $m' = \frac{2{,}000}{80} = 25$ g. As 1 g of ice occupies a volume $\bar{v} = \frac{1}{0.9}$ cm^3 and 1 g of water occupies volume $\bar{v}'$ = 1 cm^3, the total volume varies by: $\Delta V = m'(\bar{v}' - \bar{v}) = 25\left(1 - \frac{1}{0.9}\right) = -2.75$ cm^3, having therefore decreased. Therefore, the index will move toward the left. With the area of the capillary being 4×10^{-2} cm^2, the displacement will be: $l = \frac{2.75}{0.04} = 68.7$ cm.

3) The tool is therefore rather sensitive and is a very interesting way of measuring.

PROBLEM 1.3.– McLeod gauge

1) The experiment carried out using the McLeod gauge can be schematized as shown in Figure 1.6 by the two states.

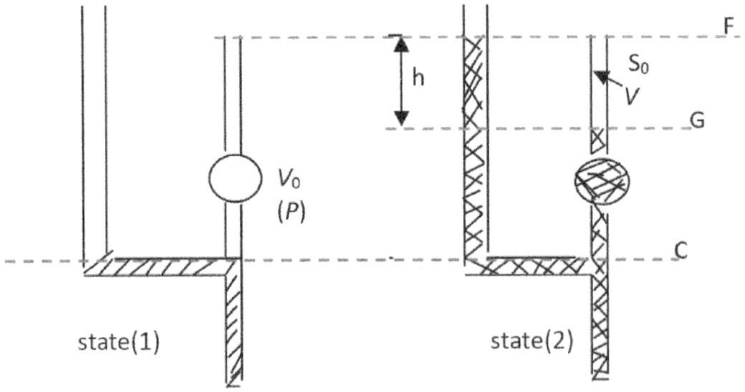

Figure 1.6. *McLeod experiment*

In state(1), the gas occupies a volume V_0 under pressure P. In state(2), it occupies volume V under pressure $(P + \text{h})$. Therefore, $V = \text{S}_0\text{h}$. If we apply Mariotte's law to both states, we obtain:

$$V_0P = V(P + \text{h}) = \text{S}_0\text{h}\,(P + \text{h})$$

So, $P(V_0 - \text{S}_0\text{h}) = \text{S}_0\text{h}^2$, therefore disregarding[1] S_0h we obtain: $P \cong \dfrac{\text{h}^2\text{S}_0}{V_0}$.

2) The accuracy of the measurement is given by $\dfrac{\Delta P}{P}$, which we obtain from expression P: $\dfrac{dP}{dh} = \dfrac{2\text{hS}_0}{V_0}$, i.e. $\Delta P = \dfrac{2\text{hS}_0}{V_0}\Delta\text{h}$ and so $\dfrac{\Delta P}{P} = \dfrac{2\Delta\text{h}}{\text{h}}$.

PROBLEM 1.4.– U-shaped tube

1) Schematically, the experiment can be described as shown in Figure 1.7.

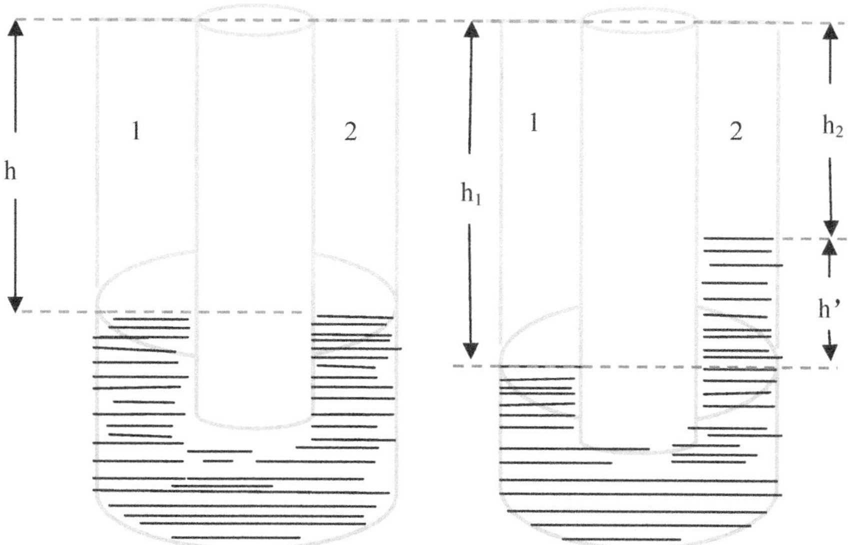

Figure 1.7. *Diagram of the experiment*

1 V_0 is large with respect to the volume of part of the capillary, which is the case for the McLeod gauge.

2) Initially, the number of moles of air in each branch is the same, since the two branches have the same volume and are at the same P and T. If n is the number of moles, the equation of states for ideal gases is applied to each branch gives: $P(\mathrm{S.h}) = nRT$, $P_1(\mathrm{S.h_1}) = nRT_1$ and $P_2(\mathrm{S.h_2}) = nRT_2$.

On the other hand, $\mathrm{h_1 - h_2 = h'}$ and volume of mercury is maintained. Therefore, $\mathrm{h_1 + h_2 = 2h}$. The pressure in each branch at $\mathrm{h_1}$ are, respectively, P_1 and $P_2 + \mathrm{h}^{(*)}$, i.e. $P_1 - P_2 = \mathrm{h}$.

We obtain the following relations:

$$\mathrm{h_1 = h_2 + h'} \text{ and } \mathrm{2h_2 + h' = 2h}$$

i.e. $\mathrm{h_2 = h - \dfrac{h'}{2}}$ and so: $\mathrm{h_1 = h + \dfrac{h'}{2}}$.

Now let $P \cdot \mathrm{S} \cdot \mathrm{h} = P_2\ \mathrm{S} \cdot \mathrm{h_2}$, i.e. $P \cdot \mathrm{h} = P_2 \cdot \mathrm{h_2}$, and so: $P_2 = \dfrac{P \cdot \mathrm{h}}{\mathrm{h_2}} = \dfrac{P \cdot \mathrm{h}}{\left(\mathrm{h} - \dfrac{\mathrm{h'}}{2}\right)}$.

This results in:

$$P_1 = \frac{P \cdot \mathrm{h}}{\mathrm{h} - \dfrac{\mathrm{h'}}{2}} = \frac{nRT_1}{S \cdot \mathrm{h_1}}, \text{ i.e. } T_1 = P_1 \mathrm{h_1} \frac{S}{nR} \text{ with } \frac{S}{nR} = \frac{T}{P \cdot \mathrm{h}}$$

hence

$$T_1 = T\frac{P_1 \cdot \mathrm{h_1}}{P \cdot \mathrm{h}} = T\frac{\left(\dfrac{P \cdot \mathrm{h}}{\mathrm{h} - \dfrac{\mathrm{h'}}{2}} + \mathrm{h'}\right)\left(\mathrm{h} + \dfrac{\mathrm{h'}}{2}\right)}{P \cdot \mathrm{h}} = \frac{\mathrm{h} + \dfrac{\mathrm{h'}}{2}}{\mathrm{h} - \dfrac{\mathrm{h'}}{2}}\left(1 + \frac{\mathrm{h'}}{P} \cdot \frac{\mathrm{h} - \dfrac{\mathrm{h'}}{2}}{\mathrm{h}}\right)$$

3) The calculation gives: $T_1 = 300.15\dfrac{110}{90}\left(1 + \dfrac{20}{76}\dfrac{90}{100}\right) = 453.75\ \mathrm{K}$.

(*) the pressure exerted by a mercury column of height h in mm is equal to h/mm Hg, by definition of the unit itself, and it is independent of the tube area, therefore:

$$P = \frac{\mathrm{m} \cdot \mathrm{g}}{S} = \frac{\rho \cdot V \cdot \mathrm{g}}{S} = \frac{\rho \cdot S \cdot \mathrm{h} \cdot \mathrm{g}}{S} = \rho \cdot \mathrm{g} \cdot \mathrm{h} = \mathrm{h}\ (\mathrm{mm\ Hg})$$

PROBLEM 1.5.– Density by the Meyer method

1) The experimental device can be schematized as per Figure 1.8.

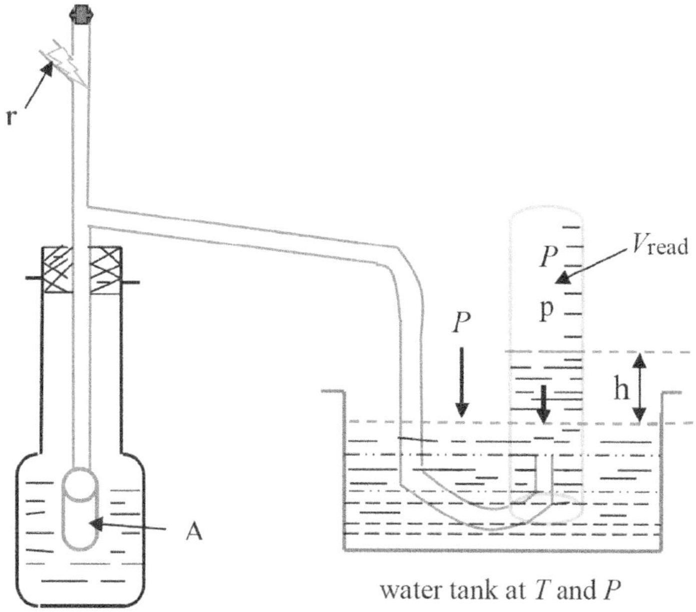

Figure 1.8. *Meyer densitometer*

2) To solve the problem, the pressure of the gas collected in the tank of water(*) needs to be calculated; it is equal to: $P - \left(p + \frac{\mathrm{h}}{13.6}\right)$; we must calculate the mass of air expelled, knowing that:

$$\rho = \frac{\mathrm{m}}{V} = \frac{\mathrm{m}}{n} \cdot \frac{P}{RT} \text{ and } \rho_0 = \frac{\mathrm{m}}{n} \frac{P_0}{RT_0}$$

So: $\dfrac{\rho}{\rho_0} = \dfrac{P}{P_0}\dfrac{T_0}{T}$, i.e. $\rho = \rho_0 \dfrac{P}{P_0}\dfrac{T_0}{T}$.

Therefore, $m = \rho_0 V \dfrac{P}{P_0}\dfrac{T_0}{T}$ with $\rho_0 = 1.293$, $d_{air} = 1.293$. We obtain:

$$m = 1.293\, V \frac{P - p - \frac{h}{13.6}}{760}\frac{273.15}{291.15} = 1.293 \times 0.242 \frac{765 - 15.8 - \frac{30}{13.6}}{760}\frac{273.15}{291.15} = 0.2885 \text{ g}$$

with the mass of the test sample being: $m' = 0.421$ g, we obtain: $d_{/air} = \dfrac{m'}{m} = \dfrac{421}{288.5} = 1.459$.

3) The density of the mixture with respect to air is associated with its molar mass: $d_{/air} = \dfrac{M}{29}$, i.e. $M = 29 \times 1.459 = 42.32$ g/mol so, if x_1 is the molar fraction of methanol in the mixture, we obtain $32\, x_1 + 46\,(1 - x_1) = 42.32$, therefore:

– for methanol $-14\, x_1 = -3.68$, i.e. $x_1 = 0.263$;

– for ethanol $x_2 = 1 - 0.263 = 0.737$.

By weight, we obtain: $\%X_1 = \dfrac{0.263 \times 32}{42.32} = 19.89$ and $\%X_2 = \dfrac{0.737 \times 46}{42.32} =$ 80.11.

4) In (A), we have $P = Cte = 765$ torr both at $T_1 = 18$ °C and $T_2 = 100$ °C, and so: $PV = nRT_1 = n'RT_2$, therefore $\dfrac{n}{n'} = \dfrac{T_2}{T_1} = \dfrac{373.15}{291.15} = 1.282$. Now the difference $(n - n')$ corresponds to 392 cm^3 in air that escaped from (A) under STP conditions.

Therefore, $n - n' = \dfrac{392}{22{,}400} = 0.0175$ mol, and so: $1.282\, n' - n' = 0.0175$, i.e. $n' = 0.062$ mol. Hence: $V = 0.062 \dfrac{760 \times 22.4}{273.15}\dfrac{373.15}{765} = 1.89$ L.

5) We have just established the number of moles remaining in (A) at 100 °C, which is $n' = 0.062$. The mixture (1)/(2) with a mass of 421 mg has a molar mass of 42.32 g/mol. This corresponds to 0.01 moles. Therefore, the number of moles of air in (A) is $n_{air} = 0.062 - 0.01 = 0.052$ mol.

Therefore, $n_{O_2} = 0.052 \times 0.2 = 0.0104$ and $n_{N_2} = 0.052 \times 0.8 = 0.0416$. For the mixture (1)/(2): $n_1 = 0.01 \times 0.263 = 0.0026$ mol and $n_2 = 0.01 \times 0.737 = 0.0074$ mol; we obtain: $x_{O_2} = \frac{104}{620} = 0.168$; $x_{N_2} = \frac{426}{620} = 0.671$; $x_1 = \frac{26}{620} = 0.042$ and $x_2 = \frac{74}{620} = 0.119$.

In a volume of 1.89 L, we obtain the following molarities (in mol/L): $m_{O_2} = \frac{0.0104}{1.89} = 5.5 \times 10^{-3}$; $m_{N_2} = \frac{0.0416}{1.89} = 22 \times 10^{-3}$; $m_1 = \frac{0.0026}{1.89} = 1.39 \times 10^{-3}$ and $m_2 = \frac{0.0074}{1.89} = 3.92 \times 10^{-3}$.

In (A), total P is 765 torr, and the partial pressures (defined by $P_i = x_i P$) are as follows: $P_{O_2} = 0.168 \times 765 = 128.5$ torr, $P_{N_2} = 0.671 \times 765 = 513.3$ torr, $P_1 = 0.042 \times 765 = 32.13$ torr and $P_2 = 0.119 \times 765 = 91.3$ torr.

(*) Calculating the pressure of a gas collected in a water tank, as a function of height h:

– if h = 0, then P in the measuring cylinder is equal to atmospheric P; in volume V, the collected gas exerts pressure P' and a certain quantity of water vapor that exerts pressure p, known as the saturation vapor pressure, measured at ambient T. Therefore: $P' = P - \text{p}$;

– if h $\neq$ 0 (water remaining in the recipient[2]), the following must be considered, for volume V of P of the water column (it acts as a piston) equal to the mass of water in the section of the tube:

$$p' = \frac{\text{m}_{\text{water}} \rho_{\text{water}} \times \text{g}}{S} = \frac{\text{h} \times S \times \rho_{\text{water}} \times \text{g}}{S} = \text{h}.\rho_{\text{water}} \cdot \text{g} \text{ (in cm } H_2O\text{)}$$

It must be expressed in mm Hg. For this, we need to calculate the height h′ of mercury corresponding to the same pressure, so: $\frac{\text{h} \times S \times \rho_{\text{water}} \times \text{g}}{S} = \frac{\text{h}' \times \text{S} \times \rho_{\text{Hg}} \times \text{g}}{S}$,

2 If the water is below the level of the tank, we obtain in this case: $P - \left(p - \frac{\text{h}}{29}\right)$.

i.e. $h' = h\frac{\rho_{water}}{\rho_{Hg}}$ with $\rho_{water} = 1$ g/cm^3 and $\rho_{Hg} = 13.6$ g/cm^3. We obtain the equivalence: $h = 13.6\,h'$, which gives a gas pressure of: $P' = P - (p + p') = P - \left(p + \frac{h}{13.6}\right)$.

PROBLEM 1.6.– Yield, composition and partial pressure

1) In this problem, we study the following at equilibrium: $2SO_2 + O_2 \rightleftarrows 2SO_3$.

Operation (I) allows the iodine solution to make it possible to determine the unreacted SO_2, i.e. remaining in the gaseous mixture:

$SO_2 + 2H_2O + I_2 \rightarrow H_2SO_4 + 2HI$.

Operation (II) allows simultaneous calculation of the SO_3 formed and the unreacted SO_2, i.e. $\begin{cases}(SO_2 + SO_3) + H_2O_2 + H_2O \rightarrow 2\,H_2SO_4 \\ 2\,H_2SO_4 + 2\,BaCl_2 \rightarrow 2\,BaSO_4 + 4\,HCl\end{cases}$.

2) For operation (I), the deci-normal solution of iodine, $\left(\frac{1}{2}0.1\right)$ mol of I_2 in 10 cm^3, we obtain $\left(\frac{1}{2}0.1\right)\frac{1}{100} = \frac{1}{2{,}000}$ mol of I_2, which correspond, according to the reaction, to $\frac{1}{2{,}000}$ mol of SO_2: this is contained in 640 cm^3 of gas. In 1 L of gas, we have: $\frac{1}{2{,}000} \times \frac{1{,}000}{640} = 0.781 \times 10^{-3}$ mol of SO_2.

For operation (II), for 1 L of gas, we obtain 0.685 g of $BaSO_4$, i.e. $\frac{0.685}{233} = 2.94 \times 10^{-3}$ mol, corresponding to 0.294×10^{-3} mol of $(SO_2 + SO_3)$ and so in 1 L of gas, we obtain: $(2.94 - 0.781) \times 10^{-3} = 2.159 \times 10^{-3}$ mol of SO_3 per liter of gas.

If no SO_2 remained in solution, we would have obtained 2.94×10^{-3} mol of SO_3. This gives a yield of: $\rho = \frac{n_{SO_3 obtained}}{n_{SO_3 theoretical}} = \frac{2.159}{2.94}100 = 73.4\%$.

3) Given the initial composition of the gas, the reaction therefore corresponds to the equilibrium: $2SO_2 + O_2 + N_2 \rightleftarrows 2SO_3 + N_2$. Assuming that initially there were 100 moles of gas, the variation of composition once in equilibrium is thus written as:

$2SO_2$	+	O_2	+	N_2	$\rightleftarrows$	$2SO_3$	+	N_2
$7-2x$		$10-x$		83		$2x$		83

So, at equilibrium, the total number of moles:

$$n = 83 + (10 - x) + (7 - 2x) + 2x = 100 - x$$

As ρ represents the ratio $\frac{n_{SO_3 \text{obtained}}}{n_{SO_2 \text{initial}\ (*)}}$, $\rho = \frac{2x}{7} = \frac{73.4}{100}$, therefore $x = 2.57$ and so $n = 100 - 2.57 = 97.43$ mol, which gives the following percentages[3] at equilibrium:

	N_2	O_2	SO_2	SO_3
x%	85.19	7.64	1.91	5.28

4) The partial pressure of a constituent (i) is defined with respect to the total P of the system and its composition by: $P_i = x_i P$. Since $P = 1 \Rightarrow P_i = x_i$, from the previous table we immediately obtain the following results:

	N_2	O_2	SO_2	SO_3
P_i/atm	0.8519	0.0764	0.0191	0.0528

5) With these values of P_i, by replacing the expression of K_P given in the statement, we obtain: $K_P = \frac{0.0528^2}{0.0191^2 \times 0.0764} = 99.86\ \text{atm}^{-1}$.

The concentration and partial pressure of a constituent(i) are linked by the relation: $P_i = c_i.RT^{(**)}$ giving the expression: $K_P = \frac{\left(c_{SO_3} RT\right)^2}{\left(c_{SO_2} RT\right)^2 \left(c_{O_2} RT\right)} = \frac{c_{SO_3}^2}{c_{SO_2}^2 c_{O_2}} \frac{1}{RT}$

$= K_C \frac{1}{RT}$, i.e. $K_C = K_P RT$ with: $R = 0.082$ L.atm/K and $T = 873$ K we obtain $K_C = 99.86 \times 0.082 \times 873 = 7{,}148.4$ L.

---

3 The percentage is defined by: $x\,\% = \frac{n_i}{n} \cdot 100$.

6) For equilibrium without N_2, we have the following reaction:

$2SO_2 + O_2 \rightarrow 2SO_3$, whose variation in composition is written as:

	$2SO_2$	+	O_2	$\rightarrow$	$2SO_3$
At $t = 0$	2		1		0
At $t > 9$	$2 - 2x$		$1 - x$		$2x$
If reaction is complete	0		0		2

Giving the yield: $\rho = \frac{2x}{2} = x$; so, the total number of moles at equilibrium is: $n = 2 - 2x + 1 - x + 2x = 3 - x$, giving the following partial pressures:

$$P_{SO_2} = \frac{2(1-x)}{3-x} P \text{ , } P_{O_2} = \frac{1-x}{3-x} P \text{ and } P_{SO_3} = \frac{2x}{3-x} P$$

Hence

$$K_P = \frac{\left(\frac{2x}{3-x}\right)^2}{\left(\frac{2-2x}{3-x}\right)^2 \left(\frac{1-x}{3-x}\right)} \frac{1}{P} = \frac{x^2 (3-x)}{(1-x)^3 P}.$$

$K_P = 99.86$; we obtain the third-degree equation: $99.86(1 - x)^3 P = (3 - x)x^2$, which must be solved for $P = 1$ and 100 atm.

The equation can be solved by successive approximations. We look for the order of magnitude of x, then we try successive approximations until we obtain 99.86. The below table presents the results for the case $P = 1$; we will then use $x = 0.765$.

x	x^2	$3 - x$	$(1 - x)^3$	K_P
0.75	0.562	2.25	15.2×10^{-3}	81 (too weak)
0.80	0.640	2.20	8×10^{-3}	176 (too strong)
0.76	0.578	2.24	13.8×10^{-3}	93 (good)
0.77	0.593	2.23	12.2×10^{-3}	108 (good)

The same trial and error for $P = 100$ atm gives $x = 0.94$ and hence the following yields at equilibrium: $\rho_1 = 76.5\%$ and $\rho_{100} = 94\%$.

(*) If the reaction is complete, the theoretical number of moles of SO_3 would be equal to that of SO_2 present at the beginning prior to reaction, since the synthesis is as follows: $1\,SO_2 \xrightarrow{(O)} 1\,SO_3$.

(**) From $PV = nRT$, we obtain $P_iV = x_inRT = n_iRT$, i.e. $P_i = \frac{n_i}{V}RT = c_iRT$.

2

Closed Systems without Chemical Reactions

2.1. Exercises

EXERCISE 2.1.– The housewife in the kitchen

The system consists of the housewife, a closed kitchen, a box of matches, a heater with a bottle of butane and a stove fueled with mains gas.

The heating and cooking will take place in several stages:

– at t_0: the housewife grabs the box of matches;

– at t_1: she strikes the match;

– at t_2: she ignites the heating;

– at t_3: she lights the stove.

All this is done in a very short time.

For each interval, the variation in energy of the system, e.g. increases and decreases, remains constant.

Answer: see corrections.

EXERCISE 2.2.– Work against the exterior

Ten kilograms of alumina is heated, at atmospheric pressure, from 20 °C to 660 °C; knowing that the densities of solid aluminum are 2.70 g/cm^3 and 2.38 g/cm^3

at 20 °C and 660 °C, respectively, calculate the work against the exterior during heating.

Answer: $W = -50$ J.

EXERCISE 2.3.– Heat of reaction and internal energy

The formation reaction of alumina by oxidation of 54 g pure aluminum, under Standard conditions of T and P (STP), releases 400 kcal as heat.

1) What is the variation in internal energy of the system (Al/O_2)?

2) What are the observations?

Answers: (1) $\Delta U = -1.669 \times 10^3$ kJ; (2) see corrections.

EXERCISE 2.4.– Differential expression of *U* as a function of *T*, *P* and C_v

Consider the function: $U(T, V)$.

1) Express the variation in coordinates (T, V).

2) Deduce the variation as a function of *T*, *P* and C_v.

Responses: (1) see corrections; (2) $d = C_V.\, dT + \left[T\left(\frac{\partial P}{\partial T}\right)_V - P\right] dV$.

EXERCISE 2.5.– Internal energy and temperature

Part I: One liter of water undergoes a variation in temperature of 10 °C when it is provided with 10^4 cal. The expansion work is disregarded.

1) What is the variation in internal energy of 3 L of water with a temperature increase of 10 °C?

2) What is the molar variation of internal energy?

Part II: We study the variation in internal energy of 1 mole of ideal gas going from state A (273 K; 22.4 L) to state B (546 K; 44.8 L).

1) Show that the variation in internal energy depends only on T.

2) Calculate this variation in the following cases:

a) a reversible transformation at constant P;

b) a reversible transformation at constant T followed by a reversible transformation at constant V.

3) Is the internal energy a state function?

Data: $C_p = 5$ cal/mol.K; $C_v = 3$ cal/mol.K and $R = 1.9872$ cal/mol.K.

Answers: Part I – (1) $\Delta U = 3 \times 10^4$ cal; (2) $\Delta U = 180$ cal; part II – (1) $dU = C_v.dT$; (2) $\Delta U = 819$ cal/mol; (3) see corrections.

EXERCISE 2.6.– Calorific capacity

How much heat energy is required, at constant pressure, to heat a mole of H_2 from 300 to 1,500 K, knowing that in this temperature range, its calorific capacity is given by the expression:

$$C_P = 6.9469 - 0.1999 \times 10^{-3}\, T + 4.808 \times 10^{-7} T^2$$

Answer: $\Delta Q = 8{,}659.72$ cal.

EXERCISE 2.7.– Enthalpy

The molar heat capacity of water vapor at a pressure of 1 atm between 373 K and 1,000 K is given by:

$$C_P = 7.17 + 2.56 \times 10^{-3} T + 8.0 \times 10^3 T^{-2} \text{ cal/mol.K}$$

1) Calculate the variation in enthalpy of gas when T increases from 100 to 900 K.

2) Why is the pressure defined by giving C_P?

Answers: (1) $\Delta H = 4.428$ kcal; (2) see corrections.

EXERCISE 2.8.– Entropy

The molar heat capacity, at atmospheric pressure, of solid magnesium in the temperature domain [0,600 °C] is given by the following relation: $Cp = 6.20 + 1.30 \times 10^{-3}T - 6.00 \times 10^{4}T^{-2}$ cal/mol.K.

Calculate the variation in entropy corresponding to the reversible heating of 1 mole of magnesium, from 27 °C to 227 °C under 1 atm.

Answer: $\Delta S = 3.22$ cal/mol.K.

(*) $\int \frac{dx}{ax+b} = \frac{1}{a}\ln(ax+b)$ and $\int \frac{dx}{(ax+b)^n} = -\frac{1}{a(n-1)(ax+b)^{n-1}}$.

EXERCISE 2.9.– Free enthalpy or Gibbs free energy

Part I: Consider the transformation of water from ice to liquid at atmospheric pressure:

1) give the sign of the variation in free enthalpy at –10 °C, 0 °C and 10 °C;

2) state the nature of the transformation.

Part II: Consider the opposite transformation of liquid water at 10 °C into ice at –10 °C, at atmospheric pressure, assuming that water completely transforms into ice. Calculate the variation in free enthalpy:

1) by numerically evaluating ΔH° and ΔS°;

2) by calculating $\Delta G^{\circ} = f(T)$ and using $T = 263$ K in this equation.

Data: At 0 °C, $\Delta H_{f ice} = 1{,}439$ cal/mol; at $-10 < T < 0$ °C, $C_{P ice} = 9$ cal/mol.K; $C_{P liquid} = 18$ cal/mol.K.

Answers: *Part I* – (1) $\Delta G_{-10} > 0$; $\Delta G_0 = 0$; $\Delta G_{10} < 0$; (2) see corrections; *part II* – (1) $\Delta G = -50.8$ cal; (2) $\Delta G = -50.4$ cal.

EXERCISE 2.10.– Isothermal transformations

A system at atmospheric pressure undergoes successive isothermal transformations:

– 100 calories are released to the exterior;

– it expands to 4 L;

– with a resistance of 10 Ω, an electric current of 1 A flows for 5 min;

– a mass of 2 kg located 15 m away vertically falls on the system.

What is the variation in internal energy of this system?

Answer: $\Delta U = 591$ cal.

EXERCISE 2.11.– Reversible transformation

A system (σ) reversibly changes from an initial state to a final state by exchanging a given quantity of heat $Q = a + bT$ with the external environment (σ').

1) What is the variation in entropy of the system?

2) What is the total variation in entropy experienced by the system and the external environment?

Answers: (1) see corrections; (2) $\Delta S = 0$.

EXERCISE 2.12.– Irreversibility of heating

Note that 100 g of water is heated from 20 °C to 80 °C in an enclosure maintained at this temperature. The heat capacity of water is 1 cal/g.K.

1) What are the variations in entropy of water and the enclosure?

2) How much entropy is created?

Answers: (1) $\Delta S_{water} = 18.63$ cal/K and $\Delta S_{enc} = -17.00$ cal/K; (2) $\Delta S = 1.63$ cal/K.

EXERCISE 2.13.– Heat of vaporization of water

One kilogram of water is vaporized under atmospheric pressure at 100 °C, whose latent heat of vaporization is 540 cal/g. Assuming that water vapor is an ideal gas, calculate:

1) the amount of heat exchanged with the exterior;

2) the work exchanged with the exterior;

3) the variation in internal energy of the system.

Answers: (1) $Q = 540$ kcal; (2) $W = -41$ kcal; (3) $\Delta U = 499$ kcal.

EXERCISE 2.14.– Vaporization of ice

Consider the transformation of 1 mole of ice at 0 °C into water vapor at 200 °C under a pressure of 1 atm:

1) calculate the variation in molar enthalpy;

2) calculate the variation in molar entropy.

Data:

	ΔH_f (kcal/mol)	ΔH_v (kcal/mol)	C_P (cal/mol.K)
Ice	1.44	/	/
Liquid	/	9.7	18
Vapor	/	/	6

Answers: (1) $\Delta H = 13.540$ kcal; (2) $\Delta S = 38.26$ cal/mol.K.

EXERCISE 2.15.– Mixture at different temperatures

Note that 100 g of water at 20 °C is mixed with an equal amount of water at 80 °C and we wait until the mixture is uniform.

1) By how much has the entropy of the universe increased?

If a heat engine is reversibly operated between the sources of two quantities of water at 20 °C and 80 °C, the temperatures of the two sources will vary during the operation of the engine, which will stop when the temperatures of the two sources are equal.

2) What will be the final temperature?

3) Calculate the work recovered.

Answers: (1) $\Delta S = 0.866$ cal/K; (2) $T_f = 48.6$ °C; (3) $W = -280$ cal.

EXERCISE 2.16.– Adiabatic expansion of nitrogen

Nitrogen relaxes adiabatically from an initial state at 100 atm and 200 °C against normal atmospheric pressure. Knowing that the C_P of nitrogen is 7 cal/mol.K, calculate the variation in entropy of 20 L of nitrogen during the process.

Answer: $\Delta S = 351.45$ cal/mol.K.

EXERCISE 2.17.– Case of three isolated bodies.

A system is made of three identical bodies, each with a heat capacity C_p and respective temperatures $T_1 = T_2$ and $T_3 > T_1$. Assuming there is no heat or work exchanged between this system and the external environment, what would be the minimum temperature each body could reach with the help from appropriate machines?

Data: $C_P = 1$ J/K; $T_1 = 300$ K and $T_3 = 800$ K.

Answers: $T_1 = 200$ K and $T_2 = T_3 = 600$ K.

EXERCISE 2.18.– Solidification of water

A system (σ) composed of 1 mole of liquid water at a temperature (T) of 0 °C come into contact with a thermostat (system σ') whose temperature (T') is determined.

Assuming the C_p of water and ice are equal, and the enthalpy of fusion of 1 mole of ice at constant pressure is 1,440 cal, complete the following table:

T (°C)	ΔS_σ	$\Delta S_{\sigma'}$	ΔS	Conclusion
–10				
0				
10				

Answers:

T (°C)	ΔS_σ	$\Delta S_{\sigma'}$	ΔS	Conclusion
–10	} – 5.27	5.47	0.20	See corrections
0		5.27	0	
10		5.09	–0.18	

EXERCISE 2.19.– Reversible isothermal expansion of an ideal gas

An ideal gas absorbs 2,250 cal relaxing at $T = 25$ °C reversibly from 1.5 to 10 L.

How many moles of gas are there?

Answer: $n = 2$.

EXERCISE 2.20 .– Joule–Kelvin coefficient

The variation in molar volume of ammonia under a pressure of 40 atm as a function of temperature is given in the following table:

T (°C)	225	250	275	300	325
v (cm^3)	962	1,017	1,076	1,136	1,186

The experimental value obtained for the J-K coefficient at 300 °C, under a pressure of 40 atm, is 0.370 K/atm.

Compare this value to that calculated from the results hereafter.

Data: C_p of NH_3 (at 300 °C and 40 atm) = 11.00 cal/mol.K.

Answer: $k_{J\text{-}K} = 0.344$ K/atm.

EXERCISE 2.21.– Thermoelastic coefficients

At 20 °C, copper has a coefficient of thermal expansion $\alpha = 4.9 \times 10^{-5}\ K^{-1}$, a coefficient of compressibility $\chi = 0.78 \times 10^{-6} atm^{-1}$ and a molar volume $v = 10.4$ cm^3/mol. The behavior of copper follows the Dulong–Petit law.

1) Calculate the difference $C_p - C_v$.

2) Find the value of C_v.

Data: $C_P = 6.4$ cal/at.K.

Answers: (1) $C_P - C_v = 0.23$ cal/at.K; (2) $C_v = 6.17$ cal/at.K.

EXERCISE 2.22.– Irreversibility when two pressures equalize

Two containers of the same volume V, perfectly insulated, contain the same ideal gas, at the same temperature T_0 but under two different pressures P_1 and P_2. We connect the two containers and wait until equilibrium is reached.

1) What are the final temperatures and pressures?

2) What is the variation in entropy overall?

Data: $V = 22.4$ m^3; $T_0 = 273$ K; $P_1 = 9$ atm and $P_2 = 1$ atm.

Answers: (1) $T_f = T_0$ and $P_f = \frac{1}{2}(P_1 + P_2)$; (2) $\Delta S = 21.086$ cal/K.

EXERCISE 2.23.– Joule expansion

Two liters of hydrogen, considered an ideal gas, under STP conditions – expand irreversibly but isothermally up to 4 L.

1) Calculate the variations in U, H, F and G.

2) Under these conditions, what is the function that confirms that this process is irreversible?

Answers: (1) $\Delta U = \Delta H = 0$ and $\Delta F = \Delta G = -140$ J; (2) Free energy function (F).

EXERCISE 2.24.– Condensation of water vapor

The condensation of 1 mole of water at 100 °C and 1 atm releases 9.72 kcal; knowing that the density of liquid water at 100 °C is 0.958 g/cm^3:

1) determine the work done against the pressure, and the variation in internal energy;

2) calculate the variations in entropy, free energy and free enthalpy;

3) which variable whose variation makes it possible to determine whether the process is reversible or not?

Answers: (1) $W = 740.865$ cal/mol and $\Delta U = 8.979$ kcal/mol; (2) $\Delta S = -26.05$ cal/mol. K; $\Delta F = 740.865$ cal/mol and $\Delta G = 0$; (3) see corrections.

EXERCISE 2.25.– Internal energy of a real gas

We consider 1 mole of a real gas to obey the following state equation: $PV = RT + B(T).P$, in which $B(T)$ is a linear function of T given by: $B(T) = aT + b$. It undergoes a reversible isothermal expansion of P_1 to P_2.

1) Using the results obtained in the generic example, express the calorimetric coefficients l and h as a function of the variables P and T and the constants R and a.

2) Express, as a function of P_1, P_2, T, R and a, the values of work done W and the heat Q exchanged with the external environment; hence, the variation in internal energy and entropy of the gas.

For argon, the experiment gave: $B(T) = -21.1$ cm^3/mol at 0 °C and -11.1 cm^3/mol at 50 °C.

3) Calculate: a, W, Q, ΔS and ΔU at $T = 25$ °C when P decreases from 10 atm to 1 atm. What conclusions can be drawn?

Answers: (1) $l = P + a\frac{P^2}{R}$; $h = -aT\frac{RT}{P}$; (2) $W = RT\ln\frac{P_2}{P_1}$; $Q = -aT(P_2-P_1) - RTln\frac{P_2}{P_1}$; $\Delta S = -a(P_2-P_1) - Rln\frac{P_2}{P_1}$; $\Delta U = -aT(P_2-P_1)$; (3) a = 0.2 cm^3/mol.K; $W = -5.704$ kJ; $Q = 5.758$ kJ; $\Delta S = 0.019$ kJ and $\Delta U = -0.054$ kJ; see corrections.

EXERCISE 2.26.– Adiabatic and reversible expansion of a real gas

One mole of gas, with the following state equation: $PV = RT + bP$, whose initial temperature is 700 °C, expands adiabatically and reversibly from 8 to 1 atm.

Calculate the work, knowing that for this gas $\gamma = \frac{5}{3}$.

Answer: $W = -1{,}637.97$ cal.

EXERCISE 2.27.– Reech Formula

1) Show that the Reech formula is valid for any fluid that obeys the following state equation: $f(P, V, T) = 0$.

2) A gas with the equation: $PV = RT + bP$ expands adiabatically and reversibly. Establish the relations between P and V, T and V, and P and T.

Answers: (1) see corrections; (2) $P(V{-}b)^{\gamma} = \text{Cte}$; $T(V{-}b)^{\gamma-1} = \text{Cte}$; $P^{\frac{1}{\gamma}}T^{\frac{1}{1-\gamma}} = \text{Cte}$.

EXERCISE 2.28.– Difference to ideal gas for entropy

1) Establish the expression of the residual function $(S - S^{*})$, for a real gas whose equation of state is: $PV = RT + BP$, where B is a function of T according to the following table:

T (°C)	290	300	310
B (cm^3/mol)	–125	–119	–113

2) Deduce $(S - S^{*})$ for H_2S at $P = 0.505 \times 10^6$ Pa and $T = 300$ °C.

Answers: (1) $(S - S^{*}) = -\frac{dB}{dT}P$; (2) $(S - S^{*})$ # 0.14 J/mol.K.

EXERCISE 2.29.– The Van der Waals Equation

In a 250 L recipient, there is 550 g of NH_3 under pressure P and a temperature of 50 °C.

1) Calculate the pressure, assuming that under these conditions, ammonia obeys the Van der Waals equation: $\left(P + \frac{a}{v^2}\right)(v - b) = RT$.

2) Compare the value obtained to the experimental value of 3 atm.

Data: $P_c = 111.5$ atm and $T_c = 132.4$ °C.

Answers: (1) $P = 3.37$ atm; (2) $\Delta P = 0.37$ atm.

EXERCISE 2.30.– Virial equation of state

The second coefficient of the virial equation (B), for an equimolar mixture of methane and n-hexane vapors, at $T = 50$ °C, is –517 cm^3/mol.

Calculate this coefficient at the same temperature for a mixture formed of 25% moles of methane and 75% moles of n-hexane.

Data: At 50 °C: $B_{11} = -33$ cm^3/mol and $B_{22} = -1{,}512$ cm^3/mol.

Answer: B = –951 cm^3/mol.

EXERCISE 2.31.– A vertical piston cylinder

A vertical cylinder, plunged into a thermostat and filled with fluid, is closed by a piston held in an initial position (A) by a screw. The piston with a mass of 1 kg and cross-sectional area of 10 cm^2 is in contact with the atmosphere whose pressure is 760 torr. The screw is released, the piston then rises under the action of the fluid to a height of 10 cm; during this transformation, the fluid receives 10 cal from the thermostat.

Calculate the variation in internal energy of the fluid (disregarding the friction of the piston and the energy of unscrewing the screw).

Answer: $\Delta U = 7.34$ cal.

EXERCISE 2.32.– A horizontal piston cylinder

A piston moves horizontally without friction in a thermally isolated cylinder; both sides of the piston has a volume of 20 L, filled with ideal gas at a temperature of 25 °C and a pressure of 1 atm. The gas in the left part of the cylinder is slowly heated until P in the right part reaches 2 atm.

1) Determine the volume and the temperature of the gas following compression in the right part.

2) Calculate the work required to compress the gas.

3) Calculate final temperature in the left part.

4) What is the amount of heat transferred to the gas in the left part?

Data: $C_v = 5$ cal/mol.K and $\frac{C_p}{C_v} = 1.4$.

Answers: (1) $V_2 = 12.2$ L and $T_2 = 90$ °C; (2) $W = 266.6$ cal; (3) $T_1 = 555.25$ °C; (4) $Q = 2{,}440$ cal.

2.2. Problems

PROBLEM 2.1.– Variation in internal energy of a system

Calculate the variation in U of a system when it undergoes the following transformations:

1) it releases 100 cal to the external environment;

2) its volume increases to 4 L with P remaining constant at 1 atm, both inside and outside;

3) a mass of 1 kg falls on the system, from a fixed point in the external environment and it drops 30 m;

4) an electric current of 2 A passes it for 20 seconds, with a resistance of 20 Ω inside the system;

5) the system successively undergoes the previous four transformations.

Answers: (1) $\Delta U = -100$ cal; (2) $\Delta U = -96.8$ cal; (3) $\Delta U = 70.3$ cal; (4) $\Delta U = 382.5$ cal; (5) $\Delta U = -256$ cal.

PROBLEM 2.2.– Calculation of γ – Clément and Desormes' method

A balloon with a capacity of several tens of liters contains air at ambient temperature and a pressure of $P_1 = 1.0204$ atm, slightly higher than the external pressure $P_0 = 1$ atm.

The balloon is in contact with the exterior via a valve. The valve is opened and immediately closed: P_1 quickly becomes equal to P_0. We consider the adiabatic (the heat exchanges did not have time to occur) and reversible (the ratio $\frac{P_1 - P_0}{P_0}$ is very small) expansion and the loss of matter to be negligible.

The air in the balloon is cooled during the expansion. It is allowed to heat up to a constant volume until it reaches the initial temperature; the pressure reaches $P_2 = 1.008$ atm.

1) Determine the expression of $\gamma = \frac{C_p}{C_v}$ as a function of pressure.

2) Calculate C_P and C_v, assuming the air is an ideal gas.

Answers: (1) $\gamma = \frac{P_0 - P_1}{P_2 - P_1}$; (2) $C_p = 5.07$ cal/mol.K and $C_v = 3.08$ cal/mol.K.

PROBLEM 2.3.– Maxwell Equations

1) Write the differential expressions of $S(T, P)$ and $S(T, V)$ for 1 mole of a pure body and show that:

a) $\left(\frac{\partial S}{\partial T}\right)_V = \frac{C_V}{T}$ and $\left(\frac{\partial S}{\partial T}\right)_P = \frac{C_P}{T}$;

b) $\left(\frac{\partial S}{\partial P}\right)_T = -\left(\frac{\partial V}{\partial T}\right)_P$ and $\left(\frac{\partial S}{\partial V}\right)_T = \left(\frac{\partial P}{\partial T}\right)_V$.

2) Deduce the expression of $C_p - C_v$ and apply it to the ideal gas.

3) Show that $\frac{C_P}{C_V} = \left(\frac{\partial V}{\partial P}\right)_T \left(\frac{\partial P}{\partial V}\right)_S$ and apply it to the ideal gas.

4) Demonstrate that $\left(\frac{\partial C_P}{\partial P}\right)_T = -VT\left(\alpha^2 + \left(\frac{\partial \alpha}{\partial T}\right)_P\right)$.

5) The partial derivative $\left(\frac{\partial T}{\partial P}\right)_H$ is called the Joule–Thomson coefficient. Using the relation $\left(\frac{\partial S}{\partial P}\right)_T = -\left(\frac{\partial V}{\partial T}\right)_P$, establish the expression of this coefficient as a function of C_P, T, V and $\left(\frac{\partial V}{\partial T}\right)_P$.

Answers: (1) see corrections; (2) $C_P - C_V = \alpha\beta PVT$ and $C_P - C_V = R$; (3) see corrections, $\frac{C_P}{C_V} = \gamma$; (4) see corrections; (5) $\left(\frac{\partial T}{\partial P}\right)_H = \frac{1}{C_P}\left[T\left(\frac{\partial V}{\partial T}\right)_P - V\right]$.

PROBLEM 2.4.– Effect of pressure on U, H and C_p of water

Let us study the effect of pressure on the internal energy, volume and heat capacity of water.

1) Show that for a pure body: $\left(\frac{\partial H}{\partial P}\right)_T = V(1 - \alpha T)$ and $\left(\frac{\partial U}{\partial V}\right)_T = P(\beta T - 1)$.

2) Deduce the expression of $\left(\frac{\partial U}{\partial P}\right)_T$.

3) Calculate the variation in internal energy for 1 m^3 of liquid water during the following transformation:

State 1: $H_2O_{liq}(10^3 \text{bars}, 303\text{ K}) \rightarrow$ State 2: $H_2O_{liq}(1\text{ bar}, 303\text{ K})$

4) Calculate the volumic ratio ($V_{2/}V_1$) for water, during the following transformation:

State 1: $H_2O_{liq}(1\text{ bar}, 283\text{ K}) \rightarrow$ State 2: $H_2O_{liq}(10^3 \text{bars}, 303\text{ K})$

5) Determine the effect of P on C_p for water at 303 K and 1 bar.

6) What conclusions can be drawn?

Data:

T (K)	283	293	303	303
$P.10^{-5}$ (Pa)	1	1,000	1	1,000
$\alpha.10^{+6}$ (K^{-1})	208	208	304	304
$\chi.10^{+12}$ (Pa)	458	348	446	338

Answers: (1) see corrections; (2) $\left(\frac{\partial U}{\partial P}\right)_T = V(\chi P - \alpha T)$; (3) $\Delta U = 7.24 \times 10^3$ kJ; (4) $\frac{V_2}{V_1} = 1.005$; (5) $\left(\frac{\partial C_p}{\partial P}\right)_T = -2.9 \times 10^{-3}$ J/K.Pa; (6) conclusion: see corrections.

PROBLEM 2.5.– Units and equation of state of an ideal gas

A recipient with a volume $V = 1{,}500\text{ cm}^3$ contains 0.1 g of hydrogen, considered to be ideal, at $T = 20$ °C.

1) Calculate the pressure inside the recipient, in atm, torr, bar, Pascal and kg/cm^2.

2) Calculate this pressure assuming the equation of state of the ideal gas is not known.

3) Deduce the equation of state of the ideal gas.

Answers: (1) $P = 0.8$ atm = 609 torr = 0.81 bar = 8.1×10^4 Pa = 0.83 kg/cm^2; (2) and (3) see corrections.

PROBLEM 2.6.– Compressions of an ideal gas

If we consider 1 mole of ideal gas initially contained in a cylinder with a piston at a temperature of 27 °C and pressure of 1 atm.

Part I: The gas is compressed adiabatically and reversibly up to a pressure of 10 atm.

1) Calculate the variation in entropy.

2) What is the final temperature of the gas?

3) Calculate the variation in internal energy of the gas.

Part II: The gas is compressed isothermally and reversibly to a pressure of $P = 10$ atm.

1) What is the variation in entropy?

2) Compare the heat dissipated during the transformation to the work done by the gas.

3) What can be deduced about the variation in U?

Data: $C_P = \frac{5}{2}R$ and $R = 1.9872$ cal/mol.K.

Answers: Part I – (1) $\Delta S = 0$; (2) $T_2 = 480.8$ °C; (3) $\Delta U = 1{,}360.82$ cal; part II – (1) $\Delta S = -4.6$ cal/mol.K; (2) $|\Delta Q| = |\Delta W| = 1{,}380$ cal; (3) $\Delta U = 0$.

PROBLEM 2.7.– Equation of state of a real gas

In the temperature domain 0 °C < T < 400 °C, and close to atmospheric pressure, the experimental equation of state of nitrogen is: $PV = RT(1+\frac{B}{V})$, in which B is a coefficient that depends on T according the equation: $B = b - \frac{a}{RT}$ whose experimental values are: a = 1.91×10^6 atm and b = 36.3 cm^3.

1) Show that this equation of state is derived from a Van der Waals equation where certain approximations are introduced.

2) Calculate the density ρ_0 of N_2 at 0 °C using the proposed equation of state.

3) What would be the relative error $\Delta^{(*)}$ for ρ_0 if we compare N_2 to an ideal gas?

4) At what T does nitrogen behave like an ideal gas?

Data: $n = 1$; $P = 1$ atm; $M_{N_2} = 14$ g/mol.

(*) By definition: $\Delta = \frac{(\rho_{exp} - \rho_{theo})}{\rho_{theo}}$.

Answers: (1) see corrections; (2) $\rho_0 = 1.253$ g/L; (3) $\Delta = 0.2\%$; (4) $T = 368.9$ °C.

PROBLEM 2.8.– Van der Waals Equation

A system is composed of 1 mole of carbon dioxide, considered to be a real gas and obeys the Van der Waals equation of state: $(P + \frac{a}{V^2})(V - b) = RT$; it undergoes a reversible isothermal expansion during which its volume changes from V_1 to V_2.

1) Calculate, as a function of a, b, R, T, V_1 and V_2, the expressions:

a) of the calorimetric coefficient l of CO_2;

b) of the work exchanged with the external environment;

c) of the variation in internal energy of the system;

d) of the amount of heat exchanged with the external environment.

2) One mole of CO_2 undergoes joule expansion in a recipient immersed in a melting ice calorimeter, thermally isolated from the external environment, initially containing 100 g of ice:

a) briefly describe the process in the calorimeter following this expansion;

b) what is the mass of ice in the calorimeter when equilibrium is re-established?

3) a) From the previous expressions, find the corresponding expressions for the ideal gas.

b) What conclusions can be drawn for the variation in internal energy of a gas, in the case of an isothermal transformation, depending on whether it is considered a real or ideal gas?

Data: a = 3.59 l^2 atm; b = 0.04 l; $V_1 = 1$ l; $V_2 = 50$ l and $\Delta H_{f,ice} = 80$ cal/g.

Answers: (1) (a) $l = \frac{RT}{V-b}$; (b) $W = -RTln\frac{V_2-b}{V_1-b} + a\left(\frac{1}{V_1} - \frac{1}{V_2}\right)$; (c) $\Delta U = -a\left(\frac{1}{V_1} - \frac{1}{V_2}\right)$; (d) $Q = RTln\ \frac{V_2-b}{V_1-b}$; (2) (a) see corrections; (b) m = 101.07 g; (3) see corrections.

PROBLEM 2.9.– Non-isolated thermostat

A thermostat containing water is maintained at 97 °C in a chamber where it is 27 °C. After a moment, 10^3 cal are released through the walls to the external environment.

1) Calculate the variation in entropy of water in the thermostat.

2) Calculate the variation in entropy of ambient air in the chamber.

3) Determine if this process is reversible or not.

Answers: (1) $\Delta S_{water} = -2.70$ cal/K; (2) $\Delta S_{air} = 3.33$ cal/K; (3) $\Delta S = 0.63$ cal/K, irreversible.

PROBLEM 2.10.– Transformation in an isolated system

In a completely isolated tank containing 50 g of water at a temperature of 40 °C, 10 g of ice at a temperature of 0 °C is added.

1) Determine the temperature of water prior to equilibrium.

2) What is the temperature at equilibrium?

3) Calculate the variation in total entropy; what conclusions can be drawn?

Data: $Cp_{water} = Cp_{ice} = 1.00$ cal/g.K and $\Delta H_f = 80.00$ cal/g.

Answers: (1) $T = 24$ °C; (2) $T_{eq} = 20$ °C; (3) $\Delta S = 0.34$ cal/g.K and see corrections.

PROBLEM 2.11.– Condensation of carbon sulfide

Consider the condensation of CS_2 vapor, deemed an ideal gas, at boiling point under a pressure of 1 atm.

1) Determine for 1 mole of sulfide, the work and amount of heat involved.

2) Determine for 1 mole of sulfide, the variation in U, H, S, F and G.

3) What is the relative error in the saturated volume of the vapor, if it is considered to be an ideal gas?

4) What is the relative error in the variation in volume if we disregard the V of the liquid, but not that of the vapor? What conclusions can be drawn?

Data: $T_{boil} = 46$ °C and $\Delta H_v^\circ = 85.0$ cal/g; $v_{specific}$ of the saturated vapor at 46 °C = 0.335 m^3/kg; density of liquid sulfur at 46 °C = 1.25.

Answers: (1) $W = 633.37$ cal/mol and $Q = -6\,460$ cal/mol; (2) $\Delta U = -5{,}826.63$ cal/mol; $\Delta S = 20.24$ cal/mol.K; $\Delta H = -\Delta H_v^\circ = 6{,}460$ cal/mol; $\Delta F = 615.10$ cal/mol and $\Delta G = 0$; (3) $\frac{\Delta V}{V} = 2.78\%$; (4) $\Delta(\Delta V) = 0.24\%$; very small.

PROBLEM 2.12.– Adiabatic expansion of a diatomic gas

1) Calculate the final volume, final temperature and the work done when expansion of 10 L of a diatomic gas, (initially at 0 °C) occurs by means of a piston, adiabatically and reversibly from 10 atm to 1 atm.

2) The same question if the expansion is sudden, with a constant P of 1 atm.

3) What is the variation in entropy of the gas during the sudden expansion?

Answers: (1) $V_f = 51.8$ L; $T_f = 141.4$ K; $W = -2{,}920$ cal; (2) $T_f = 203$ K; $V_f = 74.4$ L and $W = -1{,}561$ cal; (3) $\Delta S = 11.2$ cal/mol.K.

PROBLEM 2.13.– Reversible adiabatic expansion of an ideal gas

In a cylinder with a piston, there are n moles of ideal gas at 317 °C and 3 atm. A reversible adiabatic expansion changes the pressure of the gas to atmospheric pressure.

1) What is the final temperature following this expansion?

2) Calculate the variation in U and H for the gas. Deduce the work done by the gas.

3) Determine:

a) the initial and final volumes of the gas;

b) the number of moles in the cylinder.

4) What is the variation in entropy of the system?

Data: C_P = 10.5 J/K and C_v = 6.35 J/K.

Answers: (1) T_2 = 382.21 K; (2) ΔU # –1,320.40 J and ΔH # –2,183.33 J; W = –1,320.40 J; (3) (a) V_1 # 8.17 × 10^{-3} m^3 and V_2 # 15.87 × 10^{-3} m^3; (b) n # 0.5 mol; (4) ΔS_{syst} # 3.455 J/K.

PROBLEM 2.14.– Hydrogen expansion

Calculate the work done, the heat absorbed and the variations in entropy of the gas and total entropy, when 2 L of hydrogen (considered an ideal gas) under STP conditions expand to 4 L:

1) in a reversible isothermal manner;

2) according to the Joule expansion;

3) in a reversible isothermal manner and a constant pressure of 0.5 atm;

4) show graphically (P, V) the different work exchanged during these three expansions and relatively estimate the three values for the total variation in entropy.

Answers: (1) W = –140.6 J; Q = 140.6 J; $\Delta S\ H_2 = 0.123$ cal/K and ΔS_{tot} = 0; (2) W = 0; Q = 0; $\Delta S\ \ H_2 = 0.123$ cal/K; ΔS_{tot} = 0.123 cal/K; (3) W = –101.3 J; Q = 101.3 J; $\Delta S\ H_2 = 0.123$ cal/K and ΔS_{tot} = 0.035 cal/K; (4) see corrections.

PROBLEM 2.15.– Difference to ideal gas in reduced coordinates

At 20 °C, carbon dioxide is introduced into a container. A technical failure instantly causes the system to reach 197.5 °C. The behavior of CO_2 is satisfactorily described by equation: $V = \dfrac{RT}{P} + \text{A}$, where A is an independent constant of P and T.

1) Express the compressibility factor (z) as a function of A, P, R and T.

2) Establish the equation of $(H - H^*)$ as a function of z and express it as a function of A and P.

3) Show that $\left(H - H^*\right) = -RT_c T_r^2 \int_0^{P_r} \left(\frac{\partial z}{\partial T}\right)_{Pr} \frac{dP_r}{P_r}$.

4) Determine P_1 and P_2 using the general diagram from the chart given in the Appendix.

5) Deduce the variation in enthalpy during this incident.

Data: $C_p = 32.22 + 22.18 \times 10^{-3}T + 3.35 \times 10^{-6}T^2$ cal/mol.K; $P_c = 72.8$ atm; $T_c = 304.2$ K; A = –74.9 cm^3/mol and z = 0.69 at 373.15 K and 0.76 at 410.65 K.

Answers: (1) $z = \frac{AP}{RT} + 1$; (2) see corrections and $(H - H^*) = $ A; (3) see corrections; (4) $P_1 = 115.02$ atm and $P_2 = 159.12$ atm; (5) $H_2 - H_1 = -1.732$ kcal/mol.

PROBLEM 2.16.– Application of the first principle to ideal gases

An ideal gas contained in a cylinder with a piston, without friction, occupies a volume of 1 L at 1 atm and 25 °C. At equilibrium with the external environment at 1 atm, it is subjected to the following transformations:

– it is heated to 100 °C blocking the piston;

– it is gently heated to 100 °C allowing the piston to move freely.

1) How many moles of gas are there?

2) a) Calculate, in both cases, the amount of heat and work transferred to the gas, as well as the variation in internal energy.

b) Compare the results.

The gas is heated to 100 °C blocking the piston. Then it is left to expand slowly, releasing the piston gradually by hand, so as to accumulate all the work of the internal pressure forces until the system is in equilibrium with the external environment. This transformation can occur in two different ways: (i) by isothermal expansion at 100 °C; (ii) by adiabatic expansion without transferring any heat to the gas.

3) a) Determine the variation in internal energy of the gas, the amount of heat and work transferred.

b) Calculate the final temperature of the gas, the amount of heat and work transferred, as well as the variation in internal energy.

Data: $C_p = 7$ cal/mol.K and $C_v = 5$ cal/mol.K.

Answers: (1) $n = 4.09 \times 10^{-2}$ mol; (2) (a) (i) $\Delta Q = 15.3$ cal; $\Delta W = 0$; $\Delta U = 15.3$ cal; (ii) $\Delta W = -6.1$ cal; $\Delta Q = 21.4$ cal; $\Delta U = 15.3$ cal; (b) see corrections; (3) (a) $\Delta U = 0$; $\Delta W = -6.75$ cal; $\Delta Q = -6.75$ cal; (b) $T_f = 350.10$ K; $\Delta U = -4.71$ cal $= \Delta W$.

PROBLEM 2.17.– Cylinder with piston

A cylinder with a cross-section σ (vertical axis) is closed at the bottom by a fixed partition and at the top by a sealed piston with mass m. Initially, the piston is held by a pin which restricts the volume V_0 filled with an ideal gas in equilibrium with the external environment at P_0 and T_0.

The pin is removed, the piston descends under the effect of the gravitational field g and quickly reaches an equilibrium position without any heat exchange with the external environment: all the heat from friction is transferred to the internal gas, which then occupies a volume V at the temperature T:

1) Determine, between the initial state and final state, the expressions that give the variation in:

a) energy of the piston;

b) internal energy of the external air;

c) internal energy of the inner gas and deduce the expression of T as a function of T_0, P, $p = \dfrac{mg}{\sigma}$ and $\gamma = \dfrac{C_p}{C_v}$.

From the same initial state, a technician removes the pin, securing the piston by hand, so a state of equilibrium is reached very slowly and without any heat being exchanged with the external environment.

2) Determine the new expressions of the variation in state functions (a), (b) and (c) above, and deduce the expression of energy provided by the technician to secure the piston.

3) If during one of these operations, thermal equilibrium is re-established between interior and exterior, determine as a function of P_0, V_0 and p the variation in:

a) internal energy of the internal gas;

b) internal energy of the external environment.

Answers: (1) (a) $\Delta Q = p(V - V_0)$; (b) $\Delta U = P_0(V - V_0)$; (c) $\Delta U = (P_0 + p)(V - V_0) = nC_v(T - T_0)$ and $T = T_0\left(1 + \frac{\gamma - 1}{\gamma}\frac{p}{P_0}\right)$; (2) (a) and (b) results of (1); (c) $\Delta U = \frac{1}{\gamma - 1}.P_0V_0\left[\left(\frac{P_0 + p}{P_0}\right)^{\frac{\gamma-1}{\gamma}} - 1\right]$; $\Delta Q = (p + P_0)(V_0 - V) - \frac{1}{\gamma - 1}.P_0V_0$ $\left[\left(\frac{P_0 + p}{P_0}\right)^{\frac{\gamma-1}{\gamma}} - 1\right]$; (3) a) $\Delta U = 0$; (b) $\Delta U = \frac{p^2V_0}{P_0 + p}$.

PROBLEM 2.18.– Hollow piston cylinder

A cylinder with a cross-section σ (vertical axis), closed at both ends by fixed partitions, is split into two cavities (1) and (2), by a mobile piston. This piston comprises a sealed cylindrical casing, whose thickness and mass are negligible, and height h filled entirely with mercury. The internal gas is air, considered as an ideal gas.

At the initial state, both cavities (1) and (2) are the same height l and at the same temperature T_0 as the mercury in the piston; the pressure in cavity (1) is P and P_0 in cavity (2). Suddenly, the lower part of the piston casing is loosened, all the mercury falls to the bottom of the cylinder; the upper part of the piston is then at a height x above the mercury.

We make the following assumptions:

– all walls of the cylinder and piston casing are adiabatic;

– all heat generated from shocks or friction is transferred to the gas in (1). At equilibrium, the pressure becomes P_1 in (1) and (2); the mercury initially remains at temperature T_0.

1) After schematically representing the initial state and final state, write the expressions giving:

a) the variation in potential energy of mercury as a function of P_0, P, σ and l;

b) the variation in internal energy of gas in cavity (1) as a function of P_1, P, x, l, σ and γ;

c) the variation in internal energy of gas in cavity (2) as a function of P_1, P_0, x, l, σ and γ.

2) Show that these three expressions give rise to a relation giving P_1 as a function of P, P_0 and γ.

3) Give the expressions of T_2 and x as a function of P, P_0, T_0 and l.

4) From the following data:

$P_0(\mathrm{N/m^2})$	P(atm)	T(°C)	l(m)	R(J/mol.K)	$n_{(2)}$(mol)	γ
101,325	1.25	22.0	1	8.3143	0.4	1.40

where $n_{(2)}$ is the number of moles of air in compartment (2), determine:

a) the pressure P_1 in atm; height x and temperatures T_1 and T_2;

b) the number of moles of air $n_{(1)}$ in compartment (1).

5) Determine the variations in entropy:

a) of gas in each compartment and of mercury;

b) of the environment external to the cylinder and the universe.

The temperatures of mercury and gas in compartment (1) equalize to T_1' and the temperature in compartment (2) stabilizes to T_2'.

Assumptions: The piston moves so slowly that friction is negligible; the cylinder and piston remain perfectly adiabatic; the expansion of mercury is negligible.

6) Determine:

a) the new pressure P_2 of both sides parts of the piston; the new height x' between the mercury and the upper part of the piston; the temperature T_1';

b) the variations in entropy of gas in each cavity, and the environment external to the cylinder.

7) Knowing that mercury has a $C_p = 4376$ J/K, can the minimum difference between T_0 and T_1 be determined so that the second principle of thermodynamics is met?

Answers: (1) (a) $\Delta E_p = (P_0 - p)\sigma l$; (b) $\Delta U_1 = \frac{\sigma}{\gamma - 1}(P_1.\mathrm{x} - P.l)$;

(c) $\Delta U_2 = \frac{\sigma}{\gamma - 1}[(2P_1 - P_0)l - P_1 x]$; (2) $P_1 = \frac{1}{2}[(2-\gamma)P_0 + \gamma.P]$; $T_2 = T_0$

(3) $\left[\frac{(2-\gamma)P_0+\gamma.P}{2P_0}\right]^{\frac{\gamma-1}{\gamma}}$; $x=l\left[2-\left(\frac{2P_0}{(2-\gamma)P_0+\gamma.P}\right)^{\frac{1}{\gamma}}\right]$; (4) (a) $P_1=1.175$ atm. ; x = 1.109 m; $T_2=35.9°C$ and $T_1=34.5°C$; (b) $n_1=0.5$ moles; (5) (a) $\Delta S_{(2)}=0$; $\Delta S_{(1)}=0.86$ J/K and $\Delta S_{Hg}=0$; (b) $\Delta S_{ext}=0$; $\Delta S_{universe}=S_{created}=0.86$ J/K; (6) (a) $P_2=1.44$ atm; $x'=1.092$ m and $T_1'=295.15$ K; (b) $\Delta S_{(2)}=0$; $\Delta S_{(1)}=-0.49$ J/K and $\Delta S_{ext}=0$; (7) $T_1'-T_0\geq 3.3\times10^{-2}$.

PROBLEM 2.19.– Cylinder with a piston of negligible mass

A cylinder with a vertical axis, cross-section σ and height l, perfectly insulated is sealed: at the bottom by a fixed partition resting on a horizontal surface and its upper part by a piston of negligible thickness and mass whose segments rub against the inner surface of the cylinder, causing it to eventually move, with a force proportional to the speed of movement.

Above the piston, at height h from the bottom of the cylinder, a mass m is suspended whose center of gravity is on the axis of the cylinder filled with air, deemed an ideal gas, whose temperature T_0 is equal to the external T and the pressure is equal to the external pressure P_0. The mass is left to fall freely onto the upper end of the piston. When the equilibrium pressure state is reached, the h occupied by the air becomes x and the temperature becomes T_1.

1) Represent schematically the initial and final states of the cylinder and write the expression for the pressure P in the cylinder at equilibrium.

2) If all heat generated by friction and shocks is transferred to the internal gas, give the following expressions as a function of m, h, l, x, P_0, T_0, σ and g:

a) the variation in potential energy of mass m;

b) work received by the external environment;

c) the total variation in energy of the external environment (ΔU_{ext});

d) the variation in internal energy of internal air as a function of n and T_1;

e) show that writing the equation of state allows us to obtain the expression of this variation as a function of parameters l, x, P_0, P, σ and γ.

3) From the results in the following table:

P_0(atm)	T_0(°C)	m(kg)	h(m)	l(m)	σ(cm^2)	γ	G(m/s^2)	R(J/mol.K)
1	22	50	2	0.5	48.4	1.4	9.808	8.3143

a) Write an equation that allows the determination of x.

b) Determine P, n, x and T_1.

c) Calculate the variation in entropy of internal air, of the external environment and entropy created.

4) Assume that all the heat generated by shocks and friction is transferred to the external environment.

a) Calculate the temperature T_2 reached by the internal air at pressure equilibrium.

b) Calculate the new height x_2 and the variation in internal energy of internal air.

c) Calculate the work received by the external environment, the variation in potential energy of the mass m and heat transferred to the external environment.

d) Calculate the variations in entropy of air inside, of the external environment and entropy created.

5) Assume that after a long time, internal air, by heat loss, returns to the temperature T_0 at the initial state.

a) Determine the new height x_2 and the variation in internal energy of air inside.

b) Determine the work received by the external environment, the variation in potential energy of mass m and the heat transferred to the external environment.

c) Determine the variations in entropy of air inside, the external environment and the entropy created.

6) What conclusions can be drawn by comparing these three proposed equilibrium states?

Answers: (1) $P = P_0 + \frac{mg}{\sigma}$; (2) (a) $\Delta E_p = -mg(\mathrm{h} - \mathrm{x})$; (b) $W_{\mathrm{ext}} = -P_0\sigma(l - \mathrm{x})$; (c) $\Delta U_{ext} = -mg(\mathrm{h}-\mathrm{x}) - P_0.\sigma(l-\mathrm{x})$; (d) $\Delta U_{int} = \frac{\sigma}{\gamma-1}(P.\mathrm{x} - P_0.l)$; (e) see corrections;

(3) (a) $x = l + \left(\frac{\gamma-1}{\gamma}.h - l\right)\frac{mg/\sigma}{P_0 + mg/\sigma}$; (b) $P = 2$ atm; $n = 0.1$ mol; x = 0.536 m and $T_1 - 632.0$ K; (c) $\Delta S_{int} - nR\left(\frac{1}{\gamma-1}ln\frac{T_1}{T_0} + ln\frac{x}{l}\right) = 1.640 J/K$; $\Delta S_{ext} = 0$; $S_{created} - \Delta S_{universe}$ = 1.640 J/K; (4) (a) $T_2 = 86.65°C$; (b) $x_2 = 0.305$ m; $\Delta U_{int} = 134.3$ J; v (c) $W_{ext} = -95.6$ J; $\Delta E_p = -831.2$J; $Q_{ext} = 792.6$J; (d) $\Delta S_{int} = 0$; $\Delta S_{ext} = 2.685 J/K$ $S_{created} = \Delta S_{universe}$ =2.685 J/K; (5) (a) $x_3 = 0.25$ m; $\Delta U_{int} = 0$; (b) $W_{ext} = -122.6$ J; $\Delta E_p = -858.2$ J and $Q_{ext} = 980{,}8$ J; (c) ΔS_{int} = – 0.576 J/K; ΔS_{ext} = 3.323 J/K; $S_{created} = \Delta S_{universe}$ = 2.747J/K.; (6) We find that: $S_{created}(5) > \begin{cases} S_{created}(3) \\ S_{created}(4) \end{cases}$; in fact:

$$2.747 > \begin{cases} 1.640 \\ 2.685 \end{cases}.$$

CONCLUSION.– It could have arrived at this final state by the irreversible transformation described in (2), followed by another irreversible transformation, which is accompanied by creation of entropy; the same step is followed during the transformation (4).

PROBLEM 2.20.– Horizontal cylinder supplied with heating resistance

A very thin, mobile vertical and completely waterproof piston divides a cylinder with a horizontal axis length 0.70 m into two compartments (A) and (B) of the same volume.

In (A), there is heat resistance in negligible volume and (B) has a valve (r), which can be linked to the external environment at constant P_0.

All the sides and the piston are well-insulated, compartment (A) contains 1 g of hydrogen at 0 °C, the valve is open and compartment (B) is maintained at a constant temperature of 0 °C. This brings (A) to a temperature of 100 °C.

1) Represent this device in the form of a diagram.

2) Calculate the distance d between the piston equilibrium position and the bottom of the compartment (A).

3) Calculate the mass of air released into the external environment via the valve, knowing that P_0 = 1 atm.

At the end of the experiment, the valve is closed (r) and compartment (A) returns to its initial temperature of 0 °C.

4) Calculate the distance d' between the piston equilibrium position and the bottom of (A).

Data: Air and hydrogen are considered ideal gases; atomic mass: $H = 1$ g; volumic mass of air under STP conditions. = 1.293 g/L; normal volume of 1 mole of gas = 22.4 L.

Answers: (1) see corrections; (2) d = 47.8 cm; (3) m = 5.3 g; (4) d' = 42.8 cm.

PROBLEM 2.21.– Piston-mounted pump

Consider a system σ (1 mole of ideal gas) in state (A): $P_A = 1$ atm and $T_A = 409.50$ K, placed in a pump mounted onto a piston fixed by two stoppers b_1 and b_2. The whole thing is placed into a thermostatically controlled bath (source of heat σ_1) at 546 K until thermal equilibrium is reached. The system σ reaches state (B).

1) a) Schematically represent this device.

b) Calculate the volume V_A and the pressure P_B.

2) Calculate the heat Q_1, exchanged by σ with the external environment (disregarding the thermal influence on the pump, piston and stoppers).

3) a) Calculate the variations $(\Delta S_\sigma)_1$ and $(\Delta S_{\sigma 1})_1$.

b) Deduce the nature of the transformation A-B, physically justifying the answer.

With the gas in state B, the piston is subject to several overloadings, at once, at temperature T_B, the new equilibrium state (C) reaches a pressure of $P_C = 4$ atm and he stoppers pay no role in this.

4) a) Represent the new device schematically.

b) Calculate V_C and W_2, work exchanged by σ and the external environment during transformation B-C; and deduce Q_2.

c) Calculate the variations $(\Delta S_\sigma)_2$ and $(\Delta S_{\sigma 1})_2$ during this transformation.

d) Deduce the nature of the transformation B-C, physically justifying the answer.

5) Using an appropriate device, source σ_1 in the same state (C) is thermally isolated from σ; then, one by one, the small loads are placed on the piston, to instigate a reversible expansion C-D. The system σ reaches a new equilibrium state (D) at pressure $P_D = 1$ atm.

a) Calculate the volume V_D and temperature T_D.

b) Calculate W_3, the work exchanged during this expansion.

c) What are the values of $(\Delta S_\sigma)_3$ and $(\Delta S_{\sigma 1})_3$?

d) What would happen if the overloading was removed at once?

During transformation C-D, the bath σ_1 is cooled to temperature $T_A = 409.5$ K and the system σ is thermally linked to σ_1, with the gas changing from state (D) to its initial state (A).

6) a) Calculate W_4 and Q_4, work and heat exchanged during transformation D-A.

b) Calculate the variations $(\Delta S_\sigma)_4$ and $(\Delta S_{\sigma 1})_4$.

c) Is this phenomenon possible? Explain why.

7) a) Complete a table with the work, amount of heat and variations in entropy of the system σ for the four transformations.

b) Calculate the values for total W and total Q and show that U is a function of state.

c) Show that entropy is also a function of state.

8) Assume that the closed transformation subjected to σ is reversible during the successive stages: A–B (isochoric), B–C (isothermal), C–D (adiabatic) and D–A (isobaric).

a) Represent on the diagram (P, V) in Figure 2.2, cycle A–B–C–D–A taken by the gas, knowing that adiabatic C–D passes through points E, F and G of the diagram.

b) Represent on the same diagram (T, S), the corresponding cycle A'–B'–C'–D'–A' (the origin of entropy is arbitrary, 1 cal/K = 4 cm on axis S).

c) What do you think of the slopes of curves A'B' and D'A'?

9) a) What do the absolute values of the area defined by these two cycles represent as a function of the algebraic values of W_{rev} and Q_{rev}, amount of work and heat exchanged during the reversible cycle?

b) With the choice of unit on the axes, the same amount of energy, expressed in the same units, corresponds to the same area in diagrams (P, V) and (T, S). What is the absolute value of the ratio of this area? Explain why and verify this.

c) Highlight the area that determines the total value of W to be calculated in (7) (b) and try to once again determine it from the graph.

Data: $C_p = 29.12$ J/mol.K and $C_v = 20.80$ J/mol.K.

Answers: (1) (a) see corrections; (b) $V_A = 33.58$ L and $P_B = 1.33$ atm; (2) $Q_1 = 2.839$ kJ; (3) (a) $(\Delta S_\sigma)_1 = 5.984$ J/mol.K; $(\Delta S_{\sigma 1})_1 = -5.20$ J/mol.K; (b) irreversible; (4) (a) see corrections; (b) $V_C = 11.19$ L; $W_2 = 9{,}074.5$ J and $Q_2 = -9{,}074.5$ J; (c) $(\Delta S_\sigma)_2 = -9.131$ J/K and $(\Delta S_{\sigma 1})_2 = 16.62$ J/K; (d) irreversible; (5) (a) $V_D = 30.12$ L and $T_D = 367.30$ K; (b) $W_3 = -3{,}716.9$ J; (c) $(\Delta S_\sigma)_3 = (\Delta S_{\sigma 1})_3 = 0$; (6) (a) $W_4 = -350.58$ J and $Q_4 = 1{,}228.86$ J; (b) $(\Delta S_\sigma)_4 = 3.167$ J/K; $(\Delta S_{\sigma 1})_4 = -3.001$ J/K; (c) yes since $\Delta S' > 0$; (7) (a) see corrections; (b) $W_{tot.} = 5{,}007.02$ J and $Q_{tot.} = -5{,}006.64$ J; $\Delta U = 0.38$ J # 0; (c) $\Delta S_\sigma = 0.02$ J # 0; (8) the representation of the cycle by calculating intermediate points for each curve; (9) (a) the areas of the cycles are equal; (b) ratio ≈ 25; (c) see corrections.

2.3. Tests

TEST 2.1.– Variation in entropy

Part I – The molar heat capacity at constant pressure of a solid is given by the expression: $C_P = 5.0 + 0.5 \times 10^{-3}T$ cal/mol.K, valid between 300 and 1,000 K.

Calculate the variation in entropy of this solid when it is heated from 400 to 800 K.

Part II – A system reversibly moves from state (A) to state (B), exchanging Q = a + bT amount of heat with the external environment.

1) What is the variation in entropy of the system?

2) What is the variation entropy of the external environment?

TEST 2.2.– Variations in internal energy, enthalpy and entropy

Find the expressions of W, Q, ΔU, ΔH and ΔS in a reversible isothermal expansion leading gas from state A to state B, in the case where the gas obeys:

1) the Van der Waals equation;

2) the equation $PV = RT + bP$.

TEST 2.3.– Reversible isothermal expansion

Nitrogen, deemed a real gas, has the equation of state: $(P + \frac{a}{v^2})(v - b) = RT$ in which a and b are constants.

One mole of N_2, initially at 10 atm and 273.15 K, undergoes a reversible isothermal expansion with the final pressure being 1 atm.

1) Express the calorimetric coefficient l and the amount of heat δQ_{rev} in the case of isothermal expansion.

2) Calculate δQ_{rev} using the values of initial and final volumes that would be occupied, under the same conditions of T and P, a mole of N_2, deemed an ideal gas.

TEST 2.4.– Volumetric piston compressor

A volumetric piston compressor, whether an industrial machine or a bike pump, is always formed on the same principle: a mobile piston, in a cylinder with two valves, with access to atmospheric air at the pressure P_1 via valve S.

On return it is compressed to a pressure P_2, with both valves closed. Air occupies a volume V_2 at temperature T_2. Then, valve S' is opened and the piston retains the air at constant P_2 in reservoir R, as shown in Figure 2.1.

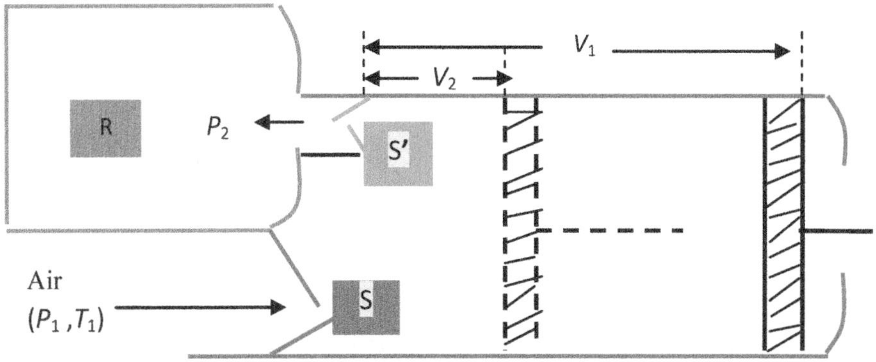

Figure 2.1. *Volumic compressor*

Air is heated during compression. If the heat created is released, the compression is isothermal; if the heat exchanged with the walls is null, it is adiabatic; in practice, compression is always between the two.

The purpose of compression is to change the pressure of air, at T_1 from P_1 to P_2. It is therefore useful to know what type of compression is most economical from an energy point of view.

At first instance, the work exchanged at suction and compression is disregarded and the compression is assumed to be reversible from P_1 to P_2.

1) From the expression $\delta W_{rev.} = -P.dV$, find the expressions of compression work of an ideal gas according to whether it is:

a) isothermal ($P.V = P_1V_1 = P_2V_2 =$ Cte);

b) adiabatic ($PV^\gamma = P_1V_1^\gamma = P_2V_2^\gamma =$ Cte);

c) polytropic ($PV^x = P_1V_1^x = P_2V_2^x =$ Cte; $1 < x < \gamma$); γ and x are deemed constant in the T domains studied, the expressions of $W_{rev.}$ are given as a function of P_1, P_2, V_1, V_2, γ and x.

2) From the expressions $\delta Q_{rev.} = C_v dT + PdV$ and $PV^\gamma =$ Cte, show that for polytropic compression of an ideal gas:

a) δQ_{rev} can be written in the form: $\delta Q_{rev} = \frac{\gamma - x}{\gamma - 1} P_1 . V_1^x \frac{dV}{V^x}$;

b) dS can be written as: $dS = \frac{\gamma - x}{\gamma - 1} P_1 . V_1^x \frac{dV}{V^x}$;

c) give the expression of $Q_{rev.}$ and ΔS as a function of P_1, P_2, V_1, V_2, T_1, γ and x.

3) If $P_1 = 1$ atm, $P_2 = 10$ atm, $V_1 = 20$ L, $T_1 = 273$ K, $\gamma = 1.40$ and $x = 1.30$, calculate the following and tabulate the results:

a) the volume final V_2 and final temperature T_2; reversible compression work $W_{rev.}$ and the amount of heat $Q_{rev.}$;

b) the variation in entropy of the gas (ΔS).

4) On a graph (P, V), plot:

a) the isothermal curve AB: A(V_1, P_1), B(V, P_2);

b) the adiabatic curve AC: A(V_1, P_1), C(V, P_2);

c) the polytropic curve AD: A(V_1, P_1), D(V, P_2).

5) Draw the lines representing the work $W_{rev.}$ in all three cases?

In reality, the work of transfer must also be considered (entry and compression).

It is easy to show that this work is equal to $(P_2V_2 - P_1V_1)$, whereby at each turn, the compressor provides the gas with work W_m such that: $W_m = W_{rev.} + P_2V_2 - P_1V_1$, i.e., $\delta W_m = \delta W_{rev} + d(PV)$.

6) Assuming the vacuum ($P = 0$) is left of the piston, show that the transfer work equals $(P_2V_2 - P_1V_1)$.

7) Find the expression of W_m and calculate its value for the three cases above.

8) In all three cases, what should be the power required for the compressor to compress 10 m^3 of air under STP conditions per minute, up to $P_2 = 10$ atm?

9) On the graph (P, V), draw the lines representing the work W_m? Where is the minimum and maximum?

In practice, there are compressors at two stages: air at T_1 and P_1 is compressed to a pressure P_1', between P_1 and P_2, under which it is cooled to T_1 and then compressed from P_1' to P_2 before being fed back into the reservoir R under this pressure P_2.

10) a) Assuming both compressions, from P_1 to P_1' and P_1' to P_2, adiabatic and reversible, draw on the graph (P, V) the area that represents the work W_m transferred to the air.

b) What qualitative conclusions can be drawn regarding the pressure of a compressor operating adiabatically, depending on whether it has one or two stages?

11) Express W_m as a function of γ, V_1, P_1, P_1' and P_2.

a) If P_1 and P_2 are fixed, what relation should exist between P_1, P_1' and P_2 for W_m to be minimum?

b) Calculate the value of W_m in this case.

c) Using the same values as above, what is the power of the compressor?

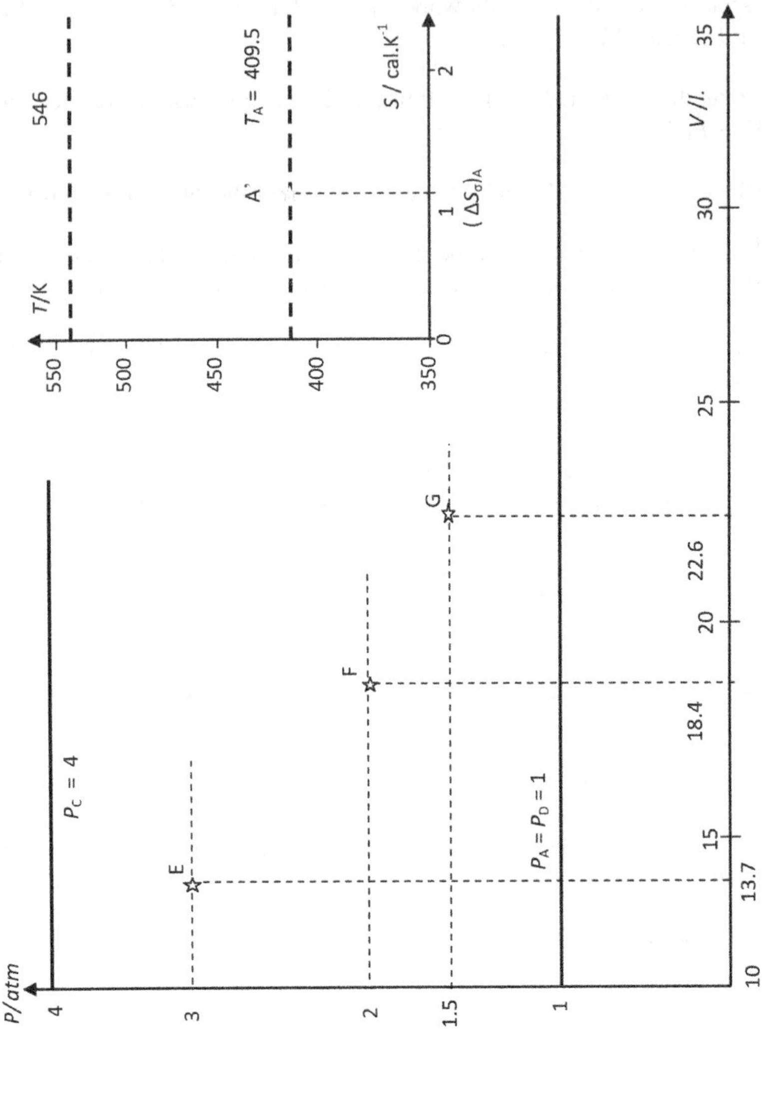

Figure 2.2. *Diagrams (P, V) and (T, S) "pump with piston"*

2.4. Detailed corrections

EXERCISE 2.1.– The housewife in the kitchen

The energy of the system is as follows:

	Increases	Decreases	Remains constant
Between t_0 and t_1			X
Between t_1 and t_2			X
Between t_2 and t_3	X		

EXERCISE 2.2.– Work against the external environment

The work done against the external environment is written as: $\delta W = -P_{ext}dV$ where P_{ext} is atmospheric pressure, against which heating occurs, and dV the variation in volume of mass of aluminum during operation, due to the variation in volumic mass with T, whereby:

$$W = -P_{ext}\Delta V = -P_{ext}(V_{660} - V_{20}) = -P_{ext}\left(\frac{\text{m}}{\rho_{660}} - \frac{\text{m}}{\rho_{20}}\right)$$

$$= -1.01312\times10^5\left(\frac{10}{2{,}700} - \frac{10}{2{,}380}\right) = -50 \text{ J}$$

EXERCISE 2.3.– Heat of reaction and internal energy

1) The formation reaction of alumina is as follows: $2Al_{2s} + \frac{3}{2}O_{2g} \rightleftharpoons Al_2O_{3s} + Q_p^{\uparrow}$. To calculate the variation in U, we use the function H, since at constant P and for a chemical reaction: $\Delta H = Q_p = \Delta U + P\Delta V$ thus $\Delta U = Q_p - P\Delta V$, where ΔV is the variation in volume during the reaction. Solids have a negligible volume compared to gases, whose variation is due to the consumption of 3/2 mol of oxygen. The reaction occurs under STP conditions, we obtain: $\Delta V = -22.4 \times 10^{-3}\times\frac{3}{2} = -33.6 \times 10^{-3}$ m^3, i.e. $\Delta U = -4 \times 10^5 \times 4.18 + 10.1325 \times 10^4 \times 33.6 \times 10^{-3} = 1.672 \times 10^6 + 3.4 \times 10^3 = 1.669 \times 10^3$ kJ.

2) In this case, we find that $\Delta U \# \Delta H \# -1.67 \times 10^3$ kJ.

EXERCISE 2.4.– Differential expression of U as a function of T, P and C_V.

1) The variation in $U(T,V)$ is given by its total differential:

$$dU = \left(\frac{\partial U}{\partial T}\right)_V dT + \left(\frac{\partial U}{\partial T}\right)_T dV$$

2) With $H = U+PV$ and then $U = H - PV$, i.e. $dU = dH - PdV - VdP = dG - TdS - SdT - PdV - VdP$; we can express the partial derivatives of $U(T, V)$ as follows:

$$\left(\frac{\partial U}{\partial T}\right)_V = \left(\frac{\partial H}{\partial T}\right)_V - V\left(\frac{\partial P}{\partial T}\right)_V = \left(\frac{\partial G}{\partial T}\right)_V + \mathrm{T}\left(\frac{\partial S}{\partial T}\right)_V + S - V\left(\frac{\partial P}{\partial T}\right)_V \text{ or } \left(\frac{\partial S}{\partial T}\right)_V = \frac{C_V}{T}$$

hence: $\left(\frac{\partial U}{\partial T}\right)_V = -S + V\left(\frac{\partial P}{\partial T}\right)_V + C_v + S - V\left(\frac{\partial P}{\partial T}\right)_V$ i.e. $\left(\frac{\partial U}{\partial T}\right)_V = C_V$.

$$\left(\frac{\partial U}{\partial V}\right)_T = \left(\frac{\partial H}{\partial V}\right)_T - P - V\left(\frac{\partial P}{\partial V}\right)_T = \left(\frac{\partial G}{\partial V}\right)_T + T\left(\frac{\partial S}{\partial V}\right)_T + T\left(\frac{\partial T}{\partial V}\right)_T - P - V\left(\frac{\partial P}{\partial V}\right)_T =$$
$$V\left(\frac{\partial P}{\partial V}\right)_T + T\left(\frac{\partial P}{\partial T}\right)_V + 0 - P - V\left(\frac{\partial P}{\partial V}\right)_T \text{ i.e. } \left(\frac{\partial U}{\partial V}\right)_T = T\left(\frac{\partial P}{\partial T}\right)_V - P$$

and finally: $dU = C_v dT + \left[T\left(\frac{\partial P}{\partial T}\right)_V - P\right]dV$.

EXERCISE 2.5.– Internal energy and temperature

Part I – (1) So, $\Delta U = Q+W$, disregarding the expansion $(\Delta V = 0)$ therefore $W = 0$, we obtain $\Delta U = Q$. For 1 L of gas, $Q = 10^4$ cal for 3 L, we obtain $\Delta U = 3 \times 10^4$cal.

2) 1 L of water weighs 1 kg with a molar mass of 18 g, this gives:

$$\Delta U = 18\frac{10^4}{10^3} = 180 \text{ cal}$$

Part II – (1) For this transformation, $dU = \delta Q_{\mathrm{rev}} + \delta W_{\mathrm{rev}}$ and according to the expression of the calorimetric coefficients δQ_{rev} is written as: $\delta Q_{rev} = C_v dT + l dV$ with by definition $l = T\left(\frac{\partial P}{\partial T}\right)_v$ and $\delta W_{rév} = -PdV$, i.e. $dU = C_v dT + (l - P)dV$. For an ideal gas, $l = \frac{RT}{V} = P$ therefore $dU = C_V dT$, which shows that for an ideal gas, (U) only depends on T, i.e. $\Delta U = C_v \Delta T = C_v(T_f - T_i)$; we can write $\Delta U = U_f - U_i$

that shows that the variation in (U) of a system depends only on the initial and final states. The calculation gives: $\Delta U = 3(546 - 273) = 819$ cal/mol.

2) This is a process: a reversible isothermal transformation (1) followed by a reversible isochoric transformation (2), for which $dU = dU_1 + dU_2$, or $dU_1 = 0$ and $dU_2 = C_v dT + (l - P)dV = C_v dT$ (since $dV = 0$), i.e. $dU = C_v dT$.

3) This is the same result as for (1) and hence U is most certainly a function of state.

EXERCISE 2.6.– Heat of combustion

At constant pressure, the heat involved is given by:

$$\delta Q = nC_P dT$$

so:

$$\Delta Q = \int_{T_1}^{T_2} C_P dT = \int_{300}^{1500} (6.9469 - 0.1999\times10^{-3} T + 4.808\times10^{-7} T^2)\mathrm{d}T =$$

$$\left[\frac{T^2}{2} + 4.808\times10^{-7}\frac{T^3}{2}\right]_{300}^{1,500} = 8.336.28 - 218.884 + 539.23 = 8,659.72 \text{ cal}$$

This is the energy required to heat 1 mole of hydrogen at constant pressure from 300 K to 1,500 K.

EXERCISE 2.7.– Enthalpy

1) By definition, at constant pressure, $dH = \delta Q_P = nC_P dT$ so $\Delta H = Q_P = \int_{T_1}^{T_2} C_P dT$ whereby:

$$\Delta H = C_p \Delta T = \left[7.17 + \frac{2.56}{2}10^{-3}T^2 - 8.0\times10^3 T^{-1}\right]_{400}^{900} = 7.17\,(900 - 400) +$$

$$1.28 \times 10^{-3}(900^2 - 400^2) + 8 \times 10^3\left(\frac{1}{400} - \frac{1}{900}\right) = 3,585 + 832 +$$

$$11 = 4,428 \text{ kcal}$$

2) We define the pressure, as C_p, since without the assumption that water vapor is an ideal gas, we would have the following expression:

$$\left(\frac{\partial C_p}{\partial P}\right)_T = -T\left(\frac{\partial^2 V}{\partial T^2}\right)_P$$

this means that $C_p = f(P, T, V) = f'(P)$.

EXERCISE 2.8.– Entropy

By definition, entropy $dS = \frac{\delta Q}{T}$ and for 1 mole, $\delta Q = C_p dT$, we obtain:

$$\Delta S = \int_{T_i}^{T_f} C_p \frac{dT}{T} = \int_{300}^{500} \frac{6.20 + 1.3\times10^{-3}T - 6\times10^4 T^{-2}}{T} dT$$

$$= \int_{300}^{500} \frac{6.2}{T} dT + \int_{300}^{500} 1.30\times10^{-3} dT - \int_{300}^{500} 6\times10^4 \frac{dT}{T^3}$$

$$= 6.2\, ln\frac{500}{300} + 1.3\times10^{-3}(500-300) + \frac{6\times10^4}{2}\left(\frac{1}{500^2} - \frac{1}{300^2}\right)$$

$$= 3.167 + 0.26 - 0.21 = 3.214 \text{ cal/mol.K}$$

EXERCISE 2.9.– Free enthalpy or Gibbs free energy

Part I – The transformation $H_2O_{ice} \rightarrow H_2O_{liquid}$ at 1 atm has a variation in free enthalpy such that: at –10 °C, $\Delta G > 0$; at 0 °C, $\Delta G = 0$ and at 10 °C, $\Delta G < 0$; so at –10 °C, it is impossible; at 0 °C, it is the ice $\rightleftarrows$ water equilibrium and at 10 °C the transformation occurs.

Part II – It involves studying the transformation: H_2O_{liq} (supercooled at –10 °C) $\rightarrow$ H_2O_{ice} (at –10 °C). For this, it must decompose via a reversible mechanism, for which thermodynamic calculations are possible, as follows:

$$\underset{1}{H_2O_{liq}(\text{at } -10°C)} \rightarrow \underset{2}{H_2O_{liq}(\text{at } 0°C)} \rightarrow \underset{3}{H_2O_{ice}(\text{at } 0°C)} \rightarrow \underset{4}{H_2O_{ice}(\text{at } -10°C)}$$

By definition of G, $\Delta G = \Delta H - T\Delta S$, with $\Delta H = \sum \Delta H_i$ which can be easily calculated from the expressions:

$$\Delta H_{(1,2)} = \int C_P . dT = Cp\Delta T = 18(273.15 - 263.15) = 180 \text{ cal}$$

$$\Delta H_{(2,3)} = -\Delta H_{(3,2)} = -\Delta H_f = -1{,}438 \text{ cal}$$

$$\Delta H_{(3,4)} = \int C_P dT = C_p\Delta T = 9(263.15 - 273.15) = -90 \text{ cal}$$

which gives $\Delta H = 180 - 1{,}438 - 90 = -1{,}348$ cal.

The variation in entropy is obtained by writing: $\Delta S = C_{p(1,2)}\int_{T_1}^{T_f}\frac{dT}{T} + \frac{\Delta H_{solid.}}{T_f} + C_{p(3,4)}\int_{T_f}^{T_1}\frac{dT}{T}$ with $\Delta H_{solid.} = -\Delta H_f = -1{,}438$ cal, whereby: $\Delta S = 18ln\frac{273.15}{263.15} - \frac{1438}{273.15} + 9ln\frac{263.15}{273.15} = 9ln\frac{273.15}{263.15} - 5.265 = 0.336 - 5.265 = -4.929$ cal/K, which gives: $\Delta G = -1{,}348 + 263.15 \times 4.929 = -50.99 \# -51$ cal.

For this transformation: $\frac{\partial}{\partial T}\left(\frac{\Delta G}{T}\right) = -\frac{\Delta H}{T^2}$ with: $\Delta H = C_{p(1,2)}\int_T^{T_f} dT + \Delta H_{sol} + C_{p(3,4)}\int_{T_f}^{T} dT = 18(273.15 - T) - 1{,}438 + 9(T - 273.15) = 1{,}020.35 - 9T$, whereby $\frac{\partial}{\partial T}\left(\frac{\Delta G}{T}\right) = \frac{9}{T} - \frac{1{,}020.35}{T^2}$, i.e. by integration: $\frac{\Delta G}{T} = \int\left(\frac{9}{T} - \frac{1{,}020.35}{T^2}\right)dT = 9\int\frac{dT}{T} - 1{,}020.35\int\frac{dT}{T^2} = 9\ln T + \frac{1{,}020.35}{T} + C$ thus $\Delta G = 9T\ln T + C.T + 1{,}020.35$.

So at $T = 273.15$ K, this is the ice/water equilibrium, i.e. $\Delta G = 0$ whereby: $273.15C = -9 \times 273.15.ln273.15 - 1{,}020.35$ and so $C = -54.23$ which gives: $\Delta G = 9T.lnT - 54.23T + 1{,}020.35$ and by replacing T with 263.15 K, we obtain $\Delta G = -52.11$ cal # -52 cal.

Exercise 2.10.– Isothermal transformations

By convention, for any system, the energy received is positive and the energy lost is considered to be negative. We obtain:

$$\Delta U_1 = Q + W = Q = -100 \ cal$$

$$\Delta U_2 = Q + W = W = -P_{ext}\int_{V_1}^{V_2} dV = -\ 4\text{L.atm}$$

i.e. $R = 8.32$ J./mol.K. $= \frac{22.4}{273.15}$ L.atm./K and so:

$$1\text{L.atm.} = 8.32\frac{273.15}{22.4} = 101.456\text{J}$$

whereby $\Delta U_2 = -4\dfrac{101.456}{4.184} = -97$ cal.

$$W = \mathcal{R}.I^2t = 10\times1\times300 = 3\times10^3\,\text{J.} = 717\text{cal} \text{ and } \Delta U_3 = Q + W = 717 \text{ cal}$$

$$W = \text{mgh} = 2\times9.81\times15 = 294.3\text{ J}, \text{ i.e. } \int_{T}^{T} C_P\frac{dT}{T} - \frac{\Delta H_f}{273.15}$$

whereby the variation in internal energy overall the system is:

$$\Delta U = \sum_i \Delta U_i = -100 - 97 + 717 + 70 = 590 \text{ cal}$$

EXERCISE 2.11.– Reversible transformation

1) The second principle is written as: $dQ_{rev} = TdS$ whereby the relation:

$$\Delta S_c = \int_{T_i}^{T_f}\frac{dQ}{T} = \int_{T_i}^{T_f}\frac{\text{b}dT}{T} = \text{b}ln\frac{T_f}{T_i}$$

2) We obtain $\Delta S = \Delta S_\sigma + \Delta S_{\sigma'}$, i.e. the transformation is reversible, and hence $\Delta S = 0$ so $\Delta S_\sigma = -\Delta S_{\sigma'}$.

EXERCISE 2.12.– Irreversibility of heating

1) For water: $\Delta S_1 = (S_f - S_i) = \int_{T_1}^{T_2}\frac{q_1}{T} = \int_{T_1}^{T_2}\text{m}C_p\frac{dT}{T} = \text{m}C_p(lnT_2 - lnT_1)$ (it is positive).

For the enclosure: $\Delta S_2 = (S_f - S_i) = \dfrac{q_2}{T_2}$ and application of the first principle overall (water + enclosure) assumed to be insulated, gives: $q_1 + q_2 = 0$ whereby $q_2 = -q_1 = -\text{m}C_p\Delta T$, i.e. $\Delta S_2 = \text{m}C_p\left(\dfrac{T_2 - T_1}{T_2}\right) = -\text{m}C_p\left(1 - \dfrac{T_1}{T_2}\right)$ (it is negative).

2) The creation of entropy in the universe is $\Delta S = \Delta S_1 + \Delta S_2$ giving the following expression: $\Delta S = mC_p\left[ln\right]\frac{T_2}{T_1} - \left(\frac{T_2 - T_1}{T_2}\right)$.

APPLICATION.–

1) $\Delta S_1 = 100 ln\frac{353}{293} = 18.63$ cal/K; $\Delta S_2 = -100\frac{60}{353} = -17.00$ cal/K.

2) $\Delta S = 18.63 - 17.00 = 1.63$ cal/K.

EXERCISE 2.13.– Heat of Vaporization of water

1) The latent heat of vaporization represents the amount of heat that needs to be transferred to a mass of substance in order to vaporize it, whereby: $Q = 540 \times 1{,}000 = 540$ kcal.

2) The work exchanged with the external environment is that against the pressure forces, i.e. $W = -P_{ext}(V_v - V_i)$, where V_1 and V_v are the volumes of 1 kg of water, in its liquid state and in its vapor state at 100 °C, respectively, meaning $V_v >> V_i$, i.e. $W \# -P_{ext}V_v$. The vapor is considered to be an ideal gas and at 1 atm, we obtain:

$$W = -nRT = -\frac{1\,000}{18}\frac{22.4}{273.15}373.15 = 1700.06\ l.\text{atm} = -41{,}14 \text{ kcal.}$$

3) The variation in U is therefore: $\Delta U = W + Q = -41.14 + 540 = 498.86$ kcal.

EXERCISE 2.14.– Vaporization of ice

This involves the transformation $H_2O_s \rightarrow H_2O_{vap}$ between 0 °C and 200 °C at constant P. In this case, ΔH of transformation is ΔQ_P, whereby the calculation for the variation in enthalpy is:

1) $$\Delta H = \Delta H_f + \int_{273.13}^{373.15} C_{p,l}dT + \Delta H_v + \int_{373.15}^{473.15} C_{p,v}dT$$

$$= \Delta H_f + C_{p,l}\Delta T + \Delta H_v + C_{p,vap}\Delta T'$$

$$= (1{,}440 + 18\times100) + (9{,}700 + 6\times100) = 13{,}540 \text{ cal;}$$

2) $$\Delta S = \Delta S_{273.15}^{373.15} + \Delta S_{373.15}^{473.15} = \frac{\Delta H_f}{T_f} + \frac{\Delta H_v}{T_v} + \int_{273.15}^{373.15} \frac{C_{p,l}}{T} dT + \int_{373.15}^{473.15} \frac{C_{p,v}}{T} dT$$

$$= 5.272 + 18 ln\frac{373.15}{273.15} + 25.995 + 6 ln\frac{473.15}{373.15} = 32.262 \text{ cal/mol.K}$$

EXERCISE 2.15.– Mixture at different temperatures

We have not obtain the following values: m = 100 g, $T_1 = 293$ K, $T_2 = 353$ K and T_f temperature when the mixture is at equilibrium. It can be calculated by applying the equation from the first principle (with $W = 0$ and $Q = 0$).

1) $\Delta U = W + Q = 2mC_pT_f - mC_p(T_1 + T_2) = 0 \Rightarrow T_f = \frac{T_1 + T_2}{2} = 323$ K, i.e. 50 °C; we calculate the variation in S for each mass of water: $\Delta S_1 = \int_{T_1}^{T_f} mC_p \frac{dT}{T} = mC_p ln\frac{T_f}{T_1}$ (it is > 0) and $\Delta S_2 = \int_{T_2}^{T_f} mC_p \frac{dT}{T} = mC_p ln\frac{T_f}{T_2}$ (it is > 0).

The variation in entropy of the universe is:

$$\Delta s_{univ} = \Delta s_1 + \Delta s_2 = mC_p ln\frac{T_f^2}{T_1T_2} = 0.866 \text{ cal/K}$$

2) The thermal machine produces work W, with the equation of the first principle becoming $\Delta U = W = 2mC_pT_f' = mC_p(T_1 + T_2)$, where T_f' is the new final temperature. It is clear that: $\Delta S' = mC_p ln\frac{T_f'^2}{T_1T_2} < \Delta S$ (since W is < 0), i.e. $T_f' < T_f$.

So for W produced by the machine to be maximum, $\Delta S'$ must be minimum: according to the second principle, it is only possible if $\Delta S' = 0$ and so $T_f'^2 = T_1T_2$ i.e. $T_f' = \sqrt{293 \times 353} = 321.6$ K = 48.6 °C.

3) $W = mC_p[2\sqrt{T_1T_2} - (T_1 + T_2)] = -280$ cal = –1,171.5 J.

EXERCISE 2.16.– Adiabatic expansion of nitrogen

The process is adiabatic and irreversible, there is therefore a creation of entropy such that: $\Delta S_{cre} = \Delta S_{gas}$ (since $\Delta S_{ext} = 0$) and so $\Delta S_{cre} = \frac{\delta Q_{rev}}{T}$. For this calculation, we need to know the T_f reached by the gas, and therefore we write:

– for an ideal gas, $dU = nC_V dT$;

– for an adiabatic expansion, $dU = \delta W_{irrev.,}$ with $\delta W_{irrev} = -P_{ext}dV$ whereby the equality $nC_v(T_f - T_i) = -P_f(V_f - V_i)$. So $-P_fV_f + P_fV_i = nC_v(T_f - T_i)$ with $P_fV_f = nRT_f$ and $P_iV_i = nRT_i$ or $P_fV_i = P_iV_i\frac{P_f}{P_i} = nRT_i\frac{P_f}{P_i} \Rightarrow nRT_i\frac{P_f}{P_i} - nRT_f = nC_v(T_f - T_i)$, i.e. $T_f(C_v + R) = T_i\ (C_v + \frac{P_f}{P_i})$. So $T_f = T_i\frac{C_v + R\frac{P_f}{P_i}}{C_v + R}$ with $\frac{R}{C_v} = \gamma - 1$, we obtain $T_f = T_i\frac{1+(\gamma-1)\frac{P_f}{P_i}}{\gamma} = 473.15\,\frac{1+0.4\frac{1}{100}}{1.4} = 339.32$ K.

We use the typical relation $dQ_{rev} = nC_p dT - VdP$ with $PV = nRT$, and so $V = nR\frac{T}{P}$, whereby $\Delta S_{gas} = \int_{T_i}^{T_f}\frac{dQ_{rev}}{T} = nC_P\int_{T_i}^{T_f}\frac{dT}{T} - nRT\int_{P_i}^{P_f}\frac{dP}{P}$.

i.e.

$$\Delta S_{gas} = nC_p ln\frac{T_f}{T_i} - nRln\frac{P_f}{T_i} = \frac{P_iV_i}{T_i}\left(\frac{C_p}{R}ln\frac{T_f}{T_i} - ln\frac{P_f}{T_i}\right)$$

$$= \frac{100\times 20}{473.15}\left(\frac{7\times 273.15}{22.4}ln\frac{339.32}{473.15} - \frac{101.325}{4.184}ln\frac{1}{100}\right)$$

$$= -19.936 + 471.413 = 351.45 \text{ u.e.}$$

EXERCISE 2.17.– Case of three isolated bodies

This involves an isolated system where $\Delta U = 0$ (first principle); so, maintain a large temperature difference between some sources, it is necessary to avoid creating entropy and for this to be optimum, the system must conserve entropy (second principle), i.e. $\Delta S = 0$, which gives $\Delta U = \Delta U_1 + \Delta U_2 + \Delta U_3 = 0$ and $\Delta S = \Delta S_1 + \Delta S_2 + \Delta S_3 = 0$.

Consider the following transformations:

– for body 1, $T_1 \rightarrow T_1'$;

– for body 2, $T_2 \rightarrow T_2'$;

– for body 3, $T_3 \rightarrow T_3'$.

Therefore, $\Delta U = mC_p[(T_1' - T_1) + (T_2' - T_2) + (T_3' - T_3)] = 0$ and $\Delta S = mC_p(ln\frac{T_1'}{T_1} + ln\frac{T_2'}{T_2} + ln\frac{T_3'}{T_3}) = 0$, i.e. $\left\{T_1 + T_2 + T_3 = T_1' + T_2' + T_3'\right\}$ and $\left\{T_1 T_2 T_3 = T_1' T_2' T_3'\right\}$. We want one of the temperatures to be minimal, T_1' for example. This suggests that $T_2' = T_3'$ (otherwise we could consider a thermal machine operating between these two sources and whose work would serve to lower T_1'). Therefore, the system will be described by T_1' and $T_2' = T_3'$ with $T_1' < T_2'$, so by writing: $T_1 = T_2 = T$ and $2T + T_3 = 2x + T_1'$, we obtain: $T_2' = T_3' = x$ and $T^2 T_3 = x^2 T_1'$; eliminating T_1' between these two relations, we obtain: $2x^3 - (2T + T_3)x^2 - T^2 T_3 = 0$; this equation describes a system where two sources are the same temperature, the third being different; it is clear that the initial state verifies this equation, i.e. x = T is a solution:

$$2x^3 - (2T + T_3)x^2 - T^2 T_3 = (x - T)(2x^2 - T_3 x - TT_3) = 0$$

whereby: $x = \frac{1}{4}\left[T_3 + \sqrt{T_3^2 + 8TT_3}\right]$, which gives the two following values: $T_1' =$ 200 K and $T_2' = T_3' = 600$ K.

CONCLUSION.– A body can be cooled from 300 to 200 K without using external energy.

EXERCISE 2.18.– Solidification of water

This involves knowing whether the water → ice transformation is possible at three proposed temperatures. So, let us imagine a reversible mechanism of changing liquid water, at its freezing point (T close to 0 °C), to ice at a given T; we can write the following:

$$H_2O_{liq}(T) \rightarrow H_2O_{liq}(273.15) \rightarrow H_2O_{liq}(273.15) \rightarrow H_2O_S(T)$$

for which: $= \Delta S = \int_T^{273} C_{P_{liq}} \cdot \frac{dT}{T} - \frac{\Delta H_f}{273.15} + \int_{273}^T C_{P_s} \frac{dT}{T} = \int_T^T C_P \frac{dT}{T} - \frac{\Delta H_f}{273.15} = \frac{-1,440}{273.15}$ $= -5.27$ cal/mol.K. This amount of heat $-\Delta H = 273.15\Delta S$ is exchanged with the thermostat such that: $T'\Delta S_{\sigma'} = \Delta H$ where for the three thermostat temperatures, we obtain (in cal/mol.K):

$$\Delta S_{\sigma.} = \begin{pmatrix} \frac{1440}{263.15} = 5.47 \\ \frac{1440}{273.15} = 5.27 \\ \frac{1440}{283.15} = 5.09 \end{pmatrix} \text{ and } \Delta S = \Delta S_{\sigma} + \Delta S_{\sigma.} = \begin{Bmatrix} -5.27 + 5.47 = +0.20 \\ -5.27 + 5.27 = 0 \\ -5.27 + 5.09 = 0.18 \end{Bmatrix}$$

CONCLUSION.– At –10 °C, the transformation is spontaneous; at 0 °C, it is reversible (equilibrium); at 10 °C, it is impossible.

EXERCISE 2.19.– Reversible isothermal expansion of an ideal gas

The variation in U is generally written as: $dU = \delta Q + \delta W = \delta Q - PdV$ and for the ideal gas, it only depends on T. Since the expansion is isothermal, $dU = 0$, i.e. $\delta Q = PdV$, with $PV = nRT$, therefore $P = nRT\frac{1}{V}$ where $\delta Q = nRT\frac{dV}{V}$ and $Q = nRT\int_{V_1}^{V_2} \frac{dV}{V} = nRT\ln\frac{V_2}{V_1}$. So $n = \frac{2250}{1.987 \times 298.15 \times 1.897} = 2.002$ # 2.

EXERCISE 2.20.– Joule–Kelvin coefficient

The J-K coefficient is defined by: $\left(\frac{\partial T}{\partial P}\right)_H = -\frac{V}{C_p}(1 - \alpha T)$ with $\alpha = \frac{1}{V}\left(\frac{\partial V}{\partial T}\right)_P$ whence we plot the graph $v = f(T/°C)$, given in Figure 2.3.

To determine α_{300}, we find $\left(\frac{\partial V}{\partial T}\right)_P = 2.246$ cm^3/°C then $\alpha_{300} = \frac{2.246}{1,136} = 1.977 \times 10^{-3}$K; on the other hand, we calculate the number of moles under these conditions: $n = \frac{PV}{RT} = \frac{40 \times 1.136 \times 273.15}{22.4 \times 573.15} = 0.967$ mol and with 1 cm^3 = 0.02423 cal/atm, we obtain: $\left(\frac{\partial T}{\partial P}\right)_H = -\frac{1,136 \times 0.02423}{11 \times 0.967}(1 - 1.977 \times 10^{-3} \times 573.15) = 0.344$ K/atm.

Comparing this to the experimental value, we obtain: $\frac{0.344}{0.370} = 0.93$.

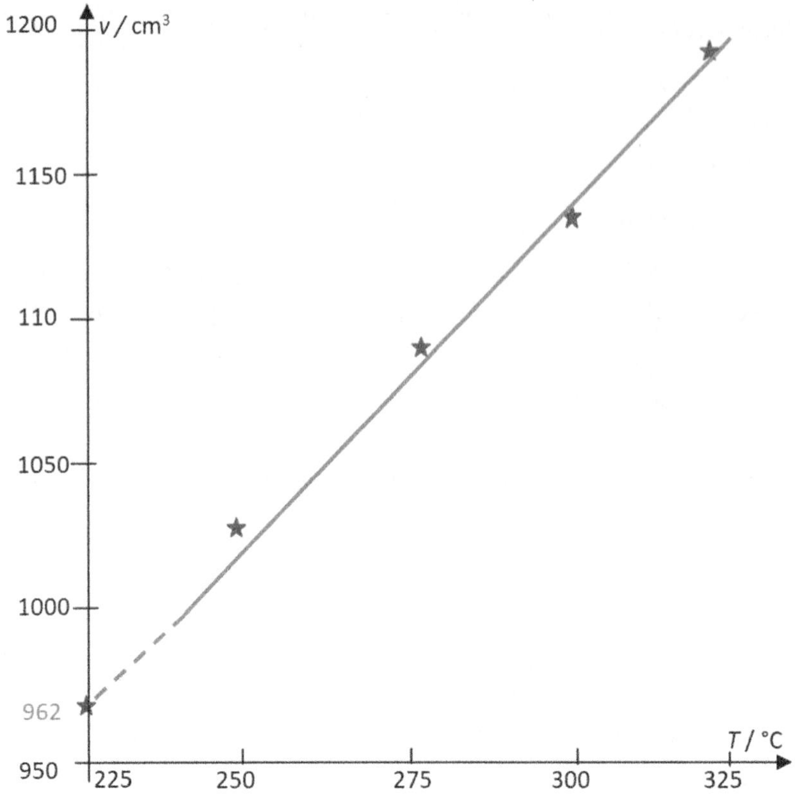

Figure 2.3. *Curve showing the molar volume as a function of T*

EXERCISE 2.21.– Thermoelastic coefficients

1) From expressions $\delta Q(T,V) = C_p dT + l dV$ and $\delta Q(T,P) = C_p\, dT + \text{h} dP$, we obtain for the calorimetric coefficients the following relations: $l = (C_P - C_v)\left(\frac{\partial T}{\partial V}\right)_P$ and $\text{h} = (C_v - C_P)\left(\frac{\partial T}{\partial P}\right)_V$. So $\delta Q = dU + PdV = \left(\frac{\partial U}{\partial T}\right)_V dT + \left(\frac{\partial U}{\partial V}\right)_T dV = C_v dT + \left[\left(\frac{\partial U}{\partial V}\right)_T + P\right] dV \Rightarrow l = \left(\frac{\partial U}{\partial V}\right)_T + P = P + T\left(\frac{\partial P}{\partial T}\right)_V - P = T\left(\frac{\partial P}{\partial T}\right)_V$, whereby:

$$C_p - C_v = T\left(\frac{\partial P}{\partial T}\right)_V \left(\frac{\partial V}{\partial T}\right)_P .$$

2) The thermoelastic coefficients α and β are defined by the expressions: $\alpha = \frac{1}{V}\left(\frac{\partial V}{\partial T}\right)_P \Rightarrow \alpha V = \left(\frac{\partial V}{\partial T}\right)_P$ and $\beta = \frac{1}{P}\left(\frac{\partial P}{\partial T}\right)_V \Rightarrow \beta P = \left(\frac{\partial P}{\partial T}\right)_V$ and they are associated with the third coefficient χ by relation $\alpha = \beta\chi P \Rightarrow \beta = \frac{\alpha}{\chi}\frac{1}{P}$, whereby replacing in the expression: $C_p - C_v = T\alpha\beta PV = T\frac{\alpha}{\chi}\frac{1}{P}P.\alpha.V = TV\frac{\alpha^2}{\chi}$, we find that $C_p - C_v = \frac{(4.9\times10^{-5})\times10.4\times10^{-3}\times293.15}{0.78\times10^{-6}} = 93.85 \times 10^{-4}$ L.atm/at.K = 0.23 cal/at.K.; the value of C_v is deduced as follows: $C_v = C_P - 0.23 = 6.40 - 0.23 =$ 6.17 cal/at.K.

EXERCISE 2.22.– Irreversibility when two pressures equalize

1) The equation of state of an ideal gas allows us to write, for each reservoir (1) and (2), from the initial state to the final state, the relation: $P_1V = n_1RT_0\,;\, P_2V = n_2RT_0$ and $P_f.2V = (n_1 + n_2)\,RT_f$, whereby: $2\frac{PV}{RT} = \left[\frac{P_1V + P_2V}{RT_0}\right]$, i.e. $P = \frac{T}{2T_0}\left[P_1 + P_2\right]$; the equation of the first principle is written as: $\Delta U = 0 = (n_1 + n_2)\,C_pT_f - (n_1C_pT_0 + n_2C_pT_0)$ whence: $T_f = T_o = 273$ K and $P = \frac{P_1 + P_2}{2} = 5$ atm.

2) The variation in S of the system is given by the expression: $\Delta S = (n_1 + n_2)S_{P,T_0} - n_1S_{P_1,T_0} - n_2S_{P_2,T_0}$ $R\left[n_1 ln\frac{P_1 + P_2}{2P_1} + n_2 ln\frac{P_1 + P_2}{2P_2}\right]$ after calculating, we obtain: $\Delta S = 21.086$ cal/K.

EXERCISE 2.23.– Joule expansion

1) The Joule expansion is defined by $W = Q = 0$, the process is isothermal and involves an ideal gas: $\Delta U = \Delta H = 0$ and $W = \Delta F + T\Delta S' = 0$ so $\Delta F = -T\,\Delta S'$. In this case, $\Delta S' = \Delta S_\sigma$, i.e. $\Delta F = -\,nRTln\frac{V_f}{V_i}$. Here, there is no work exchanged, and so:

$\Delta G + T\Delta S' = 0$, where $\Delta G = -T\Delta S' = \Delta F$. The calculation gives: $\Delta F = \Delta G = -\frac{2}{22.4}\frac{273.15\times 22.4}{273.15}ln\frac{4}{2} = -2ln2\times 101.325 = -140$ J.

2) The fact that there is a decrease in F means that the transformation is irreversible.

EXERCISE 2.24.– Condensation of water vapor

1) This involves the reversible, isothermal and isobaric transformation $H_2O_{gas} \rightarrow H_2O_{liq}$ and for which: $W = -P(V_l - V_{gas})$, where $V_{gas} = \frac{22.4\times 373.15}{273.15} =$ 30.60 L, $V_{liq} = \frac{18}{0.958} = 18.8 \times 10^{-3}$ L (negligible with respect to V_{gas}), whereby $W = PV_{gas} =$ 30.60 L.atm/mol= 3,100.55 J.

Since $\Delta U = W + Q$ and $Q = \Delta H = -9{,}720$ cal/mol, we obtain: $\Delta U = 741.05 - 9{,}720 = 8{,}978.95$ cal/mol; $\Delta S = \frac{Q}{T} = -\frac{9{,}720}{373.15} = -26.05$ cal/mol.K and $\Delta F = W =$ 741.05 cal/mol.

2) $\Delta G = \Delta U - \Delta(TS) + \Delta(PV) = -P\Delta V + T\Delta S - T\Delta S + P\Delta V = 0$, therefore the process is reversible.

EXERCISE 2.25.– Internal energy of a real gas

1) By definition, the calorimetric coefficients are expressed by the following relations: $l = T\left(\frac{\partial P}{\partial T}\right)_V$ and $h = -T\left(\frac{\partial P}{\partial T}\right)_P$. The equation proposed is written as:

$P[V - B(T)] = RT \Rightarrow P = \frac{RT}{V - B(T)}$, whereby: $\left(\frac{\partial P}{\partial T}\right)_V = \frac{R[V-B(T)] + aRT}{[V-B(T)]^2} = \frac{R\frac{RT}{P} + aRT}{\left(\frac{RT}{P}\right)^2} = \frac{\frac{R}{P} + a}{\frac{RT}{P^2}} = \frac{RP + aP^2}{RT}$ and so $l = T\left(\frac{RP + aP^2}{RT}\right) = P + a\frac{P^2}{R}$.

Similarly for h, we obtain: $V = V = \frac{RT}{P}$ B(T) $\Rightarrow \left(\frac{\partial V}{\partial T}\right)_P = \frac{R}{P} + a$, where by h = $-T\left(\frac{R}{P} + a\right) = -aT - \frac{RT}{P}$.

2) The only work involved in this case is that of the reversible expansion of the gas against the external environment, i.e. $W = -\int_{V_1}^{V_2} PdV = -\int_{V_1}^{V_2} \frac{RT}{V - B(T)} dV =$

$$-RT\ln\frac{V_2 - \mathrm{B}(T)}{V_1 - \mathrm{B}(T)} = -RT\,ln\frac{P_1}{P_2} \Rightarrow W = RT\ln\frac{P_2}{P_1}.$$

For Q, we must use the expression[(*)]: $\delta Q_{\text{rev}} = C_p.\mathrm{d}T + \mathrm{hd}P = \mathrm{hdP}$. We obtain $Q_{\text{rev}} = \int_{P_1}^{P_2} \mathrm{h.d}P = \int_{P_1}^{P_2}\left(-\mathrm{a}T - \frac{RT}{P}\right)dP = -\mathrm{a}T(P_2 - P_1) - RT\,ln\frac{P_2}{P_1}$. We simply deduce the variations in S and U by: $\Delta \mathrm{S} = \frac{Q}{T} = -\mathrm{a}(P_2 - P_1) - Rln\frac{P_2}{P_1}$ and $\Delta U = Q + W = -\mathrm{a}(P_2 - P_1)$.

3) We know two values for B at two temperatures: $\mathrm{B}(T_1) = \mathrm{a}T_1 + \mathrm{b}$ and $\mathrm{B}(T_2) = \mathrm{a}T_2 + \mathrm{b}$, whereby we deduce the value of a: $\mathrm{a} = \frac{B(T_2) - B(T_1)}{T_2 - T_1} = -\frac{11.1 + 21.1}{50} = 0.2\ \mathrm{cm^3/mol.K}$, which allows us to deduce the other values:

$$W = 0.082 \times 101.325 \times 298.15 ln\frac{1}{10} = -5.704 \text{ kJ}$$

$$Q = [-0.2.10 - 3 \times 298.15\ (1 - 10) - 0.082 \times 298.15 ln\frac{1}{10}]101.325 = -298.15\ [1.8.10 - 3 + 0.082(-2.303)]101.325 = +5.758 \text{ kJ}$$

$$\Delta S = \frac{5.758}{298.15} = 0.019 \text{ kJ/mol.K}$$

$$\Delta U = -0.2 \times 10^{-3} \times 298.15(1 - 10) = -0.054 \text{ kJ}$$

We verify that $\Delta U = W + Q = 5.758 - 5.704 = 0.054$ kJ.

CONCLUSION.– $\Delta U \neq 0$ the internal energy of a real gas does not only depend on T, as with an ideal gas, but also on P.

(*) We cannot write $\Delta U = Q + W = 0$, for the transformation to be isothermal and reversible, since the gas is not ideal.

EXERCISE 2.26.– Adiabatic and reversible expansion of a real gas

With an adiabatic and reversible expansion, from the gas equation, we have $P(V - \mathrm{b})^{\gamma} = \mathrm{Cte} = \mathrm{k}$ and during transformation, the work provided is expressed by

$W = -\int_1^2 PdV$, so by replacing: $W = -\int_1^2 k\frac{dV}{(V-b)^\gamma} = -\int_1^2 k(V-b)^{-\gamma}dV =$

$-\left[\frac{k}{1-\gamma}(V-b)^{1-\gamma}\right]_1^2 = \frac{k}{\gamma-1}\left[\left(V_2-b\right)^{1-\gamma} - \left(V_1-b\right)^{1-\gamma}\right]$, i.e. $k = P(V - b)$

$(V - b)^{\gamma-1} = RT\ (V - b\ )^{\gamma-1}$ and so $(\ V - b)^{\gamma-1} = \frac{RT}{k}$, whereby: $W =$

$\frac{k}{\gamma-1}\left(\frac{RT_2}{k} - \frac{RT_1}{k}\right) = \frac{R}{\gamma-1}(T_2 - T_1)$. Therefore, the calculation of W requires that

we know T_2. For a reversible adiabatic transformation $\frac{P_2}{P_1} = \left(\frac{T_2}{T_1}\right)^{\frac{\gamma}{\gamma-1}}$, i.e.

$ln\,P_2 - ln\,P_1 = \frac{\gamma}{\gamma-1}\left(lnT_2 - lnT_1\right)$ whose development gives:

$$lnT_2 = \frac{\gamma-1}{\gamma}(ln\,P_2 - ln\,P_1) + lnT_1 = -\frac{2}{5}ln\,P_1 + ln\,T_1$$

$$= ln\,973.15 - \frac{2}{5}ln8 = 6.049$$

which gives $T_2 = 423.59$ K, and consequently $W = \frac{3 \times 1.987}{2}(423.59 - 973.15) =$ $-1{,}637.97$ cal.

EXERCISE 2.27.– Reech formula

1) The Reech formula is: $\left(\frac{\partial P}{\partial V}\right)_Q = \gamma\left(\frac{\partial P}{\partial V}\right)_T$ for a gas of equation of state $f(P,V,T) = 0$, and by definition the calorimetric coefficients have the following relations for the different forms of the equation of state:

$V = f\,(T,P)$: $\delta Q = C_P.dT + h.dP$ (1)

$P = f\,(T,V)$: $\delta Q = C_v..dT + l.dV$ (2)

$T = f\,(P,V)$: $\delta Q = r.dP + s.dV$ (3)

which express the heat transferred to a system causing a variation in these state variables.

If dP = 0 in (1) and (3), we obtain: C_pdT = sdV and so s = $C_p\left(\frac{\partial T}{\partial V}\right)_P$ if dV = 0 in (2) and (3), we obtain: rdP = $C_v.dT$, i.e. r = $C_v\left(\frac{\partial T}{\partial P}\right)_V$, whereby if we add these values into (3), we can write the equation: $\delta Q = C_p\left(\frac{\partial T}{\partial V}\right)_P dV + C_v\left(\frac{\partial T}{\partial P}\right)_V dP$ (4).

For an adiabatic expansion, δQ = 0, equation (4) can be written as: $\left(\frac{\partial P}{\partial V}\right)_Q$ = $\frac{-\frac{C_p}{C_v}\left(\frac{\partial T}{\partial V}\right)_P}{\left(\frac{\partial T}{\partial P}\right)_V} = \frac{C_p}{C_v}\left(\frac{\partial P}{\partial V}\right)_T$ and so the equality: $\left(\frac{\partial P}{\partial V}\right)_Q = \gamma\left(\frac{\partial P}{\partial V}\right)_T$.

2) The equation of state for the proposed gas is written as: $P(V - \mathrm{b}) = RT$ = Cte (since the expansion is isothermal) and so in the differential form, we can write: $(V - \mathrm{b})\, dP + PdV = 0$ i.e. $\left.\frac{dP}{dV}\right|_T = -\frac{P}{V-\mathrm{b}}$, therefore if we apply the Reech formula: $\left(\frac{\partial P}{\partial V}\right)_Q = \gamma\frac{P}{V-\mathrm{b}} \Rightarrow \frac{dP}{P} + \gamma\,\frac{dV}{V-\mathrm{b}} = 0$, which is integrated in: $lnP + \gamma ln\,(V - \mathrm{b})$ = Cte, which can be written as: $P\,(V - b)^{\gamma}$ = Cte, equivalent to $P\,(V-b)(V - b)^{\gamma-1} = RT\,(V - b)^{\gamma-1}$ = Cte and therefore $T(V - b)^{\gamma-1}$ = Cte.

These two expressions allow us to establish, between the initial and final states of the gas, the following expression: $\frac{V_1-\mathrm{b}}{V_2-\mathrm{b}} = \left(\frac{P_2}{P_1}\right)^{\frac{1}{\gamma}} = \left(\frac{T_2}{T_1}\right)^{\frac{1}{\gamma-1}}$, whereby $\frac{P_2}{P_1} = \left(\frac{T_2}{T_1}\right)^{\frac{\gamma}{\gamma-1}}$.

EXERCISE 2.28.– Difference to ideal gas for entropy

1) The residual function of entropy is given by the following relation: $(S - S^*) = \int_0^P\left[\left(\frac{\partial S}{\partial P}\right)_T - \left(\frac{\partial S^*}{\partial P}\right)_T\right]dP$, well $\left(\frac{\partial S}{\partial P}\right)_T = \left(-\frac{\partial V}{\partial T}\right)_p \Rightarrow \left(\frac{\partial S^*}{\partial P}\right)_T = \left(-\frac{\partial V^*}{\partial T}\right)_p$ whereby $(S - S^*) = \int_0^P\left[\left(\frac{\partial V^*}{\partial T}\right)_p - \left(\frac{\partial V}{\partial T}\right)_p\right]dP$. The equation of state proposed gives: V =

$\frac{RT}{P}+B(T)$ and for the ideal gas: $V=\frac{RT}{P}$ and so $\left(\frac{\partial V^*}{\partial T}\right)_p=\frac{R}{P}$ and $\left(\frac{\partial V}{\partial T}\right)_p=$ $\frac{R}{P}+\left(\frac{\partial B}{\partial T}\right)_p=\frac{R}{p}+\frac{dB}{dT}$, whence the final expression for entropy is: $(s-s^*)=$ $\int_0^p\left(\frac{R}{P}-\frac{R}{P}-\frac{dB}{dT}\right)dP=-\int_0^p\frac{dB}{dT}dP=-\frac{dB}{dT}\int_0^p dP=-\frac{dB}{dT}P$.

2) For H_2S : $\frac{dB}{dT}$ is the slope of the curve B = f(T) obtained from the table of data and presented in Figure 2.4. We find that: $\frac{dB}{dT}=\frac{-113+125}{310-290}=0.275$ cm^3/mol.K, and the difference is: $(S-S^*)=0.275\times10^{-6}\times0.505\times10^6\ \#\ 0.14$ J/mol.K.

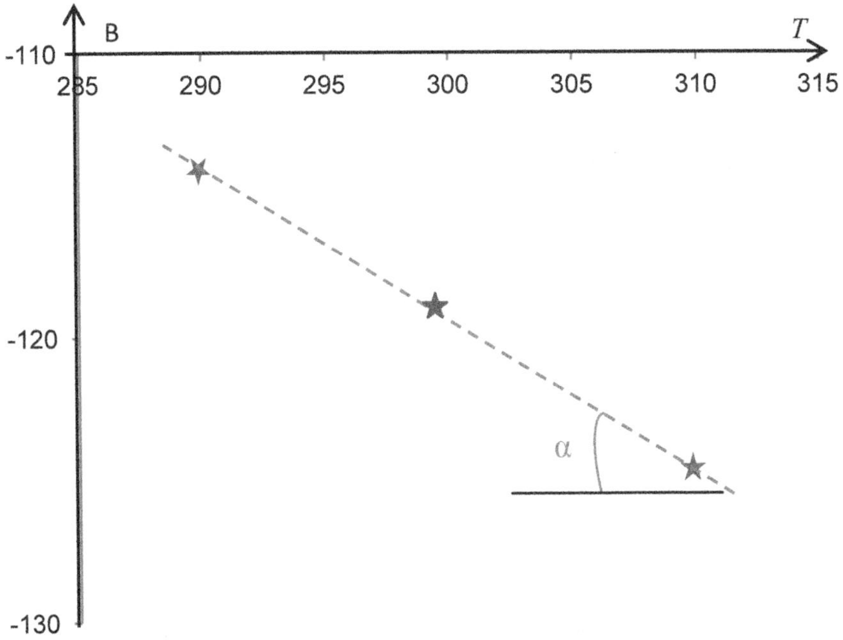

Figure 2.4. *Curve B = f(T)*

EXERCISE 2.29.– Van der Waals equation

1) The Van der Waals equation $\left(P+\frac{a}{v^2}\right)(v-b)=RT$ is expressed as a function of the molar volume $v=\frac{V}{n}$. We must determine the number of moles of NH_3 contained in the recipient.

If m = 550 g and M_{NH_3} = 17 g/mol, then $n=\frac{550}{17}$ = 32.35 mol and $v=\frac{V}{n}$ where $\left(P+\frac{a}{\left(\frac{V}{n}\right)^2}\right)\left(\frac{V}{n}-b\right)=RT$ and so $P=\frac{nRT}{V-nb}-\frac{an^2}{V^2}$.

In this new equation, we know *n, R, T* and *V*, but in order to determine *P* we need to calculate parameters a and b at the critical point, since $\left(\frac{\partial P}{\partial V}\right)_{T_c}=0$ and $\left(\frac{\partial^2 P}{\partial V^2}\right)_{T_c}=0$. We can write the differential of *P*(*T*, *V*) as follows: $dP=\left(\frac{\partial P}{\partial T}\right)_V dT+\left(\frac{\partial P}{\partial V}\right)_T dV$ and at *T* Cte of 50 °C: $dP=\left(\frac{\partial P}{\partial V}\right)_T dV=\left[\frac{\partial}{\partial V}\left(\frac{nRT}{V-nb}-\frac{an^2}{V^2}\right)\right]_T dV$, i.e. $\left(\frac{\partial P}{\partial V}\right)_T=nRT\frac{-1}{(V-nb)^2}+2an^2\frac{1}{V^3}=\frac{-nRT}{(V-nb)^2}\frac{2an^2}{V^3}$. So, at the critical point, we obtain: $\frac{2an^2}{V_c^3}-\frac{nRT_c}{(V_c-nb)^2}=0$, where $\frac{2an^2}{V_c^3}=+\frac{nRT_c}{(V_c-nb)^2}$; we also find that: $\left(\frac{\partial^2 P}{\partial V^2}\right)_T=\frac{\partial}{\partial V}\left(\frac{\partial P}{\partial V}\right)_T=\frac{\partial}{\partial V}\left[\frac{2an^2}{V^3}-\frac{nRT}{(V-nb)^2}\right]=$
$2an^2\left[\frac{\partial}{\partial V}\left(\frac{1}{V^2}\right)\right]-nRT\left[\frac{\partial\left(\frac{1}{V-nb}\right)^2}{\partial V}\right]=2an^2\left(-\frac{3}{V^4}\right)-nRT\left(\frac{-2}{(V-nb)^3}\right)=2an^2$
$\left(-\frac{3}{V^4}\right)-2nRT\left(\frac{1}{(V-nb)^3}\right)$ and at the critical point: $\left(\frac{\partial^2 P}{\partial V^2}\right)_T=0$, and so $\frac{6an^2}{V_c^4}=$

$\frac{2nRT_c}{(V_c - nb)^3}$. By making a ratio of these two expressions member by member, we obtain: b $= \frac{V_c}{3n}$ and a $= \frac{9}{8n} RT_c V_c$, where by replacing the following in the initial equation: $P_c = \frac{nRT_c}{V_c - nb} - \frac{an^2}{V_c^2}$ and at the critical point, $V_c = \frac{3}{8}\frac{nRT_c}{P_c}$, where, for a and b: a $= \frac{27}{64}\frac{R^2T_c^2}{P_c}$ and b $= \frac{1}{8}\frac{RT_c}{P_c}$. This gives: a = 4.12 L^2.atm/mol^2 and b = 0.037 L/mol; P = 3.37 atm.

2) Compared to the experimental value of 3 atm, we obtain an absolute error of $\Delta P = 3.37 - 3 = 0.37$ and relative error of 0.12, i.e. 12%.

EXERCISE 2.30.– Virial equation of state

The mixture is equimolar and so $x_1 = x_3 = 0.5$ and the virial equation of state has the form: $\frac{Pv}{RT} = 1 + \frac{B}{v} + \frac{C}{v^2} + \frac{D}{v^3} + ...$, where B = $\sum_i^n \sum_j^m x_i x_j B_{ij}$ with $B_{ij} = B_{ji}$; this is a binary (n = m = 2). We obtain: B = $x_1^2 B_{11} + x_1 x_2 B_{12} + x_1 x_2 B_{21} + x_2^2 B_{22}$, and by writing $B_{11} = B_1$ and $B_{22} = B_2$, knowing that $B_{12} = B_{21}$, we obtain:

$$B = x_1^2 B_1 + 2x_1 x_2 B_{12} + x_2^2 B_2 \text{ where } B_{12} = \frac{B - x_1^2(B_1 + B_2)}{2x_1 x_2}$$

and the calculation gives:

$$B_{12} = -517 - \left[\frac{(0.5)^2(-33 - 1512}{2(0.5)^2}\right] = -261.5 \text{ cm}^3/\text{mol}$$

at constant T, the coefficients B_{ij} are constant. So, we can calculate B for a mixture of 25% of methane:

$$B = (0.25)^2(-33) + 2(0.25)(0.75)\ (-261.5) + (0.75)^2(-1512) = -951 \text{ cm}^3/\text{mol}$$

EXERCISE 2.31.– A vertical piston cylinder

1) The system can be represented as shown in Figure 2.5.

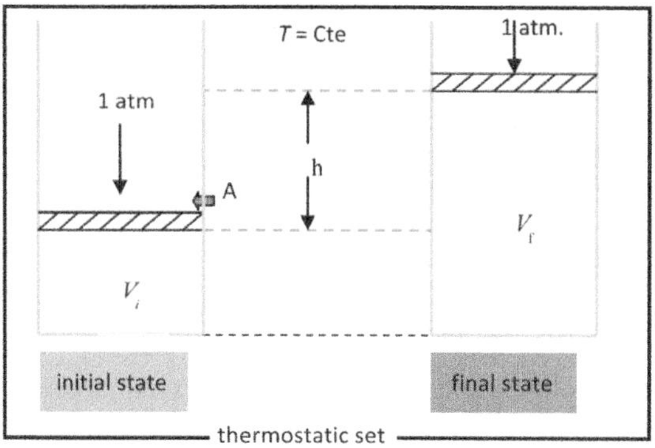

Figure 2.5. *Diagram of the device*

2) Between the initial state and final state, the system carries out work W against external pressure forces. It receives heat energy Q from the thermostat, its internal energy varies by: $\Delta U = Q + W$. So, $Q = 10$ cal and $W = -\int_{V_i}^{V_f} PdV = -P(V_f - V_i)$,

where P is the total pressure exerted on the system. Therefore, $P = P_{ext} + P_{pist}$, where $P_{pist.}$ is the pressure exerted by the piston, 0.1 kg/cm^2 whereby: $P = 1\text{ atm.} + 0.1\text{kg/cm}^2 = 101{,}325 + 0.1 \times 98{,}000 = 11.1125 \times 10^4$ Pa, and for the volume: $V_f - V_i = 10^{-3} \times 0.1 = 10^{-4}\text{m}^3$. We obtain: $W = -11.1125$ J $= -2.656$ cal and therefore $\Delta U = 10 - 2.656 = 7.344$ cal.

EXERCISE 2.32.– A horizontal piston cylinder

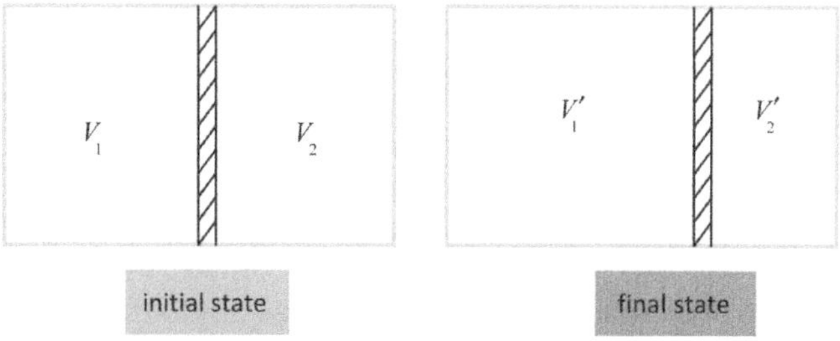

Figure 2.6. *Diagram of the instrument*

The problem can be schematized as two states, as shown in Figure 2.6.

1) The transformation is only adiabatic in compartment (2) which is thermally isolated, where the gas undergoes reversible adiabatic compression. We obtain the following relations: $P_2'(V_2')^\gamma = P_2(V_2)^\gamma$ and $T_2'(V_2')^{\gamma-1} = T_2(V_2)^{\gamma-1}$ whereby:

$V_2' = \left(\frac{P_2 V_2^\gamma}{P_2'}\right)^{\frac{1}{\gamma}}$ with $\gamma = 1.4$ we obtain: $V_2' = \left(\frac{20^{1.4}}{2}\right)^{\frac{1}{1.4}} = 12.19$ L and also:

$T_2' = T_2\left(\frac{V_2}{V_2'}\right)^{\gamma-1} = 363.45$ K.

2) The work required to compress the gas in the second compartment is equal to that received, that is to say, exchanged during the adiabatic process, which is written: $W = \frac{1}{\gamma-1}\left(P_2'V_2' - P_2V_2\right) = \frac{1}{0.4}(2\times 12.19 - 20) = \frac{4.38}{0.4} = 10.95$ L.atm $= 10.95 \times 101.325 = 1{,}109.50$ J.

3) At the end of the process, when the piston stops, the pressure in both compartments is the same, i.e. $P_1' = P_2' = 2$ atm. We can write for (1): $\begin{cases} P_1V_1 = nRT_1 \\ P_1'V_1' = nRT_1' \end{cases}$ and so $T_1' = T_1\frac{P_1'V_1'}{P_1V_1} = 40$ L, whereby $V_1' = 40 - 12.19 = 27.81$ L. Hence: $T_1' = 298.15\times 2\times\frac{27.81}{20} = 829.15$K.

4) An ideal gas whose U only depends on T has the following relation: $dU = nC_v dT$, i.e. $\Delta U = nC_v\Delta T = W + Q$ and so $Q = nC_v\Delta T - W$ where W (work received by part (2), and hence released by part(1)) $= -1{,}109.50$ with $n = \frac{P_1V_1}{RT_1} = \frac{20 \times 273.15}{22.4 \times 298.15} = 0.82$.

Therefore, $Q = 0.82 \times 5 \times (829 - 298) + 265.43 = 2{,}442.53$ cal. $= 10.21$ kJ.

PROBLEM 2.1.– Variation in internal energy of a system

The first principle has the expression $\Delta U = \Delta Q + \Delta W$; we calculate the amounts of heat and work in each case.

1) $\Delta W = 0$ and $\Delta Q = -100$, i.e. $\Delta U = -100$ cal.

2) $\Delta Q = 0$ and $\Delta W = \int_i^f \delta W$ (work by external pressure forces, equal to $-P_{ext}\, dV$) therefore:

$$\Delta W = -\int_i^f P_{ext} dV = -P_{ext} \Delta V = -4 \text{ L.atm} = -96.8 \text{ cal, whereby}$$

$$\Delta U = -96.8 \text{ cal}$$

3) $\Delta Q = 0$ and $\Delta W =$ mgh (work by force of gravity or potential energy), i.e. $\Delta W = 9.81 \times 30 = 294.3$ J, whereby $\Delta U = 70.34$ cal.

4) $\Delta W = 0$ and $\Delta Q = \mathcal{R}\, I^2\, t$ (electrical energy entirely transformed into heat by the joule effect), i.e. $\Delta Q = 20 \times 4 \times 20 = 1{,}600$ J, whereby $\Delta U = 382.41$ cal.

5) U is a function of state. As a result $\Delta U = \sum \Delta U_i$ and so we find a total variation of $\Delta U = 255.95$ cal.

PROBLEM 2.2.– Calculation of γ – Clément and Desormes' method

1) The experiment can be schematically summarized by the three following states.

$$\begin{pmatrix} P_1, T_1 \\ V_1 \end{pmatrix} \xrightarrow[\text{expansion}]{\text{adiabatic}} \begin{pmatrix} P_0, T_2 \\ V_2 \end{pmatrix} \xrightarrow[\text{constant } V]{\text{Heating at}} \begin{pmatrix} P_2, T_1 \\ V_2 \end{pmatrix}$$

After the experiment, there are many methods to calculate the ratio $\frac{C_p}{C_v}$ including:

– considering each stage applying the known relations, e.g. adiabatic and reversible expansion: $\frac{P_0}{P_1} = \left(\frac{T_2}{T_1}\right)^{\frac{\gamma}{\gamma-1}}$; heating at constant V:

$$\left.\begin{matrix} P_0 V_2 = RT_2 \\ P_2 V_2 = RT_1 \end{matrix}\right\}, \text{ whereby: } \frac{P_0}{P_2} = \frac{T_2}{T_1}$$

The gas is deemed ideal and these two relations give: $\frac{P_0}{P_1} = \left(\frac{P_0}{P_2}\right)^{\frac{\gamma}{\gamma-1}}$ where the next step is to determine γ;

– representing these transformations as a diagram (P, V):

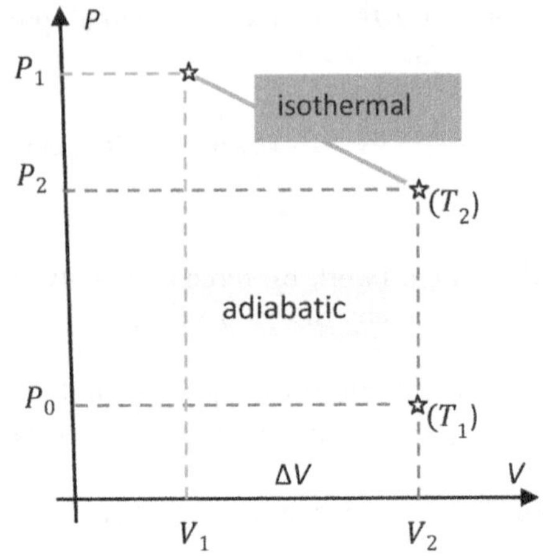

Figure 2.7. *Curve P = f (V)*

V is large and the variations in P are small, ΔV is therefore very small, since the isothermal and adiabatic are straight lines. We can therefore calculate their respective slopes as follows: Q = Cte: $\left(\frac{\partial P}{\partial V}\right)_Q = \frac{P_0-P_1}{V_2-V_1}$ and T = Cte: $\left(\frac{\partial P}{\partial V}\right)_T = \frac{P_2-P_1}{V_2-V_1}$ and the Reech formula $\left(\frac{\partial P}{\partial V}\right)_Q = \gamma\left(\frac{\partial P}{\partial V}\right)_T$ allows us to write: $\gamma = \frac{P_0-P_1}{V_2-V_1}\frac{V_2-V_1}{P_2-P_1} = \frac{P_0-P_1}{P_2-P_1}$;

– the complete calculation from the expression of δQ: during adiabatic expansion $\delta Q = C_p dT + l dP = 0$, i.e. $C_p dT - T\left(\frac{\partial V}{\partial T}\right)_P dP = 0$ whereby: $C_p = T\left(\frac{\partial V}{\partial T}\right)_P\left(\frac{\partial P}{\partial T}\right)_Q$. If, P, P', T and T' have similar values, we can write: $C_p = T\left(\frac{\partial V}{\partial T}\right)_P\frac{\Delta P}{\Delta T}$; upon heating, $\Delta P' = P_2 - P_0$ and $\Delta T' = T_1 - T_2 = -\Delta T$, and so: $\frac{\Delta P'}{\Delta T'} = \left(\frac{\partial P}{\partial T}\right)_V$, whereby $C_p = T\left(\frac{\partial V}{\partial T}\right)_P\frac{\Delta P}{\Delta P'}\left(\frac{\partial P}{\partial T}\right)_V$ since $C_p = T\left(\frac{\partial V}{\partial T}\right)_P\frac{\Delta P}{\Delta P'}\left(\frac{\partial P}{\partial T}\right)_V \Rightarrow$ $(C_p - C_v)\frac{\Delta P}{\Delta P'} = C_p$, i.e. $\frac{\Delta P'}{\Delta P} = 1 - \frac{1}{\gamma}$. Finally, we obtain:

$$\gamma = \frac{\Delta P}{\Delta P - \Delta P'} = \frac{P_0 - P_1}{P_2 - P_1}$$

2) With the experiment data: $P_0 = 1$, $P_1 = 1.020$ and $P_2 = 1.008$, we get: $\gamma = \frac{1-1.0204}{1.008-1.0204} = 1.646 = \frac{C_P}{C_V}$, whereby $C_p = 1{,}646 C_v$ and with $C_p - C_v = R = 1.987$ cal/mol.K, we obtain: $C_p = 5.07$ cal/mol.K and $C_v = 3.08$ cal/mol.K.

PROBLEM 2.3.– Maxwell equations

1) a) Entropy is a function of state. Its infinitesimal variation is therefore an exact total differential, whatever the variable considered, where as a function of (T,V) and (T,P) we write:

$$S(T,V):\ dS = \left(\frac{\partial S}{\partial T}\right)_V dT + \left(\frac{\partial S}{\partial V}\right)_T dV$$

$$S(T,P):\ dS = \left(\frac{\partial S}{\partial T}\right)_P dT + \left(\frac{\partial S}{\partial P}\right)_T dP$$

and so to bring us closer to $\left(\frac{\partial S}{\partial T}\right)_V$ of C_v and $\left(\frac{\partial S}{\partial T}\right)_P$ C_p, we must use the differentials of $U(S,V)$ and $H(S,P)$, since we know that: at constant V, $dU = C_v dT$; at constant P, $dH = C_p dT$; $dU = TdS - PdV$ and $dH = TdS + VdP$. Therefore, we can move straight to identifying the terms one by one:

$$dU = TdS = \left(\frac{\partial U}{\partial S}\right)_V dS = \left(\frac{\partial U}{\partial T}\right)_V dT = C_V dT$$

i.e. $\left(\frac{\partial U}{\partial S}\right)_V \frac{dS}{dT} = C_V$ whereby $\left(\frac{\partial S}{\partial T}\right)_V = \frac{C_V}{\left(\frac{\partial U}{\partial S}\right)_V} = \frac{C_V}{T}$.

$$dH = TdS = \left(\frac{\partial H}{\partial S}\right)_P dS = \left(\frac{\partial H}{\partial T}\right)_P dT = C_P dT \quad \text{i.e.} \quad \left(\frac{\partial H}{\partial S}\right)_P \frac{dS}{dT} = C_P$$

that is to say $\left(\frac{\partial S}{\partial T}\right)_P = \frac{C_P}{\left(\frac{\partial H}{\partial S}\right)_P} = \frac{C_P}{T}$.

b) To demonstrate the two Maxwell equations, we must first determine the second derivatives of functions F and G:

$$dF = -SdT + VdP \Rightarrow d^2F = -\left(\frac{\partial S}{\partial P}\right)_T dTdP + \left(\frac{\partial V}{\partial T}\right)_P dPdT$$

and the theorem of mathematical accuracy gives: $\left(\frac{\partial S}{\partial V}\right)_T = \left(\frac{\partial P}{\partial T}\right)_V$.

$dG = -SdT + VdP$ whereby $d^2G = -\left(\frac{\partial S}{\partial P}\right)_T dTdP + \left(\frac{\partial V}{\partial T}\right)_P dPdT$ and so, we obtain $\left(\frac{\partial S}{\partial P}\right)_T = -\left(\frac{\partial V}{\partial T}\right)_P$.

2) The previous relations imply that $C_P - C_V = T\left(\frac{\partial S}{\partial T}\right)_P - T\left(\frac{\partial S}{\partial T}\right)_V$ where the partial derivatives of $S(T,P)$ and $S(T,V)$ appear. We must therefore change the variables for one of the derivatives, so that we have a relation with only two variables, (T,V) or (T,P). Examples:

– in (T,P), we obtain $\left(\frac{\partial S}{\partial T}\right)_V = \left(\frac{\partial S}{\partial T}\right)_P + \left(\frac{\partial P}{\partial T}\right)_V \left(\frac{\partial S}{\partial P}\right)_T$ whereby: $C_P - C_V =$

$$T\left[\left(\frac{\partial S}{\partial T}\right)_P - \left(\frac{\partial S}{\partial T}\right)_P - \left(\frac{\partial P}{\partial T}\right)_V \left(\frac{\partial S}{\partial P}\right)_T\right] = -T\left(\frac{\partial P}{\partial T}\right)_V \left(\frac{\partial S}{\partial P}\right)_T = T\left(\frac{\partial P}{\partial T}\right)_V \left(\frac{\partial V}{\partial T}\right)_P = \alpha\beta PVT\ ;$$

- in (T,V), we obtain the same result; whose application to an ideal gas gives:

$$C_P - C_V = T\left[\left(\frac{\partial P}{\partial T}\right)_V \left(\frac{\partial V}{\partial T}\right)_P\right] = \frac{PVT}{T^2} = \frac{PV}{T} = R.$$

3) The ratio can be written as: $\frac{C_P}{C_V} = \left(\frac{\partial S}{\partial T}\right)_P \frac{1}{\left(\frac{\partial S}{\partial T}\right)_V} = \left(\frac{\partial S}{\partial T}\right)_P \left(\frac{\partial T}{\partial S}\right)_V$, i.e.

$\left(\frac{\partial S}{\partial T}\right)_P = -\left(\frac{\partial S}{\partial P}\right)_T \left(\frac{\partial P}{\partial T}\right)_S$ and $\left(\frac{\partial S}{\partial T}\right)_V = -\left(\frac{\partial S}{\partial V}\right)_T \left(\frac{\partial V}{\partial T}\right)_S$, so $\frac{C_P}{C_V} =$ $\left(\frac{\partial S}{\partial P}\right)_T \left(\frac{\partial P}{\partial T}\right)_S \left(\frac{\partial V}{\partial S}\right)_T \left(\frac{\partial T}{\partial V}\right)_S$; therefore, by grouping the terms with regard to each constant variable, we obtain: $\frac{C_P}{C_V} = \left(\frac{\partial V}{\partial P}\right)_T \left(\frac{\partial P}{\partial V}\right)_S$ and its application to an ideal gas ($PV = RT$) gives: $V = \frac{RT}{P}$. Hence $\left(\frac{\partial V}{\partial P}\right)_T = -\frac{RT}{P^2}$; and at constant S, we obtain the relation PV^{γ} = Cte, whereby $d\left(PV^{\gamma}\right) = 0$; as a differential: $d(PV^{\gamma}) = \gamma V^{\gamma-1}P + V^{\gamma}\left(\frac{\partial P}{\partial V}\right)_S$, i.e. $\left(\frac{\partial P}{\partial V}\right)_S = -\gamma\frac{PV^{\gamma-1}}{V^{\gamma}} = -\gamma\frac{P}{V}$, whereby $\frac{C_P}{C_V} = \left(-\frac{RT}{P^2}\right)\left(-\gamma\frac{P}{V}\right) = \gamma\frac{RT}{PV}$, that is to say: $\frac{C_P}{C_V} = \gamma$.

4) To demonstrate the proposed relation, we must first determine the second derivative of $S(T,P)$ knowing that: $dS = \frac{C_P}{T}dT - \left(\frac{\partial V}{\partial T}\right)_P dP$, so: $d^2S = \frac{\partial}{\partial P}\left(\frac{C_P}{T}\right)dTdP - \frac{\partial}{\partial T}\left(\frac{\partial V}{\partial T}\right)_P dPdT$ and: $\frac{\partial}{\partial P}\left(\frac{C_P}{T}\right) = -\frac{\partial}{\partial T}\left(\frac{\partial V}{\partial T}\right)_P$, therefore $\frac{1}{T}\left(\frac{\partial C_P}{\partial P}\right) = -\left(\frac{\partial \alpha V}{\partial T}\right)_P$, whereby $\left(\frac{\partial C_P}{\partial P}\right) = -T\left[\alpha\left(\frac{\partial V}{\partial T}\right) + V\left(\frac{\partial \alpha}{\partial T}\right)\right]_P = -T\left[\alpha.\alpha V + V\left(\frac{\partial \alpha}{\partial T}\right)_P\right] = -TV\left[\alpha^2 + \left(\frac{\partial \alpha}{\partial T}\right)_P\right]$.

5) The J-T coefficient is given at constant H, and then $dH = 0$. All that is required is to replace the differential of $S(T,P)$ in $dH = TdS + VdP$: $dS = \frac{C_P}{T}dT - \left(\frac{\partial V}{\partial T}\right)_P dP$, to obtain $C_P dT + \left[V - T\left(\frac{\partial V}{\partial T}\right)\right]_P dP = 0$, which shows that: $\frac{dT}{dP} = \frac{1}{C_P}\left[T\left(\frac{\partial V}{\partial T}\right) - V\right]_P = \left(\frac{\partial T}{\partial P}\right)_H$.

PROBLEM 2.4.– Effect of pressure on *U*, *H* and C_p of water

1) Seeing that $dH = TdS + VdP = \left(\frac{\partial H}{\partial T}\right)_P dT + \left(\frac{\partial H}{\partial P}\right)_T dP$, we must replace dS with $dS = \frac{C_P}{T}dT - \left(\frac{\partial V}{\partial T}\right)_P dP$, which gives $dH = C_P dT + \left[V - T\left(\frac{\partial V}{\partial T}\right)_P\right] dP$

and identifying term by term gives: $\left(\frac{\partial H}{\partial P}\right)_T = \left[V - T\left(\frac{\partial V}{\partial T}\right)_P\right] = V - \alpha VT$, that is to say: $\left(\frac{\partial H}{\partial P}\right)_T = V(1-\alpha T)$.

Similarly with: $dU = TdS - PdV$ and $dS = \frac{C_V}{T}dT + \left(\frac{\partial P}{\partial T}\right)_V dV$ gives $C_V dT + \left[T\left(\frac{\partial P}{\partial T}\right)_V - P\right]dV = \left(\frac{\partial U}{\partial T}\right)_V dT + \left(\frac{\partial U}{\partial V}\right)_T dV$ and then: $\left(\frac{\partial U}{\partial V}\right)_T = \left[T\left(\frac{\partial P}{\partial T}\right)_V - P\right] = P\left[T\frac{1}{P}\left(\frac{\partial P}{\partial T}\right)_V - 1\right]$. If $\beta = \frac{1}{P}\left(\frac{\partial P}{\partial T}\right)_V$, we finally obtain $\left(\frac{\partial U}{\partial V}\right)_T = P(\beta T - 1))$.

2) Since $H = U + PV$, then $dH = dU + PdV + VdP$ where at constant T we obtain $\left(\frac{\partial H}{\partial P}\right)_T dP - VdP = \left(\frac{\partial U}{\partial P}\right)_T dP + P\left(\frac{\partial V}{\partial P}\right)_T dP$ and therefore: $\left(\frac{\partial U}{\partial P}\right)_T = \left(\frac{\partial H}{\partial P}\right)_T - V - P\left(\frac{\partial V}{\partial P}\right)_T$ i.e. $\left(\frac{\partial V}{\partial P}\right)_T = -\chi V$ and $\left(\frac{\partial H}{\partial P}\right)_T = V(1-\alpha T)$, whereby $\left(\frac{\partial U}{\partial P}\right)_T = V(1-\alpha T) - V + PV\chi = V(\chi P - \alpha T)$.

3) The transformation is isothermal: $dU = \left(\frac{\partial U}{\partial P}\right)_T dP$, i.e. $\left(\frac{\partial U}{\partial P}\right)_T = \chi PV - \alpha TV$ and so $dU = (\chi PV - \alpha TV)dP$, whereby: $\Delta U = \int_{P_1}^{P_2} dU = \int_{P_1}^{P_2} [V(\chi P - \alpha T)]dP$. Since the pressure does not have much of an influence on liquids, we consider the transformation to also be isochoric. The equation then becomes: $\Delta U = \chi V \int_{P_1}^{P_2} PdP - \alpha VT \int_{P_1}^{P_2} dP$ and by taking the mean value for χ (α is constant at this P), we obtain: $\Delta U = \frac{1}{2}\chi_{moy} V \left(P_2^2 - P_1^2\right) - \alpha VT\left(P_2 - P_1\right)$; the values in the table give for χ $\chi_{moy} = 392$ and α = 304 where the variation in U is written as: $\Delta U = 0.5\ (392.10^3.10^6) - (304 \times 10^3 \times 303 \times 10^3) = 7.24 \times 10^3$ kJ.

4) The transformation occurs as a variation in T and P and so $V(T,P)$, whereby: $dV = \left(\frac{\partial V}{\partial T}\right)_P dT + \left(\frac{\partial V}{\partial P}\right)_T dP = \alpha V dT - \chi V dP$, which is written as:

$\frac{dV}{V} = \alpha dT - \chi dP$ whose integration between the two states gives: $ln\frac{V_2}{V_1} = \alpha(T_2 - T_1) - \chi(P_2 - P_1)$. Since $P_1 << P_2$, we obtain: $ln\frac{V_2}{V_1} = \alpha(T_2 - T_1) - \chi P_2$. In this case, α is no longer constant and the values in the table give $\alpha_{moy} = 256$ and $\chi_{moy} = 398$, which gives: $ln\frac{V_2}{V_1} \approx 0.00512$, whereby: $\frac{V_2}{V_1} \approx 1.0051$, i.e. $V_2 = 1.005 V_1$.

5) The variation in C_p with P is given by the relation obtained in the previous problem: $\left(\frac{\partial C_P}{\partial P}\right) = -TV\left[\alpha^2 + \left(\frac{\partial \alpha}{\partial T}\right)_P\right]$. We must now determine $\left(\frac{\partial \alpha}{\partial T}\right)_P$, i.e. the slope of line α $(T, 1)$ in the table. We find that $\left(\frac{\partial \alpha}{\partial T}\right)_P = 9.6 \times 10^{-6}\ K^{-2}$, where the calculation of the variation: $\left(\frac{\partial C_P}{\partial P}\right) = -303 \times 10^3\left[\left(304.10^{-6}\right)^2 + 9.6 \times 10^{-6}\right]$ $= -2.9 \times 10^{-3}\ J./KPa$.

CONCLUSION.– The variations (ΔV and ΔC_P) are negligible, which confirms that pressure has no significant influence on liquids, generally speaking.

PROBLEM 2.5.– Units and equation of state of an ideal gas

1) Since $PV = nRT$ with $n = \frac{m}{M} = \frac{0.1}{2} = 0.05$ mol and $R = \frac{P_0 V_0}{T_0}$, where P_0, V_0 and T_0 are the STP conditions (1 atm; 22.4 L and 273.15 K), then: $P = 0.05 \times \frac{1 \times 22\,400}{273.15} \times \frac{293.15}{1500} = 0.8$ atm Since 1 atm = 760 torr = 1.01325 bar = 1.033 kg/cm^2 = 10.1325×10^4 Pa, we obtain $P = 609$ torr = 0.81 bar = 0.83 kg/cm^2 = 8.1×10^4 Pa.

2) If we do not know the ideal gas equation, we consider the conditions of the problem to be a final state obtained from an initial reference state at STP, which gives the following:

	T/°C	P/atm	V/cm^3
Initial state (STP conditions)	0	1	V_0
Final state	20	P	1,500 cm^3

where V_0 is the volume occupied by 0.1 g of H_2 at 0 °C under 1 atm; to determine it, we must consider a simple intermediate stage, consisting of an intermediate state (a) or (b) as follows:

– (20 °C, 1 atm, V_1): the gas is heated to 20 °C at constant P, the volume varies and equals V_1; in this case, we can apply the Gay–Lussac law: $V_1 = V_0(1+\alpha.T)$ and the Mariotte law between the final state and this intermediate state: $PV = P_0V_1$, whereby $PV = P_0V_0(1+\alpha.T)$;

– (20 °C, P_1, V_0): the gas is heated to 20 °C at constant V_0, the pressure varies and equals P_1; in this case, we can apply the Charles law: $P_1 = P_0(1+\beta T)$ then the Mariotte law: $PV = P_1V_0$, which gives: $PV = P_0V_0(1+\beta T)$.

Now the gas is ideas and so $\alpha = \beta = \dfrac{1}{T_0}$, whereby $PV = P_0V_0\left(1+\dfrac{T/°\mathrm{C}}{273.15}\right)$; the calculation of P required the value V_0 (volume occupying 0.1 g of hydrogen at 0 °C under 1 atm) knowing that 1 mole of any ideal gas, under these conditions, occupies 22.4 L. Therefore if 2 g of hydrogen occupies 22,400 cm^3, then 0.1 g occupies $V_0 = \dfrac{22\,400}{2} \times 0.1 = 1120\ \mathrm{cm}^3$. This gives $P = \dfrac{1120}{1500}\left(1+\dfrac{20}{273.15}\right) = 0.8$ atm.

The ideal gas equation is simply deduced from the previous relation, since $PV = P_0V_0\left(\dfrac{273.15+T/°\mathrm{C}}{273.15}\right) = \dfrac{P_0V_0}{273.15}(273.15+T/°\mathrm{C})$, since $\dfrac{P_0V_0}{273.15} = R^{(*)}$ and $(273.15+T/°\mathrm{C}) = T/K$ we find that $PV = RT$, for 1 mole, and for n moles: $PV = nRT$.

(*) The product PV is homogeneous at a given energy, R will therefore be expressed as units of energy per mole and per degree temperature: 8.32 J./mol.K or 1.987 cal/mol.K or 0.082 L.atm/mol.K.

PROBLEM 2.6.– Compressions of an ideal gas

1) a) The compression is adiabatic $\delta Q = 0$, since $dS = \frac{\delta Q}{T}$ then $dS = 0$.

b) The compression is also reversible, for an ideal gas, T and P are related by the relation: $T_1 P_1^{\frac{1-\gamma}{\gamma}} = T_2 P_2^{\frac{1-\gamma}{\gamma}}$ with $\gamma = \frac{C_p}{C_v}$, $C_p - C_v = R \# 2$ and $C_p = \frac{5}{2} R$, whereby $\gamma = \frac{5}{3}$ and $T = 300 \times 10^{2/5} = 753.94$ K # 480.8 °C.

c) For an ideal gas, U only depends on T and its variation is given by: $dU = C_v.dT$, i.e. $\Delta U = C_v.(T_2 - T_1)$ with $C_v = C_p - 2 = 5 - 2 = 3$, where $\Delta U = 3(753.94 - 300) = 3 \times 453.94 = 1{,}361.82$ cal.

2) a) We always get $dS = \frac{\delta Q}{T}$, in this case: $\delta Q = dU + PdV$, i.e. $\Delta S = \int \frac{dU + PdV}{T}$ and at constant T, for an ideal gas, dU = 0 and $PV = RT$, whereby $\Delta S = \int \frac{PdV}{T} = R\int_{V_1}^{V_2} \frac{dV}{V} = R ln\frac{V_2}{V_1} = R ln\frac{P_1}{P_2} = 2 ln\frac{1}{10} = -2 \times 2.3 = -4.6$ cal/mol.K.

b) The heat dissipated corresponds to $\int \delta Q = \int TdS = T\int dS$ and so $\Delta Q = T\Delta S = 300(-4.6) = -1{,}380$ cal/mol.

The expression of W transferred to the gas is $\delta W = -PdV$, i.e. $W = -\int PdV = -RT\int \frac{dV}{V} = -RT ln\frac{V_2}{V_1} = RT ln\frac{P_2}{P_1} = 2 \times 2.3 \times 300 = 1{,}380$ cal/mol. Therefore, $\Delta Q = -W$, the heat dissipated during compression is equal to the work transferred to the gas by the piston.

c) By definition $dU = \delta Q + \delta W$, i.e. $\Delta U = Q + W$ and so $\Delta U = 0$, we verify that for an ideal gas, the internal energy only depends on T.

PROBLEM 2.7.– Equation of state of a real gas

1) The Van der Waals equation of state for 1 mole of gas is written as: $\left(P + \frac{a}{V^2}\right)(V - b) = RT$ or $PV + \frac{a}{V} - bP - \frac{ab}{V^2} = RT$, i.e. $PV = RT - \left(\frac{a}{V} - bP - \frac{ab}{V^2}\right)$, where these latter terms are corrective terms for the ideal gas equation of state written in the form: $\frac{PV}{RT} = 1 - \frac{a}{RTV} + \frac{bP}{RT} + \frac{ab}{RTV^2}$, whereby introduction of the following approximations into the right side of the equation:

– we disregard the second-degree term in V, i.e. $\frac{ab}{RTV^2} \approx 0$;

– for the second term. we can write $PV = RT$ and so $\frac{bP}{RT} \# \frac{b}{V}$.

It gives: $\frac{PV}{RT} = 1 - \frac{a}{RTV} + \frac{b}{V} = 1 + \frac{b - \frac{a}{RT}}{V} = 1 + \frac{B}{V}$, which is the proposed equation (*) with B a function of T.

2) Under the given conditions: a = 1.91×10^6 atm.cm^6, b = 36.3 cm^3 and $R = \frac{P_0V_0}{T_0} = \frac{1\times 22400}{273.15}$ = 82.01 cm^3.atm/K and so B = $36.3 - \frac{1.91.10^6}{82.01.T} = \frac{2.27.10^4}{T}$; knowing that ρ_0 of nitrogen is $\frac{M_{N_2}}{V_0}$, $\rho_0 = \frac{28}{V_0}$: we must therefore calculate V_0 with the help of the proposed equation of state: B + $V = \frac{PV^2}{RT}$, whereby $PV^2 - RTV - BRT = 0$. So $P = 1$, and so: $V^2 - RTV - BRT = 0$; the calculation of B at 0 °C gives $B_0 = 36.3 - \frac{2.327\text{x}10^4}{273.15}$ = – 48.92 cm^3, whereby the second-order equation to solve is: $V^2 - 22{,}400\ V + 10.958 \times 10^5 = 0$, which gives the positive solution: $V_0 = \frac{22{,}400 \pm \sqrt{22{,}400^2 - 4\times 10.958\times 10^5}}{2} = \frac{22{,}400 + 22{,}301.95}{2}$ = 22,350.98 cm^3 (the root with the minus sign gives an aberrant molar V), whereby: $\rho_0 = \frac{28}{22{,}350.98} = 1.2527 \times 10^{-3}$ g/cm^3 # 1.253 g/L.

3) If we consider N_2 to be an ideal gas, under STP conditions, we obtain: $\rho'_0 \frac{28}{22.4}$ =1.25 g/L, i.e. a relative error of: $\Delta = \frac{1.253 - 1.250}{1.250} = 2.2 \times 10^{-3}$ # 0.22%. It is a relatively large error given that experimentally we obtain the volumic mass of a gas with an accuracy greater than 10^{-3}.

4) According to the proposed equation, N_2 will behave like an ideal gas at T^* such that B = 0. Its equation then becomes $PV = RT^*$ and therefore $b - \frac{a}{RT} = 0$, i.e. $36.3 - \frac{2{,}327\text{x}10^4}{T^*} = 0$, whereby a temperature of: $T^* = \frac{2.327 \times 10^4}{36.3}$ = 641.05 K = 367.90 °C.

It is at this temperature that the previously defined error is canceled out to once again increase.

(*) This equation is derived from a general form commonly used for gases, known as the "Virial equation": $\frac{PV}{RT} = 1 + \frac{B}{V} + \frac{C}{V^2} + \cdots$ where B, C, etc., are known as

"Virial coefficients". These reflect the existence of interactions between the gas molecules.

PROBLEM 2.8.– Van der Waals equation

1) a) The Van der Waals equation gives the relation: $P = \frac{RT}{V-b} - \frac{a}{V^2}$, and by definition: $l = T\left(\frac{\partial P}{\partial T}\right)_V$ and therefore $l = \frac{RT}{V-b}$.

b) The transformation is reversible, therefore $\delta W_{rev} = -PdV$ whereby $W = -\int_{V_1}^{V_2} PdV = -RT\ln\frac{V_2-b}{V_1-b} + a\left(\frac{1}{V_1} - \frac{1}{V_2}\right)$.

c) By definition: $dU = \delta Q + \delta W$ with $\delta Q = C_v dT + ldV$, i.e. $dU = C_v dT + ldV - PdP$. Since $dT = 0$, we obtain $dU = (l-P)\,dV$, and so $dU = \left(\frac{RT}{V-b} - P\right)dV = \left(\frac{RT}{V-b} - \frac{RT}{V-b} + \frac{a}{V^2}\right)dV = \frac{a}{V^2}dV$. Thus, $\Delta U = a\int_{V_1}^{V_2}\frac{dV}{V^2} = -a\left(\frac{1}{V_2} - \frac{1}{V_1}\right) = a\left(\frac{1}{V_1} - \frac{1}{V_2}\right)$.

d) From $\Delta U = W + Q$, we obtain $Q = \Delta U - W = RT\ln\frac{V_2-b}{V_1-b}$.

2) a) The Joule expansion, so $W = 0$, whereby $Q = \Delta U$ and as $V_2 > V_1$, $\Delta U > 0$ therefore $Q > 0$: this means that the system receives heat from the calorimeter and therefore this leads to formation of ice.

b) The amount of heat released is: $Q = \Delta U = a\left(\frac{1}{V_1} - \frac{1}{V_2}\right) = 3.59\left(1 - \frac{1}{50}\right) = 3.59 \times 0.98$ L.atm= 356.482 J. Knowing that $Q = \Delta m.\Delta H_f$, we obtain: $\Delta m = \frac{356.482}{80 \times 4.184} = 1.065$ # 1.07 g, whereby the total mass of ice at equilibrium: $m = 100 + \Delta m = 101.07$ g.

3) For an ideal gas $PV = RT$, so for the Van der Waals equation: a = b = 0, we find that: $l = \frac{RT}{V} = P$; $W = -RTln\frac{V_2}{V_1}$; $\Delta U = 0$ and $Q = RTln\frac{V_2}{V_1}$.

CONCLUSION.– During an isothermal transformation, for an ideal gas $\Delta U = 0$ (function of T) and for the real gas:$\Delta U \neq 0$ (so a function of T and P or V).

PROBLEM 2.9.– Non-isolated thermostat

In this case, the exchange takes place between the thermostat at $T = 370.15$ K and air at $T' = 300.15$ K; with the thermostat not being well-insulated, there is contact between the two systems water/air, then air receives an amount of heat Q lost by water from the thermostat, which allows us to write:

1) for the water, the expression: $T\Delta S_{water} = Q = 1{,}000$ cal gives the following value $\Delta S_{water} = \dfrac{1{,}000}{370.15} = 2.702$ cal./K ;

2) for the air: $T'\Delta S_{air} = Q = 1{,}000 \Rightarrow \Delta S_{air} = \dfrac{1{,}000}{300.15} = 3.332$ cal/K;

3) for the water–air unit, the following amount of entropy is created: $\Delta S = \Delta S_{water} + \Delta S_{air} = -2.702 + 3.332 = 0.63$ cal/K.

The process is therefore irreversible, as with all processes, heat losses are impossible to recover without spending more energy.

PROBLEM 2.10.– Transformation in an isolated system

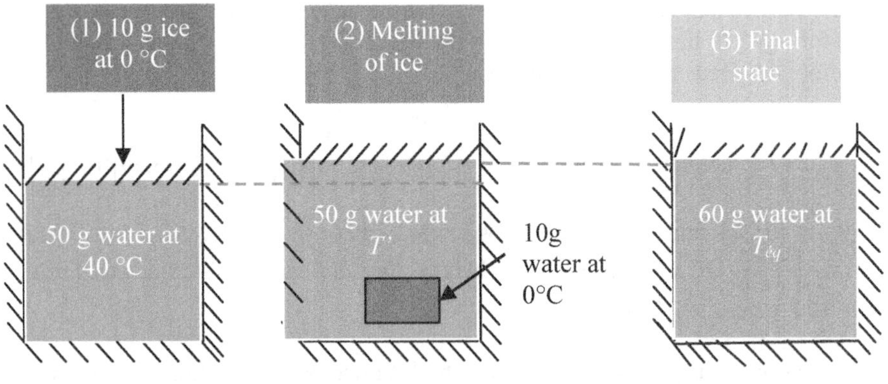

Figure 2.8. *The three successive states in the transformation*

The problem can be represented as shown in Figure 2.8.

1) The melting of ice absorbs $Q = 80 \times 10 = 800$ cal provided by water at 40 °C, which will cool down, i.e. $mC_p(313.15 - T')$, whereby $T' = 313.5\frac{800}{mC_p} = 298.15$ K.

2) The system then reaches equilibrium where 60 g of water will be at temperature $T_{éq}$, so that the heat transferred to 10 g of water at 0 °C to reach T_{eq} is equal to the heat transferred to 50 g of water at 24 °C to reach the same T_{eq}, so: $10(T_{eq} - 273.15) = 50(297.15 - T_{eq})$, whereby $T_{eq} = 293.15$ K.

3) The total variation in entropy, of the isolated system, is given by $\Delta S = \Delta S_{eau} + \Delta S_{glace}$ with, on the one hand, $\Delta S_{water} = \Delta S_{50} + \Delta S_{10}$ which corresponds to the transition of 50 g of water from 40 °C to 20 °C and 10 g of water from 0 °C to 20 °C i.e. $\Delta S_{50} = \int_{313.15}^{293.15} 50 C_p \frac{dT}{T}$ and $\Delta S_{10} = \int_{273.15}^{293.15} 10 C_p \frac{dT}{T}$ so $\Delta S_{water} = 50 \ln\frac{293.15}{313.15} + 10 \ln\frac{293.15}{273.15} = -3.30 + 0.71 = -2.59$ cal/mol.K; on the other hand, $\Delta S_{ice} = \frac{\Delta H_f}{273.15} = \frac{800}{273.15} = 2.93$ cal/mol.K, which ultimately gives: $\Delta S = 2.93 - 2.59 = +0.34$ cal/mol.K.

CONCLUSION.– The transformation is irreversible.

PROBLEM 2.11.– Condensation of carbon sulfide

1) This involves studying the reversible isothermal and isobaric transformation: $CS_{2\,gas} \rightarrow CS_{2\,liq}$ at 46 °C and 1 atm.; at constant P, W occurs against atmospheric P: $W = -P\Delta V$ with $\Delta V = V_{liq.} - V_{gas}$ and for 1 mole: $V_{liq.} = \frac{M}{d} = \frac{76}{1.25} = 60.8$ cm^3 = 0.0608 L. and $S = \frac{0.335 \times 76}{1{,}000} = 25.46 \times 10^{-3}$m^3 = 25.46 L, therefore $W = 1(25.46 - 0.0608) = 25.399$ L.atm/mol # $25.4 \frac{101.325}{4.184} = 615.10$ cal/mol. The amount of heat Q_P involved in this isobaric process is equal to the heat of condensation, since $\Delta H_P = \Delta U_P + P\Delta V$ and $\Delta U_P = Q_P + W = Q_P - P\Delta V$ and so $\Delta H_P = Q_P = -\Delta H_v = -85\ M_{CS_2} = -85 \times 76 = -6.46$ kcal/mol.

2) The calculation of the other state functions is as follows: $\Delta U = W + Q = 615.09 - 6460 = -5.845$ kcal/mol; $\Delta S = -\frac{6{,}460}{319.15} =$

20.24 cal/mol.K and $\Delta F = W - T\Delta S'$, where $\Delta S'$ is the total entropy (CS_2 + external environment). The process is isothermal and reversible, $\Delta S' = 0$, and so $\Delta F = W = 615.10$ cal/mol and $\Delta G = \Delta U - T\Delta S + P\Delta V$ $= W + Q - T\Delta S + P\Delta V = -P\Delta V + T\Delta S - T\Delta S + P\Delta V = 0$.

3) Instead of calculating V_{gas} from the specific volume, then: $\Delta V = V_f - V_i = -V_i$ (we disregard V of the liquid with respect to the gas), whereby: $W = P.V_i$; the saturated vapor of CS_2, considered an ideal gas, we can calculate the volume of 1 mole at 46 °C and 1 atm from the equation: $PV = RT$, whereby $V_i = \frac{RT}{P} =$ $\frac{22.4}{1 \times 273.15} \times 319.15 = 26.171$ L.atm/mol, which gives a relative error with respect to the true value: $\frac{\Delta V}{V} = \frac{26.17 - 25.46}{25.46} = 2.78\%$.

Disregarding V_{liq} with respect to V_{gas}, we find a relative error of ΔV of condensation equal to: $\Delta(\Delta V) = \frac{0.0606}{25.46 - 0.0608} = 0.24\%$.

CONCLUSION.– This error is very low, and so the assumption of approximation is justified.

PROBLEM 2.12.– Adiabatic expansion of a diatomic gas

This expansion process is shown in Figure 2.9.

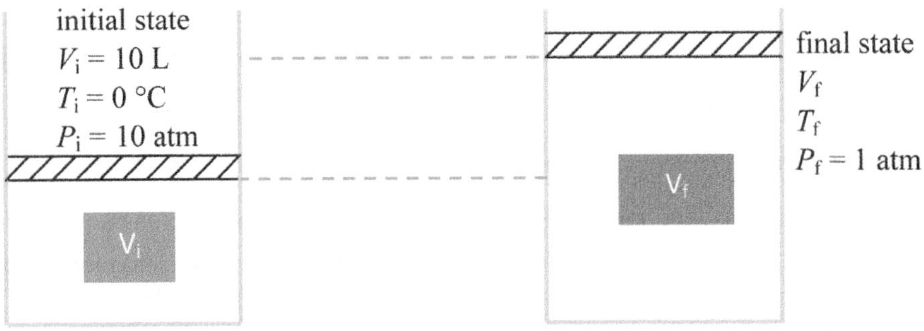

Figure 2.9. *Expansion process*

1) The piston is slowly raised until it reaches equilibrium at atmospheric pressure. In this case, we write: $P_f V_f^\gamma = P_i V_i^\gamma \Rightarrow V_f = \left(\frac{P_i V_i^\gamma}{P_f}\right)^{1/\gamma} = \left(\frac{10\times 10^{1.4}}{1}\right)^{1/1.4} =$ 51.795 L and $T_f V_f^{\gamma-1} = T_i V_i^{\gamma-1}$ and so $T_f = T_i\left(\frac{V_i}{V_f}\right)^{\gamma-1} = 273.15\left(\frac{10}{51.795}\right)^{0.4} =$ 141.48 K. The application of the first principle gives: $\Delta U = W + Q$ with $Q = 0$ and so $\Delta U = W$, and for an ideal gas, we obtain $\Delta U = nC_v(T_f - T_i) = \frac{R}{\gamma-1}(T_f - T_i)$, whereby $W = \frac{nR}{\gamma-1}(T_f - T_i)$. So, $nR = \frac{P_i.V_i}{T_i}$, i.e. $W = \frac{P_i.V_i}{T_i(\gamma-1)}(T_f - T_i) =$ $\frac{10\times 10}{273.15\times 0.4}(141.48 - 273.15)\times 101.325 = -12.211$ kJ.

2) The gas is expanded suddenly under atmospheric pressure. The work involved during this irreversible process is: $W = -P_{ext}\Delta V = -P_f(V_f - V_i)$ and $Q = 0$, where $W =$ $\Delta U = nC_v dT = nC_v(T_f - T_i)$. Thus, $P_f(V_f - V_i) = nC_v(T_f - T_i)$ and for the ideal gas, we obtain: $P_f V_f = nRT_f$, $P_i V_i = nRT_i$ and $nC_v = \frac{nR}{\gamma-1}$, whereby $T_f + T_i\frac{P_f}{P_i} = \frac{1}{\gamma-1}(T_f - T_i)$.

So $T_f\left(\frac{1}{\gamma-1}+1\right) = T_i\left(\frac{P_f}{P_i} + \frac{1}{\gamma-1}\right)$ and $\frac{T_f}{T_i} = \frac{\left(\frac{P_f}{P_i}\right) + \frac{1}{\gamma-1}}{1+\frac{1}{\gamma-1}}$, whereby: $T_f = T_i\frac{1+(\gamma-1)\frac{P_f}{P_i}}{\gamma}$ $= 273.15\frac{1+0.4\times\frac{1}{10}}{1.4} = 202.91$ K.

Not that the final T reached in this case is greater than that reached in the first case, that is to say, that the gas cools to a lesser temperature: The corresponding difference in heat is related to the creation of entropy due to the irreversibility.

The calculation of V_f is given by $P_f V_f = nRT_f$ with $nR = \frac{P_i V_i}{T_i}$, i.e. $V_f = \frac{P_i V_i T_f}{T_i P_f} =$ $\frac{10\times 10\times 202.91}{273.15} = 74.29$ L, whereby $W = -P_f(V_f - V_i) = -(74.29 - 10)\times 101.325 =$ -6.314 kJ.

3) With S being a function of state, its variation during the irreversible process will be the same for a system σ (the gas) during a reversible transformation from state (i) to the state (f): $\Delta S_\sigma = \int_i^f \frac{\delta Q_{rév}}{T}$ with $\delta Q_{rev} = nC_v dT + PdV$ and $PdV = nRT\frac{dV}{V}$. Since $n = \frac{P_i V_i}{RT_i}$, i.e. $\Delta S_\sigma = \int_{T_i}^{T_f} nC_v \frac{dT}{T} + \int_{V_i}^{V_f} nR\frac{dV}{V} = nC_v ln\frac{T_f}{T_i} + nRln\frac{V_f}{V_i} = \frac{10\times 10\times 273.15}{22.4\times 273.15}\times ln\frac{202.91}{273.15} + ln\frac{74.29}{10} 0.4609 l.\text{atm} = 0.4609\times 101.325 = +46.70\text{ J}$

PROBLEM 2.13.– Reversible adiabatic expansion of an ideal gas

The transformation of the gas can be shown in Figure 2.10.

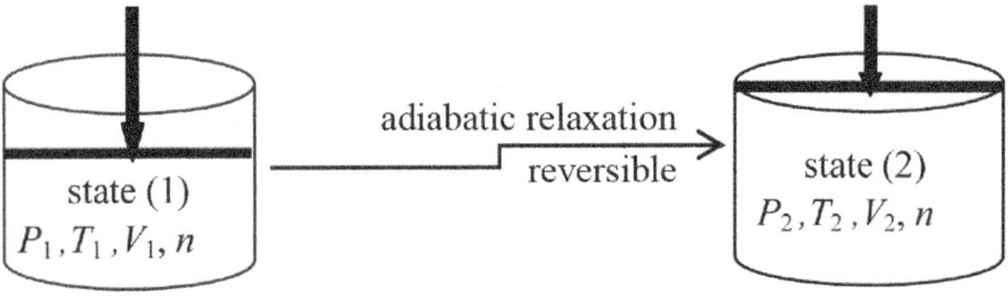

Figure 2.10. *Diagram showing the expansion*

1) Such a transformation is known as isentropic, in this case, for an ideal gas, the relations between T, P and V are $TP^{\frac{1-\gamma}{\gamma}} = \text{Cte}$ and so $T_1 P_1^{\frac{1-\gamma}{\gamma}} = T_2 P_2^{\frac{1-\gamma}{\gamma}}$, where $T_2 = T_1\left(\frac{P_2}{P_1}\right)^{\frac{\gamma-1}{\gamma}}$ with $\gamma = \frac{C_P}{C_V} = \frac{10,5}{6,35} = 1.654$ we obtain $T_2 = 590.15\left(\frac{1}{3}\right)^{\frac{0,654}{1,654}} = 382.21$ K.

2) a) One of the properties of an ideal gas is that its internal energy and enthalpy depend only T and are given by: $dU = nC_v dT$ and $dH = nC_p dT$, whereby:

$$\Delta U = C_v \Delta T = C_v(T_2 - T_1) = -6.35\ (590.15 - 382.21)\ \#\ -1{,}320.40\text{ J}$$

$$\Delta H = C_p \Delta T = C_p(T_2 - T_1) = -10.5\ (590.15 - 382.21)\ \#\ -2{,}183.33\text{ J}$$

b) Adiabatic expansion is therefore $\delta Q = 0$, whereby $U = W = -1{,}320.40$ J.

3) a) Determining volumes is possible as soon as P_i, T_i and W are known: $W = C_v \Delta T = C_v(T_2 - T_1)$ and $C_p - C_v = n$, then $(\gamma - 1) = n\frac{R}{C_V}$ and so $C_v = \frac{nR}{\gamma - 1}$ where $W =$

$\frac{nR}{\gamma-1}(T_2 - T_1)$. Here $n = \frac{P_1V_1}{RT_1} = \frac{P_2V_2}{RT_2}$, therefore $(\gamma - 1)W = \frac{P_1V_1}{T_1}(T_2 - T_1) = P_1V_1\left(\frac{T_2}{T_1} - 1\right) = \frac{P_1V_1(T_2-T_1)}{T_1}$, i.e. $V_1 = \frac{(\gamma-1)WT_1}{P_1(T_2-T_1)} = \frac{0.654\times1320.4\times590.15}{3(382.21-590.15)}$ # 8.17×10^{-3} m^3 and $V_2 = \frac{P_1}{P_2}\frac{T_2}{T_1}V_1 = \frac{3}{1}\times\frac{382.21}{590.15}$. 8.17×10^{-3} # 15.87×10^{-3} m^3.

b) The number of moles is deduced from one of the equations $P_iV_i = nRT_i$, for example: $n = \frac{P_1V_1}{RT_1} = \frac{3\times10^5\times 8.17\times10^{-3}}{9.114\times590.15}$ # 0.5 mol.

4) The system is composed of a gas and a cylinder, so: $\Delta S_{syst} = \Delta S_{gas} + \Delta S_{cyl}$; the expansion of the gas is isenthalpic, so $\Delta S_{gas} = 0$, whereby $\Delta S_{syst} = \Delta S_{cyl} = \frac{\delta Q}{T_{ext}} + S_{created}$. As the system is adiabatic, $\delta Q = 0$ then $\Delta S_{syst} = S_{created}$, with $S_{created} = \frac{-W}{T_2} = -\frac{-1,320.40}{382.21}$ # 3.455J/K.

PROBLEM 2.14.– Hydrogen expansion

For a system (σ), going from an initial state (i) to a final state (f), the transformation is isothermal if the system exchanges heat with one source (σ') at constant temperature T', and if $T_f = T' = T_i$; (σ) represents hydrogen considered as an ideal gas.

1) The transformation is reversible and so: $\delta W_{rev} = -PdV$, whereby $W = -\int_i^f PdV$ with $P = \frac{nRT}{V}$ and constant T, whereby: $W_{rev.} = nRT\int_i^f \frac{dV}{V} = nRT ln\frac{V_i}{V_f}$, and for 2 L under STP conditions, we obtain $n = \frac{2}{22,4}$. This gives: $W_{rev.} = \frac{2}{22.4}ln\frac{2}{4} = -$ 140.6 J and with $\Delta U = W_{rév} + Q_{rév} = 0$, we obtain $Q_{rev} = -W_{rev} = 140.6$ J and so $\Delta S_\sigma = \frac{Q_{rev}}{T} = \frac{-W_{rev}}{T} = \frac{140.6}{273.15} = 0.5147$ J/mol.K. Whence, $\Delta S = \Delta S_\sigma + \Delta S_{\sigma'} = 0$ and the process is reversible, so $\Delta S_{\sigma'} = -0.5147$ J/mol.K.

2) For the Joule expansion, by definition, there is no work against external P and so: $W = 0$; the expansion is isothermal, i.e. $\Delta U = 0$, where $W = Q = 0$. The Maxwell equation is written as: $\left(\frac{\partial S}{\partial V}\right)_T = \left(\frac{\partial P}{\partial T}\right)_V$, with $PV = nR$. Therefore, $\left(\frac{\partial P}{\partial T}\right)_V = \frac{nR}{V}$,

where $dS_\sigma = nR\frac{dV}{V}$ and $\Delta S_\sigma = nR\,ln\frac{V_f}{V_i}$, where $\Delta S_\sigma = 0.5147$ J/mol and $\Delta S_{\sigma'} = \frac{Q}{T} = 0$ so $\Delta S = \Delta S_\sigma + \Delta S_{\sigma'} = 0.5147$ J/mol.

The process is therefore irreversible: so there is the creation of entropy.

3) The irreversible transformation occurs against $P_{ext} = 0.5$ atm, i.e. $W_{irrev} = -P_{ext}dV = -P_{ext}\Delta V = -0.5\times 2 = -1$ L.atm $= -101.325$ J and ΔU is always $= 0 = W_{irrev} + Q_{irrev}$, therefore $Q_{irrev} = -W_{irrev} = 101.325$ J. This gives $\Delta S_{\sigma'} = -\frac{Q_{irrev}}{T} = -\frac{101.325}{273.15} = -0.371$ J.

4) The graphical representation shown in Figure 2.11 is simple and allows the variation in total entropy to be calculated as follows:

– reversible isothermal expansion: $|W| = A_1 + A_2$ and $\Delta S = 0$;

– Joule expansion: $W = 0$ and $\Delta S = \frac{(A_1 + A_2)}{T}$;

– irreversible isothermal expansion: $|W_{irrev}| = A_2$ and $\Delta S = \frac{A_1}{T}$.

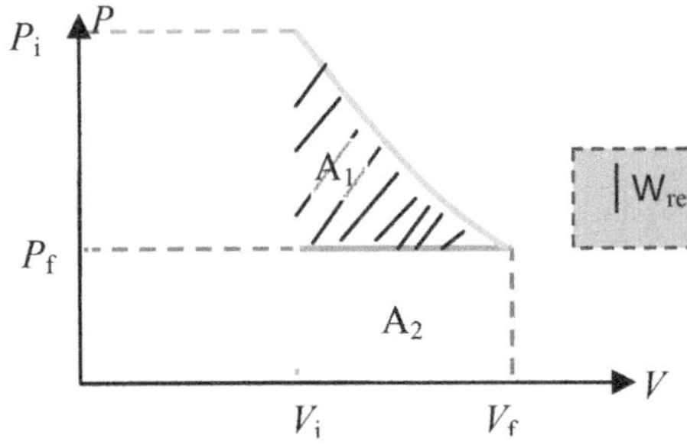

Figure 2.11. *Estimating the variation in entropy*

PROBLEM 2.15.– Difference to ideal gas in reduced coordinates

1) The compressibility factor (z) is defined by: $PV = zRT$ and the gas has the following equation $V = \frac{RT}{P} + A$, whereby its expression will be: $z = \frac{P}{RT}\left(\frac{RT}{P} + A\right) = \frac{AP}{RT} + 1$.

2) By definition, the difference to ideal gas is given by the relation: $H - H^* = \int_0^P \left[\left(\frac{\partial H}{\partial P}\right)_T - \left(\frac{\partial H^*}{\partial P}\right)_T\right] dP$, we must therefore identify the two partial derivatives for the real gas and the ideal gas:

– ideal gas: Joule's second law shows that $\left(\frac{\partial H^*}{\partial P}\right)_T = 0$;

– real gas: the differentials of $H(T,P)$ and $H(S,P)$ give the equality: $\left(\frac{\partial H}{\partial T}\right)_P dT + \left(\frac{\partial H}{\partial P}\right)_T dP = C_P dT + \left[V - T\left(\frac{\partial V}{\partial T}\right)_P\right] dP$ and the identification term gives: $\left(\frac{\partial H}{\partial P}\right)_T = \left[V - T\left(\frac{\partial V}{\partial T}\right)_P\right]$, where the expression of the difference is: $H - H^* = \int_0^P \left[V - \left(\frac{\partial V}{\partial T}\right)_P\right] dP$. We obtain:

$$V = \frac{zRT}{P} \quad \Rightarrow \left(\frac{\partial V}{\partial T}\right)_P = \quad \frac{zR}{P} + \frac{RT}{P}\left(\frac{\partial z}{\partial T}\right)_P \quad \text{so} \quad V - T\left(\frac{\partial V}{\partial T}\right)_P$$

$$= \frac{zRT}{P} - \frac{zRT}{P} + \frac{RT^2}{P}\left(\frac{\partial z}{\partial T}\right)_P = -\frac{RT^2}{P}\left(\frac{\partial z}{\partial T}\right)_P$$

where: $H - H^* = \int_0^P -\frac{RT^2}{P}\left(\frac{\partial z}{\partial T}\right)_P dP$.

Next $\left(\frac{\partial z}{\partial T}\right)_P = -\frac{\mathrm{A}P}{RT^2}$, and so $H - H^* = \int_0^P -\frac{RT^2}{P}\left(-\frac{\mathrm{A}P}{RT^2}\right) dP$, and $H - H^* = \int_0^P \mathrm{A}dP = \mathrm{A}P$.

3) The reduced coordinates are defined with respect to those of the critical point such that: $T = T_r T_c$, $P = P_r P_c$ and $V = V_r V_c$, which gives relation: $H - H^* = R\int_0^P -\frac{T_r^2 T_c^2}{P_r P_c}\left(\frac{\partial z}{\partial T}\right)_P dP_r P_c$; by removing the constants of the integral, we obtain: $H - H^* = -RT_c^2 T_r^2 \int_0^P \left(\frac{\partial z}{\partial T}\right)_P \frac{dP_r}{P_r}$.

4) The generalized diagram (see Appendix) gives the reduced isotherms for (z, P_r); it allows the determination of P_r (z, T_r) simply by reading, i.e.

– calculating P_1: at T_1 = 373.15 K, we obtain z_1 = 0.69 that corresponds to $T_{r1} = \frac{T_1}{T_C} = \frac{373.15}{304.15} = 1.226$; but in the diagram, we find that T_r = 1.2 and T_r = 1.3. An interpolation is necessary to determine P_{r1}. The calculation gives the value 1.58, where: $P_1 = P_{r1}\,P_c = 1.404 \times 72.8$, so $P_1 = 115.024$ atm;

– calculation of P_2: at T_2 = 410.50 K, we obtain z = 0.76 that corresponds to $T_{r2} = \frac{T_2}{T_C} = \frac{410{,}50}{304{,}15} = 1.35$. Therefore, it is between 1.3 and 1.4. An interpolation gives $Pr_2 = 2.19$, where: $P_2 = 2.19 \times 72.8 = 159.432$ atm.

This breakdown also induces a variation in *P*. The variation in *H*(*T,P*) will be given by: $\Delta H = H_2 - H_1$, which can be written as follows:

$$\Delta H = H_2 + \left(H_2^* - H_2^*\right) - H_1 + \left(H_1^* - H_1^*\right) = \left(H_2 - H_2^*\right) - \left(H_1 - H_1^*\right) + \left(H_2^* - H_1^*\right)$$

And each of the terms can be calculated using the appropriate equation:

a) $\left(H_2^* - H_1^*\right) = \int_{T_1}^{T_2} C_p dT$; we know the expression of C_p, we therefore obtain:

$$(H_2^* - H_1^*) = \int_{373.15}^{410.5} \left(32.22 + 22.18\times10^{-3}T + 3.35\times10^{-6}T^2\right) dT$$

$$= 32.22 + \frac{22.18}{2}10^{-3}T^2 + \frac{3.35}{3}10^{-6}\,T^3\Big]_{373.15}^{410.5} = 1.571 \text{ kcal/mol}$$

b) $\left(H_1 - H_1^*\right) = AP_1 = -74.90\times115.02 = -8.615$ kcal/mol;

c) $\left(H_2 - H_2^*\right) = AP_2 = -74.90\times159.12 = -11.918$ kcal/mol.

Here

$\Delta H = 1.571 + 8.615 - 11.918 = -1.732$ kcal/mol.

PROBLEM 2.16.– Application of the first principle to ideal gases

1) The ideal gas is characterized by its equation of state $PV = nRT$ and so $n = \dfrac{PV}{RT}$. The ideal gas constant is defined at STP conditions $R = \dfrac{P_0 V_0}{T_0}$, i.e.

$$n = \frac{P V T_0}{P_0 V_0 T} = \frac{273.15}{22.414 \times 298.15} = 4.09 \times 10^{-2} \text{ mol.}$$

2) (a) (i) This involves heating at constant V and so $dV = 0$ and therefore $\Delta W = -PdV = 0$ and $\Delta Q = nc_v \Delta T = 4.09 \times 10^{-2} \times 5 \times 75 = 15.32$ cal, whereby $\Delta U = \Delta W + \Delta Q = 15{,}34$ cal. (ii) This involves heating at constant P, $\Delta Q = nc_p \Delta T = 4.09 \times 10^{-2} \times 5 \times 75 = 21.47$ cal and $\Delta W = -\int_{V_i}^{V_f} P\,dV = -P\int_{V_i}^{V_f} dV$ and $dV = \dfrac{nRdT}{P}$, therefore $\Delta W = -\int_{T_i}^{T_f} nR\,dT = -nR\Delta T = -4.09 \times 10^{-2} \times 1.987 \times 75 = -6.095$ cal, whereby $\Delta U = 21.47 - 6.095 = 15.37$ cal. This is practically the same result, since for an ideal gas, U only depends on T.

3) It is necessary to calculate P_f of the gas at $T_f = 100$ °C and they occupy the same volume $V_i = 1$ L; in this case, we obtain: $\dfrac{P_i}{P_f} = \dfrac{T_i}{T_f} \Rightarrow P_f = P_i \dfrac{T_f}{T_i}$ with $P_i = 1$ and $P_f = 1.25$ atm, which will allow the study of two transformations proposed as follows:

a) this is an isothermal expansion, so for an ideal gas: $\Delta U = \Delta Q + \Delta W = 0$ and $\Delta W = -\int_{V_i}^{V_f} P\,dV = +\int_{P_i}^{P_f} V\,dP = nRT ln\dfrac{P_f}{P_i} = -6.75$ cal, $\Delta Q = -\Delta W = 6.75$ cal;

b) this is an adiabatic expansion; it is therefore characterized by $\Delta Q = 0$; but, moreover, this operates in a way that the state of our system can be defined at any moment. It is therefore reversible and therefore $PV^{\gamma} = \text{Cte}$ and $TP^{\frac{1-\gamma}{\gamma}} = \text{Cte}$, where $T_f = T_i \left(\dfrac{P_f}{P_i}\right)^{\frac{\gamma-1}{\gamma}}$ with $\gamma = \dfrac{c_p}{c_v} = 1.4$. We obtain $T_f = 350.10$ K and this gives: $\Delta U = nc_v \Delta T = 4.09 \times 10^{-2} \times 5 \times (350.10 - 373.15) = -4.71 \text{ cal} = \Delta W$.

PROBLEM 2.17.– Cylinder with piston

1) a) On its downwards motion, the piston will generate $W = -P_{ext}\Delta V$ where $P_{ext} = \dfrac{m.g}{\sigma} = p$, which will be provided to the internal gas in the form of friction energy: $\Delta Q = W = -p(V - V_0)$.

b) The external gas undergoes a variation of U: $\Delta U_{ext} = \Delta Q - P_0 \Delta V$. The downfall of the piston is deemed adiabatic, so $\Delta Q = 0$, where: $\Delta U_{ext} = -P_0(V - V_0)$.

c) The internal gas is ideal and so: $\Delta U_{int} = nC_v\Delta T = \Delta Q + \Delta W = -p(V - V_0) - P_0\ (V - V_0) = (P_0 + p)(V_0 - V) \Rightarrow nC_v(T - T_0) = (P_0 + p)(V_0 - V)$; to deduce the expression of T, we use the ideal gas equation between both states $P_0V_0 = nRT_0$ and $(p + P_0)V = nRT$, whereby: $V = V_0 . \dfrac{T}{T_0} . \dfrac{P_0}{P_0 + p}$. By introducing γ and with $\dfrac{C_v}{R} = \dfrac{C_v}{C_p - C_v} = \dfrac{1}{\gamma - 1}$, so $V = V_0 \dfrac{P_0}{P_0 + p}\left(1 + \dfrac{\gamma - 1}{\gamma} . \dfrac{p}{P_0}\right)$, which allows the identification of the following expression for T: $\dfrac{T}{T_0} = 1 + \dfrac{\gamma - 1}{\gamma} . \dfrac{p}{P_0}$, i.e. $T = T_0\left(1 + \dfrac{\gamma - 1}{\gamma} . \dfrac{p}{P_0}\right)$.

2) a) The variation in energy of the piston always remains equal to $\Delta Q = -p(V - V_0)$.

b) The variation in energy of the external gas also remains the same: $\Delta Q = -P_0(V - V_0)$.

c) For the internal gas, by replacing T and nC_v with the values in $\Delta U_{int} = nC_v(T - T_0)$ and with $V = V_0\left(\dfrac{P_0}{P_0 + p}\right)^{\frac{1}{\gamma}}$, we obtain the variation: $\Delta U_{int} = \dfrac{1}{\gamma\ \ 1} P_0V_0\left[\left(\dfrac{P_0 + p}{P_0}\right)^{\frac{\gamma\ \ 1}{\gamma}} - 1\right]$, whereby the energy provided by the technician corresponding to the difference between the two values of ΔU of the internal gas, i.e. $Q = (p + P_0)(V_0 - V) - \frac{1}{\gamma - 1} P_0 V_0 \left[\left(\frac{P_0 + p}{P_0} \right)^{\frac{\gamma-1}{\gamma}} - 1 \right]$.

3) If we wait for thermal equilibrium to occur between the internal and external environment:

- for the internal gas: $\Delta U_{int} = nC_v \Delta T = 0$ since $\Delta T = 0$;

- for the external environment, the variation corresponds to the variation in potential energy of the piston, i.e.

$$\Delta U_{ext} = p(V_0 - V) = p\left(V_0 - V_0 \frac{P_0}{P_0 + p} \right) = pV_0\left(1 - \frac{P_0}{P_0 + p} \right) = \frac{p^2 V_0}{P_0 + p}$$

PROBLEM 2.18.– Hollow piston cylinder

1) Schematically, we can represent this problem by the initial and final states shown in Figure 2.12.

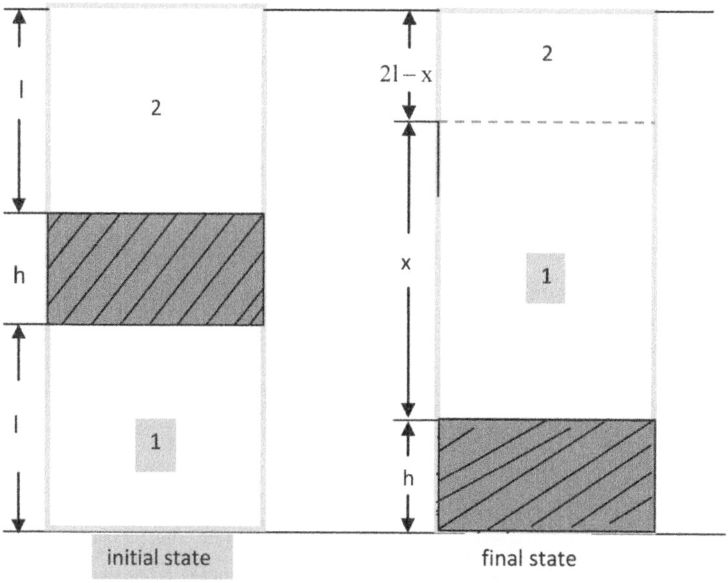

Figure 2.12. *Diagram of the transformation*

a) In the final state, the pressure at equilibrium is: $P = P_0 + \frac{\text{mg}}{\sigma}$ and the variation in potential energy of mercury between the final and initial states is given by: $\Delta E_p = -\text{mg.L}$, where $\Delta E_p = -\sigma.\text{l}.(P - P_0) = (P_0 - P)\sigma.\text{l}$.

b) $\Delta U_{(1)} = U_{\text{f}(1)} - U_{\text{i}(1)} = nC_v(T_1 - T_0) = C_v(nT_1 - nT_0)$; so, $nT_1 = \frac{P_1.\sigma.\text{x}}{R}$ and $nT_0 = \frac{P.\sigma.\text{l}}{R}$, therefore $\Delta U_{(1)} = \frac{C_v}{R}(\sigma.P_1\text{x} - \sigma.P.\text{l})$; so finally the expression is given as: $\Delta U_{(1)} = \frac{\sigma}{\gamma - 1}(P_i.\text{x} - P.\text{l})$.

c) $\Delta U_{(2)} = U_{\text{f}(2)} - U_{\text{i}(2)} = nC_v(T_2 - T_0) = C_v(nT_2 - nT_0)$, with $nT_0 = \frac{P_0.\sigma.\text{l}}{R}$ and $nT_2 = \frac{P_1.\sigma(2\text{l} - \text{x})}{R}$, which gives $\Delta U_{(2)} = \frac{C_v}{R}[\sigma.P_1(2\text{l} - \text{x}) - \sigma.P_0.\text{l}] = \frac{\sigma}{\gamma - 1}[(2P_1 - P_0).\text{l} - P_1.\text{x}]$.

2) In the equilibrium state, the sum of energies is equal to zero, i.e. $\Delta E_p + \Delta U_{(1)} + \Delta U_{(2)} = 0$, which implies that:

$$(P_0 - P)\sigma.\text{l} + \frac{\sigma}{\gamma - 1}(P_1\text{x} - P.\text{l}) + \frac{\sigma}{\gamma - 1}[P_1(2\text{l} - \text{x}) - P_0.\text{l}] = 0$$

and gives the equation: $2P_1 = (2 - \gamma)P_0 + \gamma.P$, whereby $P_1 = \frac{(2 - \gamma)P_0 + \gamma.P}{2}$.

3) In compartment (2), there is no exchange of heat, so a transformation is characterized by $PV^\gamma = \text{Cte}$, which combined with $\frac{PV}{T} = \text{Cte}$ gives $T_2 = T_0\left(\frac{P_1}{P_0}\right)^{\frac{\gamma - 1}{\gamma}}$. By replacing T with appropriate values, we obtain:

$$2\text{l} - \text{x} = \text{l}\left(\frac{P_0}{P}\right)^{\frac{1}{\gamma}}, \text{ which finally gives: } \text{x} = \text{l}.\left[2 - \left(\frac{2P_0}{(2 - \gamma)P_0 + \gamma.P}\right)^{\frac{1}{\gamma}}\right].$$

4) In each expression, we must replace the above parameters with their values and we obtain:

a) $P_1 = 1.175$ atm.; x = 1.1088.1 = 1.109 m $T_2 = 1.0472\ T_0 = 35.9$°C and $T_1 = 1.0423.T_0 = 34.5$ °C;

b) the ideal gas equation for parts (1) and (2) gives: $P_0 l\sigma = 0.4RT_0$ and $Pl\sigma = n_1 RT_0$, whereby: $n_1 = 0.4\dfrac{P}{P_0} = 0.5$ mol.

5) a) Similarly, we obtain: $\Delta S_{(2)} = 0$; $\Delta S_{(1)} = n_1 C_p\, ln\dfrac{T_1}{T_0} - n_1 R\, ln\dfrac{P_1}{P} =$

$$0.5R\left[\frac{\gamma}{\gamma-1}\ln\frac{T_1}{T_0} + \ln\frac{P}{P_1}\right] = 0.86 \text{ J/K and } \Delta S_{Hg} = 0\,.$$

b) $\Delta S_{ext} = 0$ and $\Delta S_{univ} = S_{created} = \Delta S_1 = 0.86\ \text{J/K}$.

6) a) As above, the fact that $\Delta S_{(2)} = 0$ implies that between variables T_2' and P_2 we have the relation: $\dfrac{P_2}{P_0} = \left(\dfrac{T_2'}{T_0}\right)^{\frac{\gamma}{\gamma-1}}$ with $T_2' = 306.75$ K

and $P_2 = P_0\left(\dfrac{306.75}{295.15}\right)^{\frac{1,4}{0,4}} = 1.144$ atm, whereby: $x = 1\left[2 - \left(\dfrac{P_0}{P_2}\right)^{\frac{1}{\gamma}}\right] = 1.092$ and so

$$T_1' = T_0.\frac{x}{1}.\frac{P_2}{P} = 295.15.\frac{1.092}{1}.\frac{1.144}{1.25} = 295.15 \text{ K, therefore } T_2' = T_0.$$

b) we have $\Delta S_{(2)} = 0$, where $\Delta S_{(1)} = n_1 R\left(\dfrac{\gamma}{\gamma-1} ln\dfrac{T_1'}{T_1} + ln\dfrac{P_1}{P_2}\right)$

$$= 0.5\times 8.3143\times\left(\frac{1.4}{0.4}\ln\frac{295.15}{307.60} + \ln\frac{1.175}{1.144}\right) = -0.49\ \text{J/K and } \Delta S_{ext} = 0.$$

7) According to the second principle, we always have $S_{created} \geq 0$; therefore, we can write the following: $S_{created} = \Delta S_{univ} = \Delta S_{(1)} + \Delta S_{(2)} + \Delta S_{ext} + \Delta S_{Hg} = -0.49\ \text{J/K}$,

which implies that the condition: $\Delta S_{Hg} - 0.49 \geq 0$ is $C_{P_{Hg}}\, ln\dfrac{T_1'}{T_0} \geq 0.49$, and therefore

$\ln\dfrac{T_1'}{T_0} \geq \dfrac{0.49}{4376} \approx \dfrac{T_1'}{T_0} \geq 1 + \dfrac{0.49}{4376} = 1 + 1.1\times 10^{-4}$ and $\dfrac{T_1'}{T_0} \geq 1 + 1 + 1.1\times 10^{-4}$. Hence $T_1' - T_0 \geq 3.3\times 10^{-2}\,\text{K}$.

PROBLEM 2.19.– Cylinder with piston of negligible mass

1) The initial and final states can be represented schematically by Figure 2.13 and the pressure at equilibrium has the expression $P = P_0 + \frac{\text{mg}}{\sigma}$.

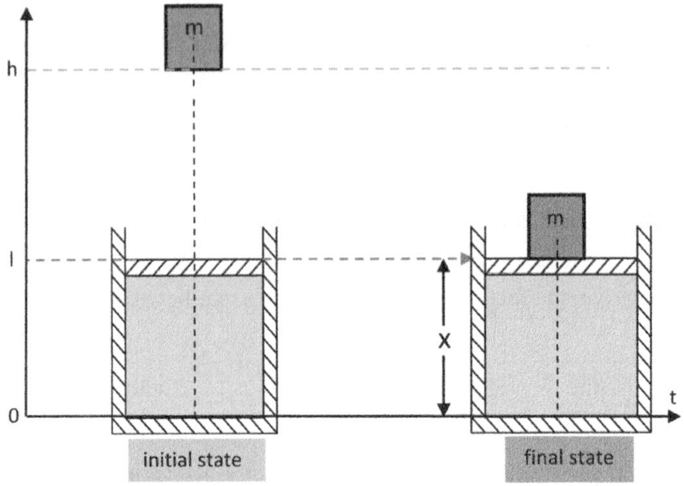

Figure 2.13. *Diagram of the transformation*

2) a) The variation in potential energy is $\Delta E_{\text{p}} = -\text{mg(h}-\text{x)}$.

b) The work is $W_{ext} = -P_0 \int dV = -P_0(V_{\text{f}} - V_{\text{i}}) = -P_0.\sigma(\text{l}-\text{x})$.

c) For U: $\Delta U_{ext} = \Delta E_{\text{p}} + W_{ext} = -\ \text{mg(h}-\text{x)} - P_0\sigma(\text{l}-\text{x})$.

d) For the interior: $\Delta U_{int} = nC_v(T_1 - T_0)$.

e) The equation of state is written as: for the final state, $nRT_1 = P\sigma.\text{x}$ and for the initial state, $nRT_0 = P_0.\sigma.\text{l}$. This gives $\Delta U_{int} = \frac{C_v}{R}.\sigma.(P.\text{x} - P_0.\text{l})$. So $\frac{C_v}{R} = \frac{1}{\gamma - 1}$, where $\Delta U_{int} = \frac{\sigma}{\gamma - 1}(P.\text{x} - P_0.\text{l})$.

3) a) At equilibrium, we have $\frac{\sigma}{\gamma-1}(P.\mathrm{x}-P_0.1)-\mathrm{mg}(\mathrm{h}-\mathrm{x})-P_0.\sigma(1-\mathrm{x})=0$. I.e. $P.\mathrm{x}-P_0.1-(\gamma-1)(P-P_0)(\mathrm{h}-\mathrm{x})-(\gamma-1)P_0(1-\mathrm{x})=0$ whose solution is: $\mathrm{x}=\frac{P_0}{P}1+\frac{\gamma\ \ 1}{\gamma}.\frac{P.P_0}{P}\mathrm{h}=1+\left(\frac{\gamma\ \ 1}{\gamma}\mathrm{h}\ \ 1\right)\frac{\mathrm{mg}/\sigma}{P_0+\frac{\mathrm{mg}}{\sigma}}$.

b) In each of the previous expressions, we need to replace the parameters with numerical values to obtain:

$$P=1+\frac{50\times 9.808}{48.4\times 10^{-4}\times 101325}=1.99997\ \#2\ \mathrm{atm}$$

$$n=\frac{101325\times 48.4\times 10^{-4}\times 0.50}{8.3143\times 295.15}\ \#0.1$$

$$\mathrm{x}=0.50+\left(\frac{4}{7}-\frac{1}{2}\right)\times 0.50=0.5357\ \#\ 0.536\ \mathrm{m}$$

$$T_1=\frac{2\times 101325\times 48.4\times 10^{-4}\times 0.5357}{0.1\times 8.3143}\ \#632\ \mathrm{K}$$

c) We obtain $\Delta S_{int}=n.\left[C_v\,ln\frac{T_1}{T_0}+R\,ln\frac{P_1}{P_0}\right]=nR\left(\frac{1}{\gamma-1}ln\frac{T_1}{T_0}+ln\frac{\mathrm{x}}{1}\right)=$ 1.6399 J/K and with $\Delta S_{\mathrm{ext}}=0$, and so $S_{created}=\Delta S_{univ}=\Delta S_{int.}=1.640$ J/K.

4) a) All heat produced is transferred to the external environment. For the gas, the reversible adiabatic compression is characterized by $PV^{\gamma}=\mathrm{Cte}$, which with $\frac{PV}{T}=\mathrm{Cte}$ gives the relation $T_2=T_0\left(\frac{P}{P_0}\right)^{\frac{\gamma-1}{\gamma}}$, whereby $T_2=359.79$ K.

b) From the equality: $\frac{P\sigma.\mathrm{x}_2}{T_2}=\frac{P_0\sigma.1}{T_0}$, we obtain the following expression:

$\mathrm{x}_2=1.\frac{T_2}{T_0}.\frac{P_0}{P}=1.\left(\frac{P_0}{P}\right)^{\frac{1}{\gamma}}$ which, after calculation, gives: $\mathrm{x}_2=0.3047\ \#\ 0.305$ m and $\Delta U_{int}=nC_v(T_2-T_0)=\frac{nR}{\gamma-1}(T_2-T_0)=134.3$ J.

c) $W_{ext} = -P_0\Delta V = -P_0.\sigma(1-x_2)$ = – 95.6 J and $\Delta E_p = -mg(h-x_2)$ = 831.2 J. So, Q received by the external environment is obtained from the energy conservation equation: $Q + W_{ext} + \Delta E_p + \Delta U_{int} = 0$, which gives: $Q = P_0.\sigma(1-x_2) + mg(h-x_2) - \frac{nR}{\gamma - 1}(T_2 - T_0) = 792.6$ J.

d) $\Delta S_{int} = 0$; $\Delta S_{ext} = \frac{Q}{T_0} = 2.685 J/K$ and $S_{created} = \Delta S_{ext} = \frac{Q}{T_0} = 2.685 J/K$.

5) a) At constant temperature T_0, $P.\sigma.x_3 = P_0.\sigma.1$, where: $x_3 = 1.\frac{P_0}{P}$, i.e. $x_3 = 0{,}25$ m and $\Delta U_{int} = 0$.

b) $W_{ext} = -P_0.\sigma(1-x_3) = -122.6$ J; $\Delta E_p = -mg(1-x_3) = -858.2J$ and $Q = -W_{ext} - \Delta E_P - \Delta U_{int}. = P_0\sigma(1-x_3) + mg(h-x_3) = 980.8J$.

c) With $\Delta S_{int} = nR\ln\left(\frac{x_3}{1}\right)$ = – 0.576 J/K and $\Delta S_{ext} = \frac{Q}{T_0} = 3.323 J/K$, then $S_{created} = \Delta S_{univ} = \Delta S_{int} + \Delta S_{ext}$ 2.747 J/K.

6) It could have been possible to reach the state proposed in (5) by the transformation described in (2), followed by another irreversible transformation accompanied by the creation of entropy. Therefore, we should find that $\Delta S_{univ}(5) > \Delta S_{univ}(2)$, which is verified (2.747 > 1.640).

Similarly, going through stage (4), we should find that $\Delta S_{univ}(5) > \Delta S_{univ}(4)$, which is verified (2.747> 2.685).

PROBLEM 2.20.– Horizontal cylinder supplied with heating resistance

1) The instrument can be represented in Figure 2.14.

2) In compartment (A), the gas passes from an initial state at (P_0, V_0, T_0) to a final state at (P_0, V and T); the ideal gas law allows us to write: $\frac{P_0V_0}{T_0} = \frac{P_0V}{T}$, i.e. $\frac{V_0}{T_0} = \frac{V}{T}$ with T_0 = 273.15 K; T = 373,15 K; $V_0 = V_A = V_B = S\frac{1}{2}$ and V = Sd.

Here, $\frac{l/2}{T_0} = \frac{d}{T}$ and d $= \frac{l.T}{2T_0} = \frac{70 \times 373.15}{2 \times 273.15} = 47.81$ cm # 0.478 m.

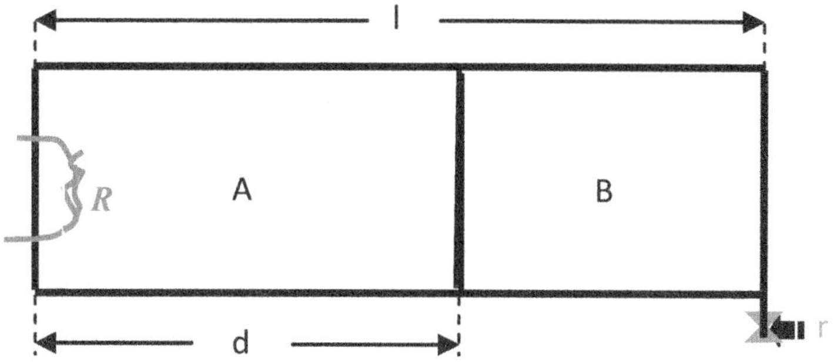

Figure 2.14. *Diagram showing cylinder with two compartments*

3) V of compartment (B) goes from $V_B = S.\frac{l}{2}$ at $V = S(l - d)$ with a variation of corresponding to V of air released to the external environment; S is then determined from (A) at the initial state, knowing that it contains 1 g of hydrogen, i.e. ½ mol under STP conditions that occupies $\frac{22.4}{2}$ L. So, $S\frac{l}{2} = \frac{22{,}4}{2}$ i.e. $S = \frac{22{,}400}{l}$ cm^2 and

$$\Delta V_B = \frac{22\,400}{l}\left(d - \frac{l}{2}\right) = \frac{22\,400\text{ d}}{l} - \frac{22\,400}{2} = \frac{22{,}400 \times 47.81}{70} - 11.200 = 4{,}096 \text{ cm}^3$$

$= 4.096$ L; with $\rho_{air} = 1.293$ g/L, giving $m_{air} = 4.096 \times 1.293 = 5.3$ g, this is the mass of air released.

4)(*) At the initial state, $V_A = V_B$ and (A) contains ½ mole and therefore (B) also contains ½ mole. The molar mass of air is equal to 29, whereas the mass of air released corresponds to $\frac{5{,}3}{29} = 0.183$ mol; the number of moles of air that remains contained is therefore: $0.5 - 0.183 = 0.317$ mol.

When the valve (r) is closed, if we return to initial STP conditions then the ratio of volumes is equal to the number of moles: $\frac{d'}{l-d'} = \frac{0.5}{0.317}$ and so d' $\frac{0.5 \times 0.7}{0.817} = 0.428$ m $= 42.8$ cm(*).

(*) We can address this question by once again writing the initial and final states for (A) and (B) in the following way:

– compartment (A):

Initial	P_0	S.d	T	$\frac{P_0 d}{T} = \frac{Pd'}{T_0}$
Final	P	S.d'	T_0	

– compartment (B):

Initial	P_0	S $(l-d)$	T_0	$\frac{P_0(l-d)}{T_0} = \frac{P(l-d')}{T_0}$
Final	P	S $(l-d')$	T_0	

Whereby: $\frac{l-d'}{l-d} = \frac{Td'}{Td}$, i.e. $l.d.T_0 - d.d'T_0 = d'.T(l-d) \Rightarrow d' = \frac{l.d.T_0}{T(l-d)+d.T_0} = \frac{70 \times 47.8 \times 273.15}{373.15\,(70-47.8) + 47.8 \times 273.15} = 42.8$ cm.

We can also express (d') as a function of l, T_0 and T, knowing that $d = \frac{l.d.}{2T_0}$ and so $d' = \frac{l.T_0\frac{l.T}{2T_0}}{l.T\left(1-\frac{T}{2T_0}\right)+\frac{l.T}{2}} = \frac{1}{2\left(1-\frac{T}{2T_0}\right)+1}\,\frac{1}{2\left(\frac{2T_0-T}{2T_0}\right)+1} = \frac{l.T_0}{3T_0-T} = \frac{0.70 \times 273.15}{3 \times 273.15 - 373.15} = 0.428$ m= 42.8 cm.

PROBLEM 2.21.– Piston-mounted pump

1) a) The diagram of this instrument can be shown in Figure 2.15.

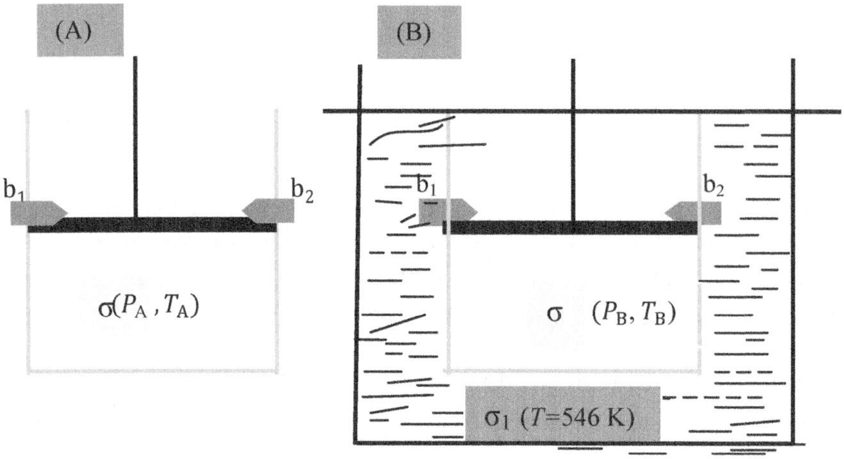

Figure 2.15. *Diagram of the body's pump*

2) According to the first principle: $\Delta U_1 = Q_1 + W_1$ and V is constant, therefore $W_1 = 0$, whereby $\Delta U_1 = Q_1$. So for an ideal gas, $\Delta U = nC_v\Delta T$, i.e. $Q_1 = C_v(T_B - T_A) = 20.8\,(546 - 409.5) = 2{,}839$ J.

3) a) Since Q_1 is the amount of heat released by σ_1 to σ, the variation in S of the external environment is therefore: $(\Delta S_{\sigma 1})_1 = -\frac{Q_1}{T_{\sigma 1}} = -\frac{2839}{546} = -5.20$ J and for the gas, we obtain $\Delta S_\sigma = S_B - S_A \Rightarrow \Delta S_\sigma = \int_A^B \frac{dQ}{T} = C_v\int_{T_A}^{T_B} \frac{dT}{T} = C_v\, ln\frac{T_B}{T_A} = 20.8\ln\frac{546}{409.5} = 5.984$ J.

b) To determine the nature of this transformation, we must calculate the total variation in entropy: $(\Delta S')_1 = (\Delta S_\sigma)_1 + (\Delta S_{\sigma 1})_1 = 5.984 - 5.20 = 0.784$ J. It is positive, and hence "irreversible", which means that, once at state (B), the gas will not be able to spontaneously release heat into the water bath to regain its initial temperature.

4) a) The diagram will be transformed as shown in Figure 2.16 where the stoppers no longer play a role.

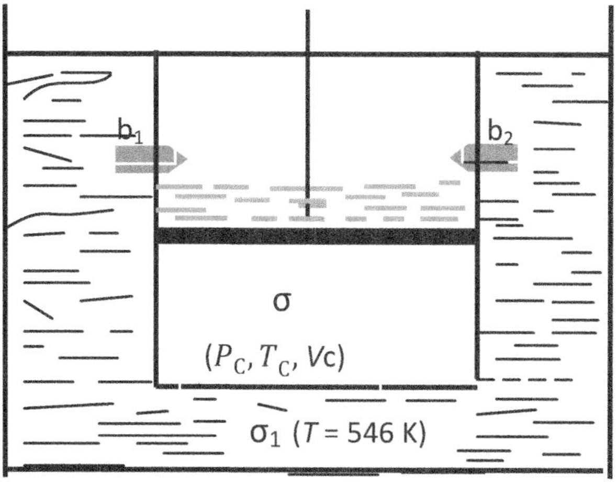

Figure 2.16. *Modified diagram of the body's pump*

b) Overloading once only causes abrupt compression of σ causing it to move from state (B) to state (C), such that $T_C = T_B$. So the transformation is isothermal, and we obtain: $P_B V_B = P_C V_C$ (Mariotte's law) whereby: $V_C = \frac{P_B V_B}{P_C} = 33.58\,\frac{1.33}{4} = 11.19$ L.

As this operation is an isothermal compression, the gas works against an external pressure of 4 atm, since this is what it reaches at equilibrium. Where $W_2 = -\int_{V_B}^{V_C} P_{ext} dV = -P(V_C - V_B) = -4\ (11.19 - 33.58) = 89.56$ L.atm; as 1 L.atm = 101.325 J(*), then $W_2 = 9{,}074.67$ J and since $\Delta U = 0 = W_2 + Q_2$, $Q_2 = -9{,}074.67$ J.

c) Temperature of the external environment $T_{\sigma 1}$ is constant, therefore: $(\Delta S_{\sigma 1})_2 = -\frac{Q_2}{T_{\sigma 1}} = \frac{9{,}074.67}{546} = 16.621$ J/mol.K, and the variation in entropy of an ideal gas during isothermal compression is given by $\int \frac{dQ}{T}$ with $dQ = dW = nRT\frac{dV}{V}$, whereby $(\Delta S_\sigma)_2 = nRTln\frac{V_C}{V_B}$, i.e. $(\Delta S_\sigma)_2 = \frac{22.4 \times 101.325}{273.15} ln\frac{11.19}{33.58} = -9.131$ J/mol.K.

d) To determine the nature of the transformation, we calculate the total variation in entropy during stage B-C: $(\Delta S')_2 = 16.621 - 9.131 = +6.490$ J. It is therefore irreversible.

In fact, the quick transition from (B) to (C) makes it impossible to define the state of the system (that is to say, the values of P, V and T) at any transitional time point.

5) The heat isolation of the gas in σ_1 prevents there from being any heat exchanged with the external environment, which makes the process adiabatic. Also, as it is reversible, we have a reversible adiabatic transformation, whereby:

a) for an ideal gas PV^γ = Cte, then $P_C V_C^\gamma = P_D V_D^\gamma$ with $\gamma = \frac{C_p}{C_v} = \frac{29.12}{20.80} = 1.4$, i.e. $V_D = V_C \left(\frac{P_C}{P_D}\right)^{1/\gamma} = 11.19 \times 4^{1/1,4} = 30.12$ L. Since $P_D V_D = RT_D$, then $T_D = \frac{P_D V_D}{R} = \frac{30.2 \times 273.15}{22.4} = 367.30$ K.

Note the significant decrease in temperature caused by this type of expansion;

b) knowing T_D, the calculation of W_3 then becomes simple(**). Knowing that for this expansion $Q_3 = 0$, then $\Delta U_3 = W_3 = C_v\ (T_D - T_C) = 20.80\ (367.30 - 546) =$ 3,716.96 J;

c) the transformation is adiabatic and reversible, i.e. isentropic, therefore $(\Delta S')_3 = 0$, and the gas does not exchange any heat with the external environment, so $(\Delta S\sigma_1)_3 = 0$, whereby $(\Delta S\sigma)_3 = -(\Delta S\sigma_1)_3 = 0$;

d) if we remove the overloading at once, the piston quickly moves back against the stoppers: the gas returns to its initial V, P and T, meaning that the system σ returns to its initial state.

However, the return trajectory will not be defined and therefore it will not be possible to determine the state of the gas at any moment (the values of P, V and T): which is precisely what characterizes irreversibility.

6) a) For the gas, the process D → A is such that: $P_D = P_A = 1$ atm, so the transformation is isobaric, where the heat exchanged at constant P, Q_p which corresponds to the ΔH of the transformation (which we now know is equal to nC_pdT), then: $Q_4 = C_p(T_A - T_D) = 29.12\ (409.5 - 367.30) = 1{,}228.86$ J.

Also, in this case, W is given by $\int(-PdV) = -P\int dV$ where $W_4 = -P\Delta V = -(V_A - V_D) = (30.12 - 33.58).101.325 = -350.58$ J.

b) As the gas exchanges Q_4 amount of heat with the external environment, the variation in entropy of σ_1 in this case is: $(\Delta S_{\sigma 1})_4 = -\frac{Q_4}{T_{\sigma 1}} = -\frac{1{,}228.86}{409.5} = -3.00$ J and for the gas it will be: $(\Delta S_\sigma)_4 = \int\frac{dQ}{T} = C_p\int\frac{dT}{T} = C_p ln\frac{T_A}{T_D} = 29.12 ln\frac{409.5}{367.30} =$ 3.167 J/mol.K.

c) To verify this process D →A, we always calculate the total variation in entropy: $(\Delta S')_4 = (\Delta S_\sigma)_4 + (\Delta S_{\sigma 1})_4 = 3.167 - 3.00 = +0.167$ J. It is > 0, and hence the transformation is certainly possible.

7) a) The table below shows the respective signs of all amounts of heat exchanged:

	A→B isochoric	B→C isothermal	C→D adiabatic	D→A isobaric
Q (J)	+2,839	−9,074.5	0	+1,228.86
W (J)	0	+9,074.5	−3,716.9	−350.58
(ΔS_σ) (J/mol K)	+5.984	−9.131	0	+3.167

b) The total values of W and Q are calculated simply by summation:

$W = 9{,}074.5 - 3{,}716.9 - 350.58 = +5{,}007.02$ J

$Q = 2{,}839 - 9{,}074.5 + 1{,}228.86 = -5{,}006.64$ J

whereby $\Delta U = W + Q = 0.38$ J # 0, which shows that for a closed transformation, U only depends on the initial and final states: this is a function of state.

c) Similarly, $(\Delta S_\sigma) = 5.984 - 9.131 + 3.167 = 0.02$ J # 0, which means that S is also a function of state.

8) The cycle can be represented as curves for each transformation and through the calculation of intermediate points, when possible, shown as follows:

a) for P, V: A → B (isochoric), equation V = Cte = 33.58 and two known points A$\begin{cases} P_A = P_D = 1 \\ V_A = 33.58 \end{cases}$ and B$\begin{cases} P_B = 1.33 \\ V_B = V_A = 33.58 \end{cases}$; B → C (isothermal), equation PV = Cte = $P_C V_C$ = 44.76 L.atm, whereby, in addition to point C{P_C = 4; V_C = 11,19}, the four intermediate points hereafter.

V (L)	15	20	25	30
P (atm)	2.98	2.23	1.79	1.49

C → D (adiabatic), equation PV^γ = Cte = $P_C.V_C^\gamma$ = 117.60 where, for example, point F{P = 2; V = 18.36} as well as point D{P_D = 1; V_D = 30,12} and those (E, F, G) obtained from the graph; D → A (isobaric), equation P = Cte = 1 atm, this is plotted on the same graph;

b) for (T, S), we have the following points: A' is given: {S_A = 1; T_A = 409.5}; B' is such that: {$S_B = S_A + 1.43$; T_B = 546}; Therefore: {$S_C = S_B - 2.18$; T_C =546}; D' is such that: {$S_D - S_C = 0$; T_D = 367.30}.

We can also plot curves B'C' and C'D' on the same graph; we find that: $S_A - S_D = S_A - S_B + S_B - S_D = -1.43 + 2.19 = +0.76$.

The equations of curves A'B' and A'D' remain to be determined, which corresponds to: A'B': isochoric, and so $S = C_v.\ln T + k$ and at point A', then: $S_A = \frac{20.8}{0.184}.ln409.5 + k = 1$ so k = 1 – 29.90 = – 28.90, where its equation: $S = \frac{20.8}{4{,}184}.lnT -$ 28.90, which allows there to be an intermediate point: M{T = 475 K, S = 4.97.ln475 – 28.90 = 30.64 – 28.90 = 1.74}; D'A': isobaric, and so $S = C_p.lnT + k'$; for A': $S_A = \frac{29.12}{4.184}.ln409.5 + k' = 1$, whereby k' = 1 – 41.90 = –40.90, and its equation is: $S = \frac{29.12}{4.184}lnT - 40.90$, which allows there to be an intermediate point: N{T = 380 K, S = 6.96 ln380 – 40.90 = 0.44}. We obtain the trace as presented on the same graph; the slopes of the curves are, respectively, $\left(\frac{dT}{dS}\right)_V = \frac{T}{C_v}$ for A'B' and $\left(\frac{dT}{dS}\right)_p = \frac{T}{C_p}$ for D'A'; at point A', the ratio of slopes is equal to $\frac{C_p}{C_v} > 1$, therefore isochoric is "faster" than isobaric, whereby the larger situation at point A', presented in Figure 2.17, giving the shape of the two curves at this point.

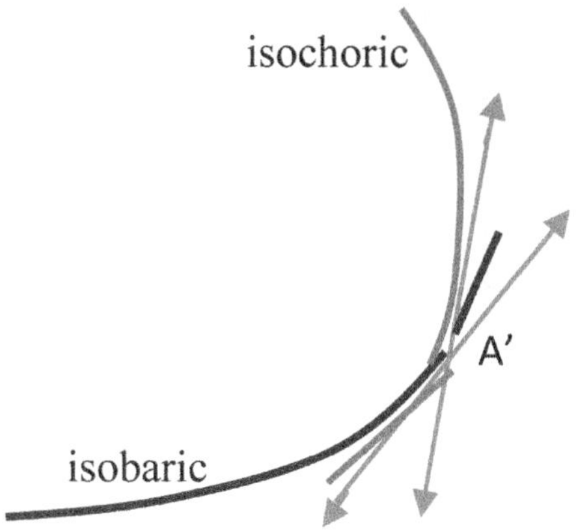

Figure 2.17. *Shape of the curves at point A'*

9) a) By definition, $\delta W = -\int PdV$, i.e. $|W\text{rev}|$ is represented by the cycle area A-B-C-D-A; similarly $\delta Q = \int TdS$, i.e. $|Q\text{rev}|$ is represented by the cycle area A'-B'-C'-D'-A'.

In fact, during the proposed closed transformation, the gas receives work and heat is released to the external environment, hence the area ABCD in $(P, V) = +W_{rev}$; area A'B'C'D'A' in $(T,S) = -Q_{rev}$. Since $\Delta U = W + Q = 0$, then $W = Q$: the cycle areas are equal.

b) The choice adopted per unit area for each system gives: in (P,V): $\begin{cases} 1 \text{ atm} \leftrightarrow 4 \text{ cm} \\ 1 \text{ L} \leftrightarrow 1 \text{ cm} \end{cases}$ so $\text{cm}^2 \leftrightarrow 0.25$ L.atm $= 0.25 \times 101.325 = 25.33$ J; in (T,S): $\begin{cases} 1 \text{ cal.K}^{-1} \leftrightarrow 4 \text{ cm} \\ 50 \text{ K} \leftrightarrow 2 \text{ cm} \end{cases}$ so 1 cal $\leftrightarrow$ 0.16 cm^2, i.e. 1 cm$^2 \leftrightarrow$ 26.15 J, where a ratio of $\frac{26.15}{25.33} = 1.03 \approx 1$ and hence the unit areas are practically the same. We should therefore verify on the graph that the cycle areas are in fact equal (≈ 36 cm^2).

c) The work calculated in (7b) is the total W, not to be confused with W_{rev} which corresponds to a reversible transformation, whereas here the isothermal part of the transformation is not reversible: W total $= W_2 + W_3 + W_4$ as shown in Figure 2.18.

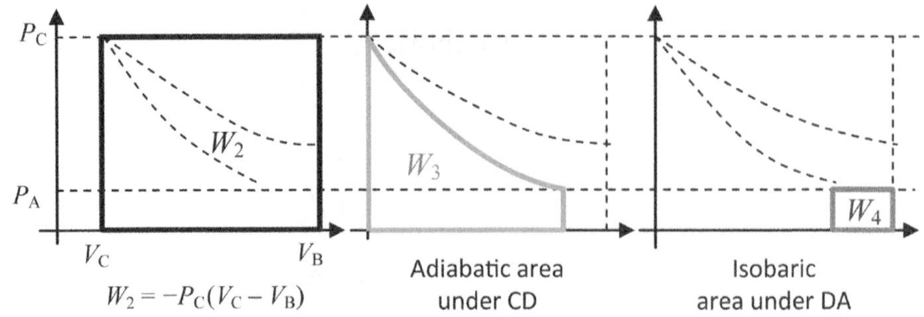

Figure 2.18. *Areas for the different works*

Hence the total work represented is shown in Figure 2.19.

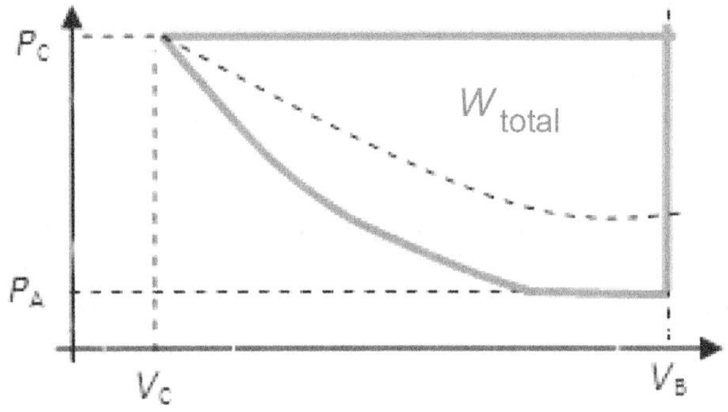

Figure 2.19. *Total work area*

(*) To go from L.atm to J, we use R in two unit systems:

$$R = \begin{cases} \dfrac{101325\text{Pa} \times 22.4 \times 10^{-3}\,\text{m}^3}{273.15\text{K}} \,/\, \text{J.K}^{-1}\text{mol.}^{-1} \\ \dfrac{1\,\text{atm} \times 22.4l.}{273.15\,\text{K}} \,/\, \text{atm.K}^{-1}\text{mol.}^{-1} \end{cases}$$

, which gives 1 L.atm = 101.325 J.

(**) If we did not previously calculate T_D, we can use the relation $W_3 = \frac{1}{\gamma - 1}(P_D V_D - P_C V_C)$, since: $T = \frac{PV}{nR}$, $C_p - C_v = R$ and $C_p = \gamma C_v$.

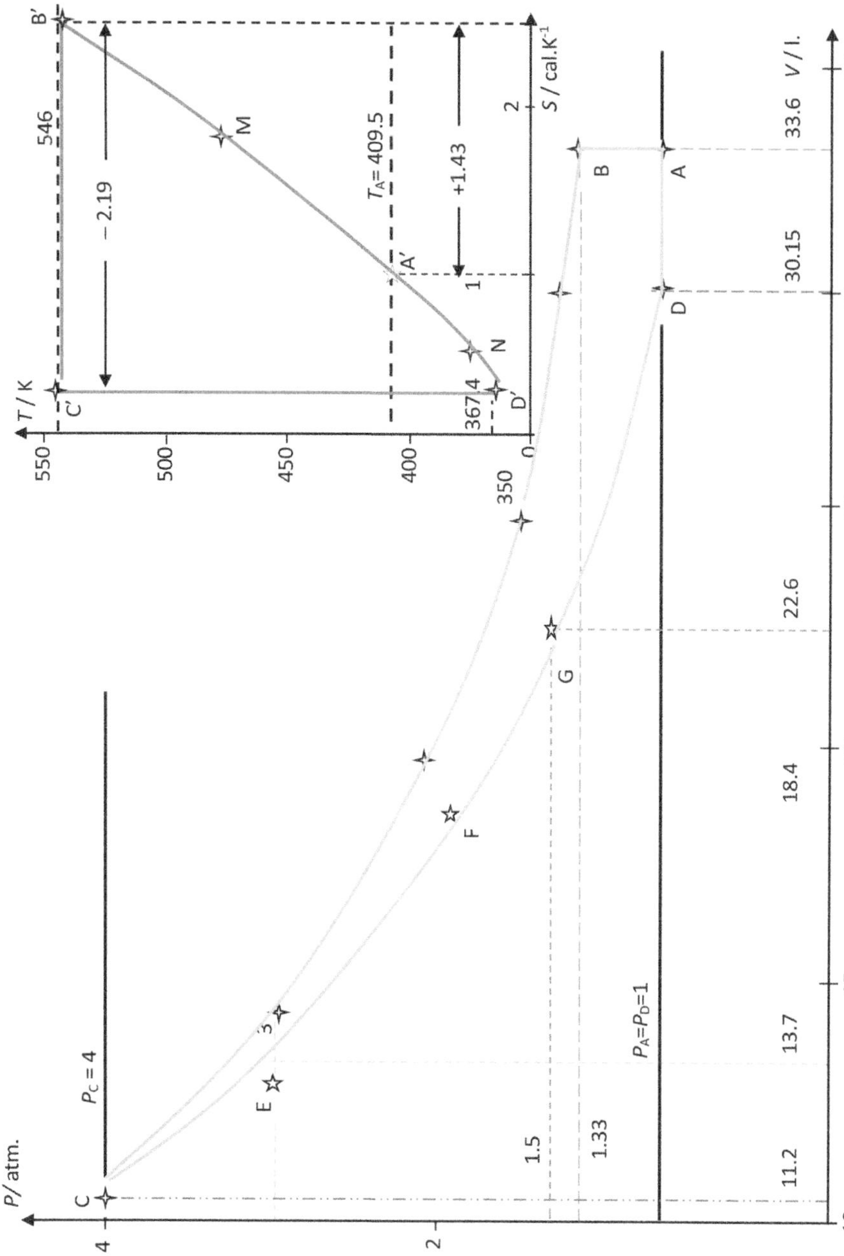

Figure 2.20. *Graphs P = f(V) and T =f(S)*

3

Open and Reacting Systems During Reaction

3.1. Exercises

EXERCISE 3.1.– Heat of reaction at constant *P* and *V*

Consider the following reactions:

(1) $H_{2liq} + ½O_{2gas} \rightarrow H_2O_{liq}$ at 25 °C;

(2) $CH_3CO_2C_2H_{5liq} + H_2O_{liq} \rightarrow CH_3CO_2H_{liq} + C_2H_5OH_{liq}$ at 25 °C;

(3) $N_{2gas} + 3H_{2gas} \rightarrow 2NH_{3gas}$ at 400 °C.

Assuming the gases obey the ideal gas law, calculate the difference in heat of reaction at constant pressure (Q_P) and at constant volume (Q_v) for each reaction.

Answers: $\Delta Q_1 = -888.64$ cal, $\Delta Q_2 = 0$ and $\Delta Q_3 = -2{,}675.10$ cal.

EXERCISE 3.2.– Enthalpy of formation of a component

The variations in enthalpy during combustion of components are as follows:

– ethyl alcohol: −326.70 kcal;

– carbon: −94.052 kcal;

– saccharose: −1,352 kcal;

– hydrogen: −68.317 kcal.

Calculate the enthalpy of formation of ethyl alcohol (1) and saccharose (2) at $T = 25$ °C.

Answers: $\Delta H_1 = -66.355$ kcal and $\Delta H_2 = -527$ kcal.

EXERCISE 3.3.– Chemical potential of water

Let us consider water in its stable state under atmospheric pressure, and its temperature will vary between 25 °C and 200 °C.

Express its chemical potential as a function of temperature from its chemical potential at 25 °C.

Data: Molar heat of vaporization at 100 °C = 9.770 cal/mol; molar specific heat of the liquid = 18 cal/mol·K; molar specific heat of the solid = 8 cal/mol·K; standard entropy of the liquid at 25 °C = 16.7 cal/mol·K.

Answers: liquid water ($T < 373$ K): $g_l\ (T) - g_l\ (298\ \text{K}) = -387.6 + 103.86T - 41.44\ T\log(T)$; water vapor ($T > 373$ K): $g_v\ (T) - g_l\ (298\ \text{K}) = 13{,}113 + 8.44T - 18.4\ T\log(T)$.

EXERCISE 3.4.– Boudouard reaction

By studying the transformation of carbon dioxide into carbon monoxide in the presence of graphite, Boudouard obtained for the standard enthalpy and the standard entropy the following values:

	CO_2	CO	C
ΔH°_{298}/ (kcal·mol^{-1})	–94.03	–26.39	0
ΔS°_{298}/ (cal·mol^{-1}K^{-1})	51.09	47.32	1.36

We propose to determine from which temperature this reaction is possible, assuming we have ideal gases and the variation in C_P is negligible.

1) Write the Boudouard reaction.

2) Determine the expression for $\ln K_p$ as a function of T.

3) Calculate $\ln K_p$ for the following values of T: 600 K, 800 K, 1,000 K, 1,200 K, 1,500 K and 2,000 K.

Answers: For (1), (2) and (3), see corrections. T should be $\geq 10^3$ K.

EXERCISE 3.5.– Calculation of chemical potential at infinite dilution

Consider a diluted solution of titanium (1) in iron (2) at constant pressure.

1) Show that the variation in molar Gibbs free energy of titanium with respect to two different references, one being a pure body (denoted*) and the other at infinite dilution (denoted $^{\infty}$), is written as: $\Delta g_1 = RT\ln\gamma_1\frac{M_2}{100M_1}$.

2) Calculate its value at T = 1,623 °C.

Data: $M_1 = 47.9$ g/mol; $T_{1f} = 1{,}660$ °C; $\Delta H_1 = 4{,}500$ cal/mol, $\gamma_1 = 0.011$, and $M_2 = 55.85$ g/mol.

Answers: (1) see corrections; (2) $\Delta g_1 = -33.76$ kcal/mol.

EXERCISE 3.6.– Chemical affinity of supercooled water

Consider supercooled water maintained at a temperature of –10 °C by or in contact with the external environment, whereas its normal melting point is $T_0 = 0$ °C, since the following transformation is induced (e.g. by shock): liquid water → water solid.

1) Express the variations in enthalpy and entropy of water during this transformation.

2) Express the variations in enthalpy and entropy of the external environment.

3) Express entropy created over the course of the transformation.

4) Calculate the affinity of the transformation at the same temperature.

Data: $L(T_0) = 80$ cal/g; $C_{p\,l} = 1$ cal/g·K; $C_{p\,S} = 0.5$ cal/g·K.

Answers: (1), (2) and (3) see corrections; (4) $A = 2.838$ cal/g.

EXERCISE 3.7.– Decomposition of ammonium chloride

In a closed recipient, of constant volume, there are 0.1 mole of solid ammonium chloride (occupying a negligible volume) and 0.3 mol of ammonia gas at 1 atm and 25 °C. It is heated until the decomposition of NH_4Cl into NH_3 and gaseous HCl.

1) Determine, as a function of T and ξ, the partial pressures and total pressure in the recipient.

2) If the partial pressures are linked between them by the law of mass action of constant K(T), determine the relation between P and T at equilibrium.

3) If the decomposition is total, what is the value of ξ?

Answers: (1) $P_{NH_3} = \frac{0.3+\xi}{0.3} \cdot \frac{T}{298.15}$, $P_{HCl} = \frac{\xi}{0.3} \cdot \frac{T}{298.15}$ $P = \frac{0.3+2\xi}{0.3} \cdot \frac{T}{298.15}$;

(2) $P^2 = \left(\frac{T}{298.15}\right)^2 + 4K$; (3) $\xi = 0.1$.

EXERCISE 3.8.– Industrial manufacture of ethanol

The industrial manufacture of ethanol occurs via the hydration of ethylene in the vapor phase, according to the reaction: $C_2H_4 + H_2O \leftrightarrows C_2H_5OH$ which depends on temperature and pressure.

In a reactor maintained at 125 °C and 101.3 kPa, a gaseous mixture is introduced composed of 25% ethylene and 75% water vapor.

1) Calculate the reaction progress.

2) Deduce the composition of each body in the reactor.

Data: $\Delta G_T^\circ = 4.530$ J/mol.

Answers: (1) ξ # 0.04; (2) $y_{C_2H_4}$# 0.219; y_{H_2O} # 0.740; $y_{C_2H_5OH}$ # 0.041.

EXERCISE 3.9.– Cracking of methane

The experiment showed that heating pure methane at 1,000 K, at a constant pressure of 1 atm, uniquely gives the following equilibrium reactions:

$CH_4 \rightleftarrows C_{graphite} + 2H_2$ (1)

$2CH_4 \rightleftarrows C_2H_6 + H_2$ (2)

$2CH_4 \rightleftarrows C_2H_4 + 2H_2$ (3)

1) Determine the partial pressures of the gas phase assuming that the gases are ideal.

2) Study the influence of total pressure of this composition at constant temperature.

Data:

At 10^3K	Δh_f°/ (kcal·mol)	S° (cal·K^{-1}·mol^{-1})
Graphite	0	5.844
CH_4	21.482	59.141
C_2H_6	25.280	79.390
C_2H_4	9.204	72.066
H_2	0	39.702

Answers: (1) $P_{H_2} = 0.920$ atm; $P_{CH_4} = 8.06 \times 10^{-2}$ atm; $P_{C_2H_4} = 5.45 \times 10^{-7}$ atm and $P_{C_2H_6} = 1.45 \times 10^{-6}$ atm; (2) see corrections.

EXERCISE 3.10.– Synthesis of methanol

The industrial synthesis of methanol via CO and H_2 occurs with a catalyst and under pressure at temperatures between 200 and 300 °C.

Let us study the reaction at 500 K and at a total pressure of 20 atm as a function of ξ.

1) Write the chemical reaction.

2) Calculate and represent graphically, as a function of ξ, the affinity and the variation in Gibbs free energy.

3) Calculate the progress at equilibrium.

Data: At t = 0, $n_{CO} = 100$, $n_{H_2} = 200$ and 0 mol of methanol; standard affinity at 500 K: $\overset{\circ}{A}_{500} = -4.755$ kcal/mol.

Answers: (1) $A(T) = 2{,}584 - 10^3 \ln \frac{\xi(300 - 2\xi)^2}{(100 - \xi)}$; $G(\xi) - G(0) = 329{,}100 - 2{,}584\xi - 3 \times 10^5 \ln \frac{300-\xi}{100-\xi} + 10^3 \xi \ln \frac{(300 - 2\xi)^2}{(100 - \xi)^3}$; (2) $\xi_e = 46.86$ mol.

EXERCISE 3.11.– Electromotive force of a battery cell

We propose the following battery cell: in the first compartment, there is a silver electrode in contact with AgCl, and in the second compartment, mercury chloride is in contact with Hg_2Cl_2. The electrical contact between the two compartments is ensured by a connection containing a solution of KCl.

1) Write the reaction of this battery.

2) Calculate the electromotive force (E) of this cell at 25 °C.

3) Determine the variation coefficient of this electromotive force with temperature.

Data:

At 25 °C	$\Delta H5_f^{\circ}$ (kcal·mol^{-1})	S (cal·mol^{-1}·K^{-1})
AgCl	–30.262	22.97
Ag	0	10.206
Hg_2Cl_2	–63.32	46.8
Hg	0	18.5

Answers: (1) see corrections; (2) E = 0.041 V; (3) $\left(\frac{\partial E}{\partial T}\right)_{P,V} = 3.4 \times 10^{-4}$ V/K.

EXERCISE 3.12.– Determining the electrode potential

The aqueous solution of NaF: $NaF_{cr} \rightleftarrows NaF_{aq}$ is characterized by the following data:

– a solubility limit at 298.15 K: 42.2 g/kg of water;

– an activity coefficient form ≈ 1: $\gamma_{\pm} = 0.60$;

– an oxidation potential of sodium: $E^{\circ}_{Na/Na^{+}} = 2.714$ V;

– a $\Delta H^{\circ}_{298} = 190$ cal/mol.

At 298.15 K and 1 atm	Na_{cr}	NaF_{cr}	$F_{2,gas}$
s° (cal·mol^{-1}K^{-1})	12.2	13.1	48.6
$\Delta H8^{\circ}_{f}$ (cal·mol^{-1})	/	−136,330	/
M (g·mol^{-1})	/	41.99	/

1) Determine the standard potential of electrode $F_{2,gaz} / F^{-}_{aq}$.

2) Determine the standard enthalpy of formation and standard entropy of NaF_{aq}.

Answers: (1) $V^{\circ}_{F_{2,gaz}/F^{-}_{aq}} = -2.87$ V ; (2) $\Delta H^{\circ}_{f,NaF_{aq}} = -136{,}140$ cal/mol and $S^{\circ}_{NaF_{aq}} =$ 11.73 cal/mol.K.

EXERCISE 3.13.– Dissociation of carbon dioxide

The measurements of the equilibrium constant of the dissociation reaction of carbon dioxide gave the following relation: $\ln K_p = -\frac{135{,}280}{RT} + \frac{41.49}{R}$.

1) Write the reaction and establish the expression for the variation in *G*.

2) Deduce the variations in enthalpy and entropy during this reaction and explain why they are independent of temperature.

3) Calculate the standard enthalpy of formation and standard entropy of the oxide of carbon at 298 K.

Data: CO_2: $\Delta H^{\circ}_{f298} = -94.0$ kcal/mol and $S^{\circ}_{298} = 51.0$ cal/mol.K; O_2: $S^{\circ}_{298} =$ 49.0 cal/mol.K.

Answers: (1) see corrections; $\Delta G^{\circ}_{T} = 153{,}280 - 41.49\,T$; (2) $\Delta H^{\circ}_{T} =$ 135,280 cal/mol^{-1}; $\Delta S^{\circ}_{T} = 41.49$ cal/mol.K; see corrections; (3) $\Delta H^{\circ}_{f298(CO)} =$ 26.36 kcal and $\Delta S^{\circ}_{298(CO)} = 47.24$ cal/K.

EXERCISE 3.14.– Preparation of α-cyanopyridine

The preparation of this molecule can occur at 25 °C and can be carried out in the laboratory according to the reaction: $C_2N_{2gas} + C_4H_{6gas} \rightleftarrows C_6H_4N_{2sol} + H_{2gas}$ and the gaseous mixture is similar to an ideal gas.

1) Perform the thermodynamic calculation to confirm this possibility.

2) Calculate the equilibrium constant of the reaction at atmospheric pressure.

Data:

	C_4H_{6gas}	C_2N_{2gas}	$C_6H_4N_{2sol.}$	$H_{2,gas}$
ΔH°_{f298}/ (cal.mol^{-1})	26,748	71,820	62,000	/
S°_{298}/ (cal.mol$^{-\circ}$.K^{-K})	66.42	57.64	77.09	31.21

Answers: (1) see corrections; (2) K_P # 2×10^{23}.

EXERCISE 3.15.– Patent for benzene synthesis

A company proposes a patent, describing the equipment and mode of operation, to synthesize benzene via the hydrogenation of carbon at a temperature of 298 K.

1) Write the reaction equation.

2) Thermodynamically, after performing the calculation, is this process possible?

Data:

At 298 K	C_6H_6	C	H_{2gas}
ΔH°_{comb}/ (kcal)	–789.1[1]	-94.05	–68.32[1]
S°_{298}/ (cal.mol^{-1})	64.34	1.36	31.21

Answers: (1) and (2) see corrections.

1 The water formed is in the liquid form.

EXERCISE 3.16.– Synthesis of iodine monobromide

The synthesis of iodine monobromide is obtained by combining iodine and bromine in the gas phase at a pressure of 1 atm. The standard data provided by tables in the literature are similar to the following table:

		ΔH_f° kcal/mol	ΔG_f° kcal/mol	S° cal/mol.K	C_P° cal/mol.K
Br_2	Liquid	0	0	36.4	17.2
	Gas	7.34	0.751	58.639	8.6
I_2	Solid	0	0	27.9	13.14
	Gas	14.876	4.63	62.28	8.81
IBr_{gas}		9.75	0.91	68.10 or 61.80	8.69

1) Write the reaction and verify the consistency of numerical data for ΔH°, ΔG° and S°. What is an appropriate value for the entropy of iodine monobromide?

2) Calculate the enthalpy of formation of bromine at 114 °C.

3) At 25 °C, what is the equilibrium constant of the reaction?

4) Calculate reaction yield at 25 °C, for an initial equimolar composition of iodine and bromine.

Answers: (1) see corrections; $S^\circ = 61.80$ cal/mol.K; (2) $\Delta H_{f114(IBrgas)} =$ 1,359.24 cal/mol; (3) $K_{P298} = 407.16$; (4) ρ # 91%.

EXERCISE 3.17.– Effect of pressure on chemical equilibrium

The reaction of formation of $COCl_{2(g)}$ occurs in the gas phase and is written as: $CO_{(g)} + Cl_{2(g)} \rightleftarrows COCl_{2(g)}$. At temperature T and total pressure of 1 atm, we find $K_p = 15$. This mixture is compressed, at constant T, so that P reaches 2 atm.

Discuss the variations in the number of moles of each component and the total number of moles.

Data: At t = 0, $n_{CO} = n_{Cl_2}$ and $n_{COCl_2} = 0$.

Answer: see corrections.

EXERCISE 3.18.– Metal/refractory interaction

Here, we examine whether it is possible to metl a rigorously pure iron, under inert atmosphere, in aluminum crucibles without contaminating it with aluminum or oxygen from these same crucibles.

Aluminum should dissolve in the iron, giving an ideal solution, and simultaneous dissolutions of O_2 and Al_2O_3 in Fe have no influence on one another as long as they remain in small amounts.

1) Write the reactions that occur in this case.

2) Determine the titers of dissolved oxygen and aluminum.

3) Draw conclusions.

Data: The affinity of the reaction from dissolution of O_2 in the iron for the expression: $A(T) = 27{,}900 - 6.08T\,4.57T\ln x_O + \frac{4.57}{2}T\ln P_{O_2}$.

At 1,600 K	Al_2O_3	Al	O_{CC}
ΔC_f° (kcal.mol^{-1})	402.12	/	/
s° (cal.mol^{-1}.K^{-1})	57.56	20.80	62.22

Answers: (1) see corrections; (2) $x_O = 1.652 \times 10^{-5}$; $x_{Al} = 1.101 \times 10^{-5}$; (3) see corrections.

EXERCISE 3.19.– Chemical variable, equilibrium constant and affinity

Gaseous SO_3 is prepared, in a reactor at 600 °C and under 2 atm, according to the reaction: $2SO_{2(g)} + O_{2(g)} \rightleftarrows 2SO_{3(g)}$. The initial gaseous mixture is composed of 100 mol of SO_2 and 25 mol of O_2.

1) Calculate the standard affinity at 600 °C and deduce the value of the equilibrium constant.

2) Give the expression for the affinity of the reaction as a function of ξ and K_p.

3) Write the equilibrium condition.

4) Estimate the value of the chemical variable at equilibrium.

Data:

At 298 K	S	O_2	SO_2	SO_3
s° (cal/K.mol)	7.62	49	59.4	61.2
Δg° (cal/mol)	0	0	−71,790	−88,520

Answers: (1) $A^{\circ}(873) = 7.36$ kcal/mol; (2) $K_p = 67.52$; (3) $A(T) = \frac{(25-\xi)(50-\xi)^2}{(125-\xi)\xi^2}$; (4) ξ # 24.35.

3.2. Problems

PROBLEM 3.1.– Fuel cell

Consider a fuel cell undergoing the oxygenation of H_2 at 1 atm by O_2 under the same pressure, to give H_2O also at 1 atm, at a temperature varying between 25 °C and 200 °C.

Knowing that the electromotive force (E) of this fuel cell is associated with the affinity (A) of the reaction of formation of 1 mole of water by relation $A = 2F.E$.

1) Explain how this relation arises.

2) Establish the expression for E as a function of T.

Data: H_2O: $C_{pl} = 18$ cal/mol.K; $C_{pv} = 8$ cal/mol.K; $\Delta h_{vap}(100\ °C) =$ 9,720 cal/mol; $s^{\circ}_{l(25°C)} = 16.7$ cal/mol.K; $s^{\circ}_{O_2\,298} = 49.0$ cal/mol.K; $s^{\circ}_{H_2\,298} = 31.21$ cal/mol.K.

Answers: see corrections.

PROBLEM 3.2.– Dismutation of iron protoxide

Consider the dismutation of FeO into Fe and Fe_3O_4, disregarding the non-stoichiometrics of the oxides and their specific heat.

Determine, graphically represent and discuss as a function of temperature the affinity of this transformation.

Data:

At 1,000 K	ΔH_f° (cal.mol^{-1})	$\Delta G1_f^\circ$ (cal.mol^{-1})
FeO	–62,800	–4,755
Fe_3O_4	–259,700	–188,900

Answer: $A°(T) = 8{,}500 - 9.8T$; see corrections.

PROBLEM 3.3.– Gasification of carbon by water

The gaseous mixture $CO + \frac{1}{4}H_2 + \frac{3}{2}N_2$, produced by the gasification of 12 g of C by H_2O, occurs at a temperature of 323 K under atmospheric pressure burning adiabatically in a quantity of air ($O_2 + 4N_2$).

The combustion is complete and the combustion products, considered as ideal gases, return to the initial temperature.

1) Assess the heat released at constant volume and constant pressure.

2) Calculate the temperatures at which adiabatic combustion at constant V, then at constant P, will give burnt gases.

Data: At constant V and 0 °C: $\Delta H_{C(CO)} = -68.20$ kcal; $\Delta H_C\,(H2) = -68.40$ kcal; diatomic gas: $C_v = 4.8 + 0.0012\,T$ (°C); water vapor: $C_v = 5.9 + 0.0043\,T$ (°C); CO_2: $C_v = 6.7 + 0.0052T$ (°C); at 0 °C: ΔH_v (H_2O) = 606.0 cal/g; at 50 °C: $P_v(H_2O) = 9.2$ cm Hg.

Answers: (1) $(Q_v)_{50} = 82.609$ kcal and $(Q_p)_{50} = 83.010$ kcal; (2) $T_v = 1{,}757$ °C and $T_p = 1{,}423$ °C.

PROBLEM 3.4.– Standard chemical potential of water

Consider water in its standard liquid or vapor state, at a temperature T between 298 K and 473 K; the standard heat capacities C°_{Pl} and C°_{pv}, of the liquid and vapor respectively, are independent of T. L°_v is the normal latent heat of vaporization.

1) Determine for $T > 298$ K the expressions:

a) standard molar enthalpy of liquid water and vapor;

b) standard molar entropy of liquid water and vapor;

c) standard molar chemical potential of liquid water and vapor.

2) Give the expression for the standard affinity of vaporization of water as a function of T, disregarding the influence of the standard heat capacities.

3) Calculate the variation in entropy of the universe for spontaneous evaporation at 105 °C under atmospheric pressure and spontaneous condensation at 95 °C under the same pressure for 1 kg of water.

Answers: (1) see corrections; (2) $A(T) \# \overset{\circ}{A}(T) \# \overset{\circ}{L_v}\left(\frac{T}{373.15}-1\right)$; (3) $S^{v}_{created} = \Delta S^{v}_{univ..} = 19.24$ cal/K and $S^{c}_{created} = \Delta S^{c}_{univ.} = 19.75$ cal/K.

PROBLEM 3.5.– Partial pressure and equilibrium constant

The reaction of formation of carbon monoxide: $C_{(g)} + CO_{2(g)} \rightleftarrows 2CO_{(g)}$ at 1,150 K occurs with a variation in enthalpy of 41 kcal/mol.

1) Express K_p as a function of the chemical variable and pressure.

2) At 1,150 K, $K_p = 22.4$: calculate P_{CO} and P_{CO_2} at equilibrium, knowing that P remains constant and equal to 1 atm.

3) In which direction will the equilibrium of this reaction move if:

a) the total pressure is increased;

b) the temperature is increased;

c) a catalyst is used;

d) an inert gas at constant volume is added;

e) carbon is added;

f) carbon dioxide is added.

4) Calculate K_p at 1,120 K.

Answers: (1) $K_p = \frac{4\xi^2}{1-\xi^2}P$; (2) $P_{CO} = 0.95$ and $P_{CO_2} = 0.04$; (3) see corrections; (4) K_{p1120} # 13.84.

PROBLEM 3.6.– Corrosion of a pipe

Reducing gases (CO and H_2) pass through a chrome pipe, at constant P and 1,500 K, to determine maximum levels of oxidant impurities (CO_2 or H_2O) to avoid oxidation of the pipe.

1) Calculate the maximum admissible level of hydrogen.

2) Calculate the maximum admissible level of carbonic anhydride, into the oxide of carbon.

3) Determine the limits in the case where the oxide of carbon contains traces of water and carbonic anhydride.

Data:

	Cr_2O_3	H_2O	CO_2	CO
$\Delta G O_f^{\circ}$(1,500 K) (cal.mol^{-1})	–178,600	–39,260	–94,710	–58,370

Answers: (1) $\tau = 10^{-3}$; (2) $\tau = 4.18 \times 10^{-4}$; (3) see corrections.

PROBLEM 3.7.– Synthesis of cyclohexane

At 500 K in the presence of a catalyst, oxygen at a partial pressure of 10 atm is reacted with benzene to form cyclohexane; under these conditions, benzene and cyclohexane are liquids and form an ideal solution.

1) Calculate the concentration of cyclohexane at equilibrium.

2) Calculate the total pressure of the mixture.

Data: At 500 K: $\log K_{f C_6H_6 gas} = -17.1521$; $\log K_{f C_6H_{12} gas} = -14.8932$; P_{vap} of the pure bodies: $\log P^* C_6H_6 = 1.336$; $\log P^* C_6H_6 = 1.295$.

Answers: (1) $x \cong 1$ (pure cyclohexane); (2) $P = 29.72$ atm.

PROBLEM 3.8.– Synthesis of ethanol

The reaction of water in the gas phase with ethylene to give ethanol occurs at $T = 370$ K and $P = 1.5 \times 10^5$Pa, in a reactor of constant volume.

1) a) Write the chemical reaction.

b) Calculate its equilibrium constant.

In this reactor, we introduce 33.6 g of ethylene and 2.4 g of water vapor.

2) a) Are they in stoichiometric proportions, or is one of the two in excess?

b) Calculate the amount of ethanol obtained when equilibrium is reached.

3) In which direction will the equilibrium displace if:

a) we increase T;

b) we increase P;

c) we add water vapor.

Data: At 420 K, $\Delta G_r^\circ = 9.875$ kJ/mol; $\Delta H_r^\circ = -40.17$ kJ/mol deemed constant at (350 K; 500 K).

Answers: (1) (a) see corrections; (b) $K_p = 0.28$; (2) (a) see corrections; (b) $m_{C_2H_5OH}$ # 6.12 g; (3) see corrections.

PROBLEM 3.9.– Preparation of sulfuric acid from gypsum

The dissociation of pure anhydrous calcium sulfate by heating at atmospheric pressure primarily gives CaO, SO_2 and O_2; but in the presence of silica in excess, we obtain the (SiO_2, 2CaO) form by the reaction with the released calcium oxide.

1) Write the reaction in the first case and calculate the decomposition temperature T_1.

2) Show that the hypothesis about the absence of formation of SO_3 hypothesis is valid.

3) Write the reaction in the second case and calculate the decomposition temperature T_2.

4) Is the quantity of SO_3 formed in this case appreciable?

Data:

At 298 K	State	ΔG_f°/ (cal)	ΔH_f° (cal)	s° (cal.mol$^{-\circ}$.K^{-K})
$CaSO_4$ CaS CaO SiO_2	Solid Solid Solid Solid	/	–342,400 –129,700 –151,800 –216,000	25.5 13.5 9.5 10.0
O_2	Gas	/	/	49.0
$(SiO_2,2CaO)$	Solid	–71,700	/	29.1
SO_2	Gas	/	–70,900	/
SO_3	Gas	–88,700	–94,600	/
S	Solid	/	/	7.6

Reaction: $SiO_2 + 2CaO \rightleftarrows (SiO_2,2CaO)$: $\Delta H^\circ_{298} = -3 \times 10^4$ cal.

Answers: (1) T_1 = 1,444.2 °C; (2) $P_{SO_3} = 10^{-2}$ atm; (3) T_2 = 1,227.85 °C; (4) $P_{SO_3} = 0.02 P_{SO_2}$.

PROBLEM 3.10.– Conversion of methane

The conversion of methane by water vapor, in the presence of a nickel catalyst, occurs according to the following reaction: $CH_{4gas} + H_2O_{gas} \rightleftarrows CO_{gas} + 3H_{2gas}$ with $\Delta H^\circ_{298} = 49,277$ cal.

1) Determine the ΔS°_{298} of this reaction.

2) From the previous values:

a) give an approximate expression for the standard affinity of the reaction as a function of T;

b) deduce an approximate value for the constant K_1 of this reaction at 1,000 K.

The reaction equilibrium occurs at a total pressure P and temperature T, for an initial mixture of 100 mol of methane and 100 n mol of water.

3) From the previous results, write the affinity of the reaction as a function of T, P and ξ.

4) Write the law of mass actions for this reaction.

5) We aim to convert 99% of the methane into carbon monoxide at a pressure of 1 atm with $n = 3$.

a) Calculate the approximate temperature T_1 at equilibrium.

b) Calculate more accurately the temperature T_1', using the values ΔH_T° and ΔS_T° of the reaction such that T is closest to T_1'.

c) Determine the total pressure P at which one would obtain a 99% conversion of methane with this new value of T_1' and for $n = 1$.

6) Compare these two results with case $P = 1$ atm and $n = 1$ for 99% conversion. What conclusions can be drawn?

Data:

	CH_4	H_2O	CO	H_2
S°_{298} (cal.K^{-1})	44.480	45.105	47.301	31.210
S°_{1200} (cal.K^{-1})	62.424	57.445	57.581	41.040

Answers: (1) $\Delta S^\circ_{298} = 51.346$ cal/K ; (2) (a) $A_1^\circ = -\ 49{,}277 + 51.346\ T$; (b) $K_1(10^3) = 2.83$;

(3) $A_1 = -\ 49{,}277 + 51.346T - RT \ln \dfrac{27P^2\xi^4}{(100n-\xi)(100-\xi)\left[100(n+1)+2\xi\right]^2}$;

(4) $\dfrac{27P^2\xi^4}{(100n-\xi)(100-\xi)\left[100(n+1)+2\xi\right]^2}$ $\exp\left(\dfrac{51.346 - \dfrac{49\ 277}{T}}{1.9872}\right)$; (5) (a) $T_1 = 1{,}115$ K ;

(b) $T_1' = 1011$ K ; (c) $P = 0.047$ atm; (6) $T_1'' = 1{,}307$ K. So $n = 3$, an excess of water vapor, is more advantageous.

PROBLEM 3.11.– Cyclization of hexane

Consider the gas phase reaction, at a constant pressure of 1 atm, of pure n-hexane to cyclohexane, which is written as: $C_6H_{14} \rightleftarrows C_6H_{12} + H_2$. This is accompanied by serious restrictions in the movements permitted by atoms, which causes a significant decrease in entropy.

This phenomenon can be verified using the Van Krevelen and Chermin method, with the tables providing the following data:

At 298.15 K	C_6H_{14}	C_6H_{12}
$\Delta G2_f$ (kcal)	–0.06	7.59
S°_{298} (cal.K^{-1})	92.4	71.27

1) Calculate the enthalpy of formation of n-hexane and cyclohexane gas.

2) Calculate the Gibbs free energy of reaction at 1,000 K and determine the proportion of cyclohexane formed.

3) The normal entropy of hydrogen at 25 °C is 31.2 cal/mol.K. Calculate the difference in entropy between n-hexane and cyclohexane.

4) Why does the Van Krevelen and Chermin method explain the lower entropy of cyclohexane?

Answers: (1) $\Delta G^{\circ}_{1f} = -58.3$ cal/mol; $\Delta G^{\circ}_{2f} = 7\,7{,}370$ cal/mol; (2) $\Delta G^{\circ}_{r} =$ 330 cal/mol and $y_2 = 0.20$; (3) $\Delta S^{\circ} = 10.03$ cal/mol.K ; (4) see corrections.

PROBLEM 3.12.– Dissociation of water

Consider water vapor at 1 atm that is heated to 2,000 °C, a temperature at which there is a partial dissociation in O_2 and H_2. Let α be the degree of dissociation and K_p the equilibrium constant of the reaction, with the effects of C_P being negligible.

1) a) Write the number of moles of all components of the expected reaction.

b) Starting with only water vapor, what is the variance in the system?

2) a) Give the expression for K_p as a function of p and α.

b) Determine from this relation the variance in equilibrium.

c) Calculate the constant K_p at 2,000 °C.

3) What is the value of α of water vapor at 1 atm (note that α is much smaller than 1)?

4) If we want to obtain a dissociation rate of 10% at a temperature of 2,000 °C, what pressure would be required?

Data:

	H_2O_{liquid}	H_{2gas}	O_{2gas}
ΔH°_{f298} (kcal.mol)	–68.3	0	0
ΔS°_{298} (cal.mol.K)	16.7	31.2	49
ΔH°_v (cal.mol)	9,800	/	/

Answers: (1) (a) see corrections; (b) v = 2; (2) (a) $K_P = \frac{\alpha^3 P}{(1-\alpha)^2(2+\alpha)}$; (b) see corrections; (c) $K_P = 2.05 \times 10^{-6}$ atm; (3) $\alpha = 1.6 \times 10^{-2}$; (4) $P = 2.65$ torr.

PROBLEM 3.13.– Synthesis of nitrogen monoxide

The mixture of one molecule of nitrogen and one molecule of oxygen leads to the synthesis of nitrogen monoxide according to an equilibrium whose variation in enthalpy during the formation of 1 mole of monoxide is equal to 21.60 kcal/mol, assuming that the calorific capacities do not vary with temperature.

1) Write the reaction equation and calculate its equilibrium constant at 10^3, 2×10^3, 3×10^3 and 4×10^3 K:

a) assuming that the diatomic gases have the same calorific capacities;

b) using 298 and 4×10^3 as average values: $C_{PN_2} = C_{PNO} = 8.5$ and $C_{PO_2} = 9$.

2) Calculate the yield of NO at 3×10^3 and 4×10^3 K.

3) Determine the equilibrium constant if the synthesis is carried out using air?

Consider that the composition of air is 4 mol of nitrogen for every 1 mole of oxygen.

4) What is the yield of NO with respect to O_2 at 3×10^3 K and 4×10^3 K?

5) Compare the values of the yields as related to O_2 or N_2.

Data:

	N_2	O_2	NO
S°_{298}/ (cal.mol^{-1}.K^{-1})	45.77	49.03	50.34

Answers: (1) see corrections; (a) $\ln K_P = \frac{21{,}741.32}{T} + 2.959$; (b) $\ln K_P = \frac{21{,}816.3}{T} = 0.2516 \ln T + 4.645$; (2) ρ_{3000} # 4.68% and ρ_{4000} # 10.52%; (3) see corrections; (4) ρ_{3000} # 9.82% and ρ_{4000} # 23.52%; (5) $\rho_{/O_2} = 4\rho_{/N_2}$.

PROBLEM 3.14.– Synthesis of alumina

In a reactor at constant V, 5.4 g of aluminum is burnt (with an excess of O_2 to obtain alumina (aluminum oxide). Approximately 39.86 kcal of heat is released to maintain a constant temperature.

1) a) Write the equation for the reaction.

b) Determine the variation in the number of moles of each component as a function ξ.

2) At time t, 0.2 mol of Al disappears. How many moles of alumina are created?

3) a) Calculate the variation in internal energy of the system contained in the reactor.

b) Deduce the quantity q_v of this reaction.

Data: $M_{Al} = 26.98$ g/mol.

Answers: (1) (a) see corrections; (b) $dn_{Al} = -2d\xi$; $dn_{O_2} = -\frac{3}{2}d\xi$ and $dn_{Al_2O_3} = d\xi$; (2) $dn_{Al_2O_3} = 0.1$; (3) (a) $\Delta U = 39.86$ kcal; (b) $q_v = -398.6$ cal/mol.

PROBLEM 3.15.– Combining hydrogen and oxygen

To experimentally study the dissociation of water, given by the reaction $2H_2O \rightleftarrows 2H_2 + O_2$, we measure the total pressure within a recipient, at constant volume V, containing water vapor at a temperature of 2,500 K with a device that sets the value for the partial pressure of the water vapor in the gaseous mixture. The gases involved in the reaction are deemed ideal gases.

1) Determine as a function of P and P_{H_2O} the molar fraction of the dissociated water.

2) Compare the partial pressures of hydrogen and oxygen.

3) Express the equilibrium constant relative to the partial pressures:

a) as a function of P and the molar fraction of dissociated water;

b) as a function of P and P_{H_2O}.

4) For a pressure $P_{H_2O} = 0.500$ atm, we find a total pressure of 0.540 atm. Calculate, under these conditions, the molar fraction of dissociated water and the K_P.

5) Calculate this same constant K_P at a temperature of 2,500 K:

a) knowing that at this temperature, the K_P values of the two reactions:

$$\begin{cases} 2CO_2 \rightleftarrows actiO_2 \\ CO_2 + H_2 \rightleftarrows actH_2O \end{cases} \text{are, respectively:} \begin{cases} K_{1P} = 1.4 \times 10^{-3} \\ K_{2P} = 6.4 \end{cases}$$

b) from the thermodynamic data below, disregarding the C_p of the components;

c) from the same data taking into account the C_P values.

Data:

	H_2O_{vapor}	H_{2gas}	O_{2gas}
ΔH°_{f298} (kcal.mol^{-1})	–57.8	/	/
(cal.mol^{-1}.K^{-1})	45.1	31.2	49.0
C_P (cal.mol^{-1}.K^{-1})	10.8	8.3	8.6

The combination of hydrogen and oxygen takes place at 298 K and a pressure of 1 atm. We disregard the dissociation of water at high temperature and we take the following value for water vapor: $C_P = (7.5 + 0.002T)$ cal/mol.K.

6) What is the maximum theoretical temperature obtained?

Answers: (1) $x = \frac{2(P - P_{H_2O})}{2P - P_{H_2O}}$; (2) $\frac{P_{H_2}}{P_{O_2}} = 2$; (3) (a) $K_P = P\frac{x^3}{(1-x)^2(2+x)}$; (b) $K_P = \frac{4}{27}\frac{(P - P_{H_2O})^3}{(P_{H_2O})^2}$; (4) $K_P = 37.93 \times 10^{-6}$; $x = 0.0506$; (5) (a) $K_P = 34.18 \times 10^{-6}$; (b) $K_P = 3.63 \times 10^{-6}$; (c) $K_P = 34.64 \times 10^{-6}$; (6) T # 4.850 K.

PROBLEM 3.16.– Manufacture of ferromanganese

The reduction of manganese monoxide by carbon at 1,500 K, and 1 atm, gives carbon monoxide.

1) a) Write the chemical reaction.

   b) Determine the affinity of the reduction giving pure *Mn*.

   c) What conclusions can be drawn?

The same reduction is carried out in the presence of liquid cast iron to obtain an ideal solution of manganese in the molten iron.

2) a) Write the new reduction reaction.

   b) Determine the expression for its affinity.

3) a) Calculate the titer *Mn* for which the affinity is cancelled out.

   b) What information does this titer give?

Data: The specific heat capacity and melting capacity are disregarded.

At 1,500 K	Δh_f°/ (cal.mol^{-1})	s° (cal.K^{-1}.mol^{-1})
MnO	92,000	14.40
CO	26,400	47.30
Mn	0	7.59
C	0	1.36

Answers: (1) (a–c) see corrections; (2) (a) see corrections; (b) $A(T) = -(6{,}005 + RT\ln x_{Mn}P_{CO})$; (3) (a) $x_{Mn} = 0.098$; (b) see corrections.

3.3. Tests

TEST 3.1.– Electromotive force and enthalpy of reaction

The electromotive force of the fuel cell Pt/H_2/HCl(0.1 N)/Hg_2Cl_2/Hg varies as a function of temperature (°C) according to the approximative law $E = E_{25}(1 + 1.3 \times 10^{-4} T)$ with $E_{25} = 0.3995$ V.

Estimate under these conditions the variation in enthalpy at 25 °C for the reaction:

$$H_2 + Hg_2Cl_2 \rightarrow 2HCl + 2Hg.$$

TEST 3.2.– Enthalpy of reaction at constant pressure

Ammonia is synthesized via the reaction of hydrogen with nitrogen. The experiment gives $T_1 = 893$ K, $K_{P1} = 6.3 \times 10^5$ and $T_2 = 973$ K, $K_{P2} = 2.16 \times 10^6$. The variation in enthalpy between these two temperatures is considered negligible.

1) Write the chemical reaction for the synthesis of NH_3.

2) Calculate the enthalpy of reaction at *P*.

TEST 3.3.– Allotropic form and enantiotropic solids

1) Draw a diagram of a pure body without an allotropic form, specifying the different domains determined by the equilibrium curves.

2) Establish the equations for the equilibrium curves.

3) Draw a diagram of the same body with two solid enantiotropic forms.

TEST 3.4.– Decomposition of hydrogen peroxide

The decomposition of H_2O_2 in the vapor state causes an explosion. Its combustion temperature depends on the pressure and a curve is obtained as shown in Figure 3.1.

1) Define combustion temperature.

2) Interpret this curve.

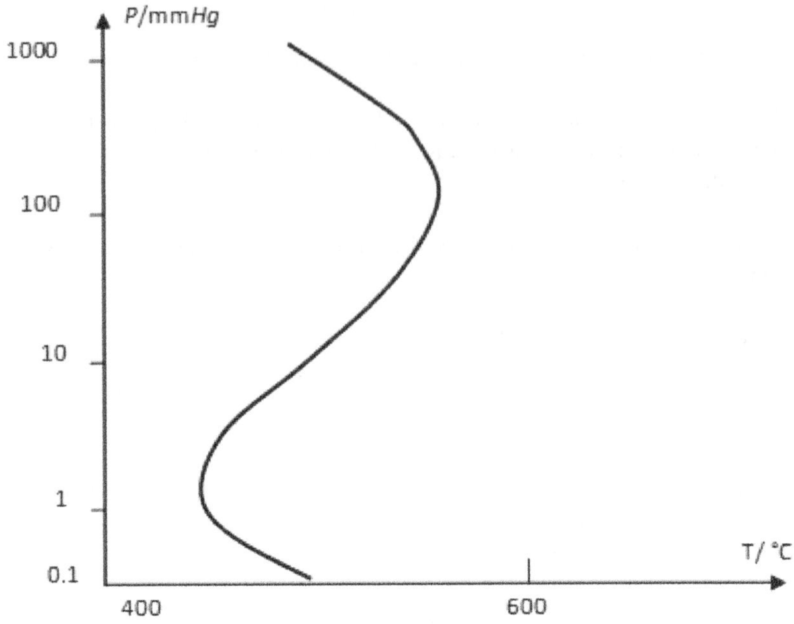

Figure 3.1. *Qualitative curve for the combustion of H_2O_2*

TEST 3.5.– Preparation of chloroform from benzene in the vapor phase

We propose to prepare C_6H_5Cl by the action of chlorine on C_6H_6. The reaction is carried out in a non-distensible recipient, with a volume of 44.8 L, at a temperature high enough for it to transition into the vapor phase.

Here, we assume that monochlorobenzene and chlorohydric acid are the only reaction products, the gases are ideal and the influence of the calorific capacities are negligible.

1) a) Express, in the case of 1 mole of C_6H_6 and 1 mole of Cl_2, the partial pressures at equilibrium as a function of the constant K_P and the total *P*.

b) Deduce the variance in equilibrium from these expressions and find the value using the phase rule. Justify your answer.

2) a) Using the saturation vapor pressure of C_6H_5Cl, show graphically that the latent heat of vaporization of this component is 9.1 kcal/mol.

b) What is the boiling point of C_6H_5Cl under atmospheric pressure?

c) What is the standard entropy of gaseous C_6H_5Cl?

3) Calculate ΔH_T° and ΔS_T° for the reaction concerned.

4) a) What is the value of K_P at 546 K?

b) At this temperature, what are the partial pressures of the reagents and products at equilibrium?

Data:

	C_6H_{6gas}	Cl_{2gas}	$C_6H_5Cl_{liq.}$ (130 °C, 1 atm)	HCl_{gas}
ΔH°_{f298} (kcal.mol^{-1})	19.8	0.0	27.8	–22.0
S°_{298} (cal.mol^{-o}.K^{-o})	64.3	53.3	47.2	44.6

3.4. Detailed corrections

EXERCISE 3.1.– Heat of reaction at constant *P* and *V*

The heat of reaction at constant *P* and constant *V* are given, respectively, by the following relations: $Q_P = nC_P\Delta T$ and $Q_v = nC_v\Delta T$, their difference being: $Q_P - Q_v = n(C_P - C_v)\Delta T = nR\Delta T$.

We assume that the gases obey the ideal gas law and that the reaction occurs at *T* Cte, therefore $Q_P - Q_v = \Delta nRT$, Δn representing the variation in the number of moles of the gaseous bodies only. For the three reactions proposed, we obtain the following:

$H_{2\,gas} + ½\,O_{2\,gas} \rightarrow H_2O_{liq}$ at 25°C

$$Q_P - Q_v = -\frac{3}{2}\,1.987 \times 298.15,\text{ i.e. } \Delta Q = -\,888.64\text{ cal}$$

$CH_3CO_2C_2H_{5\,liq.} + H_2O_{liq.} \rightarrow CH_3COOH_{liq.} + C_2H_5OH_{liq}$ at 25 °C

$$Q_P - Q_v = \Delta Q = 0\text{ cal}$$

$N_{2\,gas} + 3H_{2\,gas} \rightarrow 2NH_{3\,gas}$ at 400 °C

$$Q_P - Q_v = -2 \times 1.987 \times 673.15,\text{ i.e. } \Delta Q = -2{,}675.1\text{ cal.}$$

EXERCISE 3.2.– Enthalpy of formation of a component

The enthalpy of formation of a component corresponds to the heat of formation at constant P of this component from its elements; so for the two components proposed, we obtain:

– ethyl alcohol:

$$C_2H_5OH + 3O_2 \rightarrow 2CO_2 + 3H_2O \text{ with } \Delta H = -\,326.7 \text{ kcal}$$

$$2C + 2O_2 \rightarrow 2CO_2 \text{ with } \Delta H = -\,2 \times 94.052 = -\,188.104 \text{ kcal}$$

$$3H_2 + \frac{3}{2}O_2 \rightarrow 3H_2O \text{ with } \Delta H = -\,3 \times 68.317 = -\,204.951 \text{ kcal}$$

$$2C + 3H_2 + ½\,O_2 \rightarrow C_2H_5OH$$

Thus:

$$\Delta H_1 = -\,326.7 + 188.104 + 204.951 = -66.355 \text{ kcal}$$

– Saccharose:

$$C + O_2 \rightarrow CO_2 \text{ with } \Delta H = -94.052 \text{ kcal}$$

$$H_2 + ½\,O_2 \rightarrow H_2O \text{ with } \Delta H = -68.317 \text{ kcal}$$

$$C_{12}H_{22}O_{11} + 12O_2 \rightarrow 12CO_2 + 11H_2O \text{ with } \Delta H = -1{,}352 \text{ kcal}$$

$$12C + \frac{11}{2}O_2 + 11H_2 \rightarrow C_{12}H_{22}O_{11}$$

and so: $\Delta H_1 = -\,1{,}352 + 751.487 + 1{,}128.624 = -528.11$ kcal.

Exercise 3.3.– Chemical potential of water

To solve this problem, we use the properties of the state functions g, h and s, imagining the transition from the reference state (liquid water at 298 K and 1 atm) to the required state (water vapor at 473 K and 1 atm).

The enthalpy of vaporization of water, the normal boiling point of water and the reference state temperature are denoted as L_v, T_{eb} and T_o, respectively.

1) 298 K $<$ T $<$ 373 K: liquid water is heated, so we use the general relation $g = h - Ts$, knowing that: $h(T) = h(T_0) + \int_{To}^{T} C_l \, dT = C_l(T-T_0)$ and $s(T) = s(T_0) + \int_{To}^{T} C_l \frac{dT}{T} = C_l \ln\frac{T}{T_0}$. The combination of these three equations gives the following for the liquid: $g_l(T) = h_l(T_0) + C_l(T-T_0) - T\left[s_l(T_0) + C_l \ln\frac{T}{T_0}\right]$.

So, $h_l(T_0) = g_l(T_0) + T_0 s_l(T_0)$ therefore $g_l(T) = g_l(T_0) + \left[C_l - s_l(T_0)\right](T - T_0) - TC_l \ln\frac{T}{T_0}$.

2) For $T \geq 373$ K: the first step is the vaporization of the liquid at 373 K, and then the heating of the vapor at 373 K. The variations in h and s are successively calculated:

$$h_l(T_e) = h_l(T_0) + C_l(T_e - T_0) \text{ and } s_l(T_e) = s_l(T_0) + C_l \ln\frac{T_e}{T_0}$$

$$h_v(T_e) - h_l(T_e) = L_v \text{ and } s_v(T_e) - s_l(T_e) = \frac{L_v}{T_e}$$

$$h_v(T) - h_v(T_e) = C_v(T - T_e) \text{ and } s_v(T) - s_v(T_e) = C_v \ln\frac{T}{T_e}$$

hence:

$$h_v(T) - h_l(T_0) = C_l(T_e - T_0) + L + C_v(T - T_e)$$

$$s_v(T) - s_l(T_0) = C_l \ln\frac{T_e}{T_0} + \frac{L}{T_e} + C_v \ln\frac{T}{T_e}$$

By replacing all these expressions in the equation with $g\ (T)$, we obtain: $g_v(T) = h_l(T_0) - Ts_l(T_0) + C_l(T_e - T_0) + L_v + C_v(T - T_e) - C_l T \ln\frac{T_e}{T_0} - L_v\frac{T}{T_e} - C_v T \ln\frac{T}{T_e}$.

In this relation, we need to arrive at $g_l(T_0)$, with the final equation being:

$$g_v(T) = g_l(T_0) - (T - T_0)s_l(T_0) + C_l(T_e - T_0) + C_v(T - T_e) + L_v\left(1 - \frac{T}{T_e}\right)$$
$$-T\left(C_l \ln\frac{T_e}{T_0} + C_v \ln\frac{T}{T_e}\right)$$

APPLICATION.–

– Liquid water at $T < 373$ K: $g_l(T) - g_l(298) = -387.6 + 103.86\,T - 41.44\,T\log T$.

– Water vapor at $T > 373$ K: $g_v(T) - g_l(298) = -13{,}113 + 84.4\,T - 18.4\,T\log T$.

3) The graphical representation of the variations in these three functions *h*, *s* and *g* is given in Figures 3.2, 3.3 and 3.4.

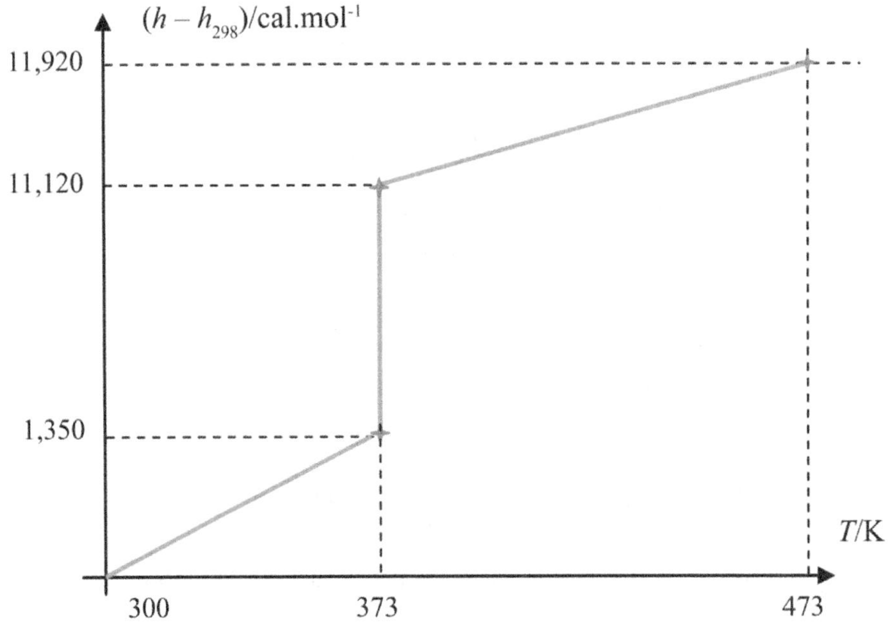

Figure 3.2. *Curve showing the variation in enthalpy as a function of T*

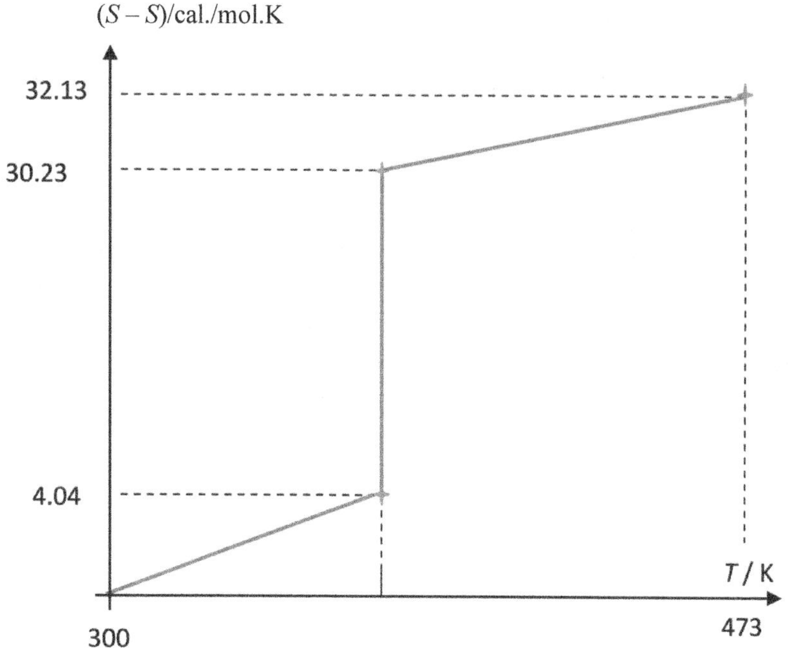

Figure 3.3. *Curve showing the variation in entropy as a function of T*

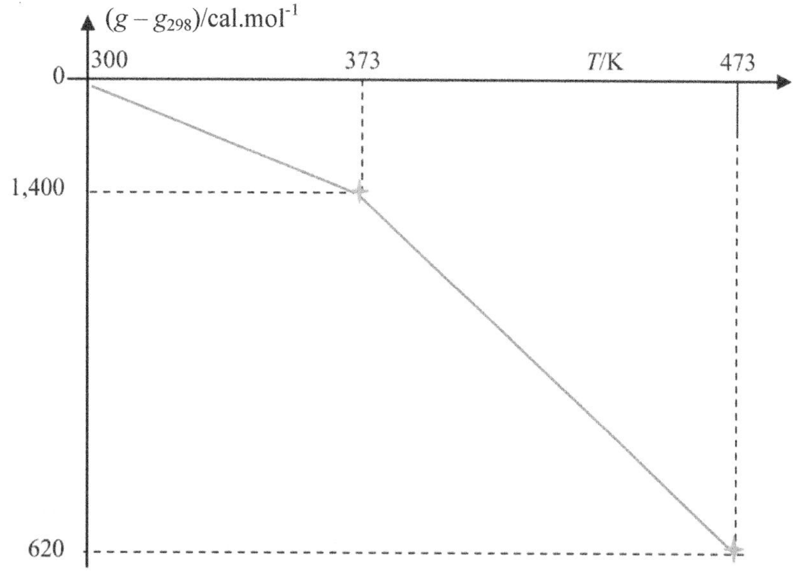

Figure 3.4. *Curve showing the variation in Gibbs free energy as a function of T*

EXERCISE 3.4.– Boudouard reaction

1) At equilibrium, the Boudouard reaction corresponds to: $C_{graphite} + CO_{2gas} \rightleftarrows 2CO_{gas}$.

2) By definition: $\Delta G^\circ_T = -RT\ln K_P = \Delta H^\circ_T - T\,\Delta S^\circ_T$, with $\Delta H^\circ_T = \Delta H^\circ_{298} + \int_{298}^{T} \Delta C_P dT$ and $\Delta S^\circ_T = \Delta S^\circ_{298} + \int_{298}^{T} \Delta C_P \frac{dT}{T}$. The assumption that $\Delta C_P \cong 0$ implies $\Delta H^\circ_T = \Delta H^\circ_{298}$ and $\Delta S^\circ_T = \Delta S^\circ_{298}$ hence: $-RT\ln K_P = \Delta H^\circ_{298} - T\Delta S^\circ_{298}$, with: $\Delta H^\circ_{298} = 2(-26{,}390) - (-94.030 + 0) = 41{,}250$ cal/mol and $\Delta S^\circ_{298} = 2 \times 47.32 - (1.36 + 51.09) = 42.19$ cal/mol.K, which gives: $\Delta G^\circ_T = -RT\ln K_P = 41.250 - 42.19T$. Therefore, $\ln K_P = -\frac{41{,}250}{RT} + \frac{42.19}{R} = \frac{42.19}{1.987} - \frac{41{,}250}{1.987}\frac{1}{T} = 21.233 + \frac{20.760}{T}$.

3) The calculations for the T proposed give the following values:

T/K	600	800	1,000	1,200	1,500	2,000
$\ln K_P$	–13.367	– 4.710	0.473	3.933	7.393	10.853
ΔG°_T	> 0	> 0	< 0	< 0	< 0	< 0

The reaction therefore can only take place from T = 1,000 K.

EXERCISE 3.5.– Calculation of chemical potential at infinite dilution

1) By affecting the sign (*) for the reference pure body and that (∞) for the reference at infinite dilution, the chemical potential of the body (1) in solution is written as: $g_1 = g_1^* + RT\ln a_1^* = g_1^\infty + RT\ln a_1^\infty$; therefore, $\Delta g_1 = g_1^\infty - g_1^* = RT\ln a_1^* - RT\ln a_1^\infty = RT\ln\frac{a_1^*}{a_1^\infty}$ and the coefficient of activity is $\gamma_i = \frac{a_i}{x_i}$. Therefore, $a_1^* = \gamma_1^* x_1^*$ and $a_1^\infty = \gamma_1^\infty x_1^\infty$; for the infinite dilution, we can express the composition in % mass, i.e. $\bar{m}_1 = \frac{m_1}{m}$ and $\bar{m}_2 = \frac{m_2}{m}$ with $\bar{m}_1 + \bar{m}_2 = 100\%$, so $\gamma_1^\infty = \frac{a_1^\infty}{\bar{m}_1}$ and $\frac{a_1^*}{a_1^\infty} = \frac{x_1^*}{\bar{m}_1}\frac{\gamma_1^*}{\gamma_1^\infty}$, then $x_1^* = \frac{\bar{m}_1/M_1}{\bar{m}_1/M_1 + \bar{m}_2/M_2} = \frac{\bar{m}_1/M_1}{\frac{\bar{m}_1}{M_1} + \frac{100-\bar{m}_1}{M_2}} = \frac{\bar{m}_1/M_1}{\frac{\bar{m}_1}{M_1} - \frac{\bar{m}_1}{M_2} + \frac{100}{M_2}}$. Both $\frac{\bar{m}_1}{M_1}$ and $\frac{\bar{m}_1}{M_2}$ are negligible with respect to $\frac{100}{M_2}$ and therefore: $x_1^* = \frac{\bar{m}_1}{M_1}\frac{100}{M_2} = \frac{\bar{m}_1.M_2}{M_1.100}$, so $\frac{a_1^*}{a_1^\infty} = \frac{\gamma_1^*}{\gamma_1^\infty}\frac{\bar{m}_1.M_2}{\bar{m}_1 M_1.100} = \frac{\gamma_1^*}{\gamma_1^\infty}\frac{M_2}{M_1.100}$, writing $\frac{\gamma_1^*}{\gamma_1^\infty} = \gamma_1$, we obtain $\Delta g_1 = RT\ln\gamma_1\frac{M_2}{100M_1}$.

2) At 1,623 K, titanium is a pure solid, whereas in solution in Fe, it is liquid at this temperature. We therefore need to take into account the melting in the form of g, such that: $\Delta g_{1f} = \Delta h_{1f} - T\Delta s_{1f} = \Delta h_{1f} - \frac{T}{T_{1f}}\Delta h_{1f} = \Delta h_{1f}\left(1 - \frac{T}{T_{1f}}\right)$. The calculation gives: Δg_{1f} = 72.16 cal/mol, so:

$\Delta g_1 = \Delta g_{1f} + RT\ln\gamma_1\frac{M_2}{100M_1} = 72.16 + 1.987 \times 1{,}896.14\ln 0.011\frac{55.85}{47.9} = -33.76$ kcal/mol.

EXERCISE 3.6.– Chemical activity of supercooled water

The transformation occurred between an initial state where water is liquid at –10 °C, and a final state where water is solid at –10 °C, therefore:

1) $\Delta H_{sys} = h(T)_s - h(T)_l$ with: $h(T)_l = h(T_0)_l + C_l(T - T_0)$; $L(T_0) = h(T_0)_l - h(T_0)_s$ and $h(T)_s = h(T_0) + C_s(T - T_0)$. This results in $h(T)_s - h(T)_l = -L(T_0) - (C_l - C_s)(T - T_0)$; also, with T_0 being the equilibrium temperature between liquid water and solid water, S of melting at T_0 is equal to $\frac{L(T_0)}{T_0}$ giving the relation: $s_s(T) - s_l(T) = -\frac{L(T_0)}{T_0} - (C_l - C_s)\ln\frac{T}{T_0}$.

2) According to the first principle, the enthalpy lost by the system is integrally gained by the external environment at constant temperature T. This results in $\Delta H_{sys} + \Delta H_{ext} = 0$, i.e. $\Delta H_{ext} = -\Delta H_{sys}$, hence: $\Delta H_{ext} = L(T_0) + (C_l - C_s)(T - T_0)$. Since the temperature of the external environment is equal to –10 °C, then: $\Delta S_{ext} = \frac{\Delta H_{ext}}{T} = \frac{L(T_0)}{T} + (C_l - C_s)\frac{T - T_0}{T}$.

3) The entropy created is obtained by summarizing the system and the external environment as follows:

$$\Delta S_{created} = \Delta S_{sys} + \Delta S_{ext} = L(T_0)\left[\frac{1}{T} - \frac{1}{T_0}\right] + (C_l - C_s)\left[\frac{T - T_0}{T} - \ln\frac{T}{T_0}\right]$$

4) The transformation can be studied as a chemical reaction: $H_2O_l \rightleftarrows H_2O_s$, whose affinity is written as: $A(T) = -\Delta H(T) + T\Delta S(T)$ with $\Delta H(T) = h(T)_s - h(T)_l$ and $\Delta S(T) = s(T)_s - s(T)_l$. This results in $A(T) = L(T_0) + (C_l - C_s)(T - T_0) - T\frac{L(T_0)}{T_0} - T(C_l - C_s)\ln\frac{T}{T_0}$, therefore $A(T) = L(T_0)\left[1 - \frac{T}{T_0}\right] + (C_l - C_s)T\left[\frac{T - T_0}{T} - \ln\frac{T}{T_0}\right]$.

This verifies that the affinity represents, at T, the entropy created over the course of the transformation, and its sign will indicate the direction in which the reaction will shift:

– if $T < T_0$: $A(T) > 0$, water solidifies;

– if $T > T_0$: $A(T) < 0$, water remains liquid.

APPLICATION.– $A(T) = 2.838$ cal/g.

Exercise 3.7.– Decomposition of ammonium chloride

The reaction that occurs is written as: $NH_4Cl_5 \rightleftarrows NH_{3gas} + HCl_{gas}$. It can be characterized as a function of the degree of process ξ as shown in the following table:

		NH_4Cl	NH_3	HCl	n	P_{total}
Initial state	$T_0 = 298.15$ K	0.1	0.3	0	0.4	$P_0 = 1$
Final state	T	0	$0.3 + \xi$	ξ	$0.3 + 2\xi$	P

1) Partial pressure is given by $P_i = x_i P = \frac{n_i}{n}.P$. Since V is constant, $P_0 = 1$ atm and $T_0 = 298.15$ K, we can determine P_i by writing the following equalities: $\begin{cases} P_0V_0 = nRT_0 = V_0 \\ PV_0 = nRT \end{cases}$ therefore $P = \frac{T}{T_0} = \frac{T}{298.15}$ and so the expressions give the partial pressures of the gaseous components: $P_{NH_3} = \frac{0.3+\xi}{0.3+2\xi}.\frac{T}{298.15}$ and $P_{HCl} = \frac{\xi}{0.3+2\xi}.\frac{T}{298.15}$.

2) At equilibrium, we obtain the equality $\Delta \overset{\circ}{G} = RT\ln K(T) = -RT\ln P_{NH_3} P_{HCl}$.

Therefore: $\left(\frac{T}{298.15}\right)^2 = 1 + 4\left(\frac{T}{298.15}\right)^2 K(T)$ and $P^2 = \left(\frac{298.15}{T}\right)^2 + 4\,K(T)$.

3) When the decomposition is total, that is to say when there is a loss of 0.1 mol of NH_4Cl, 0.1 mol of HCl will appear and so: $\xi = 0.1$.

EXERCISE 3.8.– Industrial manufacture of ethanol

1) To determine ξ of the reaction, we must establish the expression for the partial pressure of each body, as shown in the table below:

		C_2H_4	H_2O	C_2H_5OH	Total
$t_0 = 0$		0.25	0.75	0	1
$t > t_0$	n	$0.25-\xi$	$0.75 - \xi$	ξ	$1-\xi$
	P_i	$\frac{0.25 - \xi}{1 - \xi}P$	$\frac{0.75 - \xi}{1 - \xi}P$	$\frac{\xi}{1 - \xi}P$	P

The equilibrium constant K_p is given by: $RT\ln K_p = \Delta G_T^{\circ}$, therefore $\ln K_p = \frac{-\Delta G_T^{\circ}}{RT} = \frac{-4530}{1.987 \times 398.15}$, i.e. $K_p = 0.255$, which is linked to P_i by $K_p = \frac{P_{C_2H_5OH}}{P_{C_2H_4}P_{H_2O}} = \frac{\xi}{1-\xi}P.\frac{(1-\xi)^2}{(0.25-\xi)(0.25-\xi)P^2}$, therefore $K_p = 0.255 = \frac{(1-\xi)\xi}{(0.25-\xi)(0.75-\xi)P} = \frac{(1-\xi)\xi}{(0.25-\xi)(0.75-\xi)}$, since $P = 101.13$ kPa = 1 atm. The second degree equation is therefore: $0.255\xi^2 - 1.255\xi + 0.048 = 0$, so the positive root for value $\xi = 0.0396$ # 0.04, i.e. 4% mol.

The titers are given by the expression $y_i = \frac{n_i}{n}$, which gives:

$$y_{C_2H_4} = \frac{0.25-\xi}{1-\xi} = \frac{0.25-0.04}{1-0.04} \# 0.219$$

$$y_{H_2O} = \frac{0.75-\xi}{1-\xi} = \frac{0.75-0.04}{1-0.04} \# 0.740$$

$$y_{C_2H_5OH} = \frac{\xi}{1-\xi} = \frac{0.04}{1-0.04} \# 0.041$$

EXERCISE 3.9.– Cracking of methane

1) For any chemical equilibrium, K_p is linked to P_i of the reaction components by the relation: $K_p = \frac{\prod_j P_j^{\upsilon_j}}{\prod_i P_i^{\upsilon_i}}$, where υ_i and υ_j are the stoichiometric coefficients of the reagents and products, respectively. Therefore, for each equilibrium, the expressions are as follows:

$$1): K_1 = \frac{P_{H_2}^2}{P_{CH_4}}, \text{ therefore } P_{CH_4} = \frac{1}{K} P_{H_2}^2 ;$$

2): $K_2 = \dfrac{P_{H_2}.P_{C_2H_6}}{P^2_{CH_4}}$, therefore $P_{C_2H_6} = \dfrac{K_2}{K_1^2} P^3_{H_2}$;

3): $K_3 = \dfrac{P^2_{H_2}.P_{C_2H_4}}{P^2_{CH_4}}$, therefore $P_{C_2H_4} = \dfrac{K_3}{K_1^2} P^2_{H_2}$.

Since $P = \sum_i P_i \Rightarrow P_{H_2} + P_{CH_4} + P_{C_2H_4} + P_{C_2H_6} = 1$, then: $1 = \dfrac{K_2}{K_1^2} P^3_{H_2} + \dfrac{1}{K}\left(1 + \dfrac{K_3}{K_1}\right) P^2_{H_2} + P_{H_2}$. So, K_i can be determined by the relation: $RT\ln K_i = -\Delta G_i^\circ = T\Delta S_i^\circ - \Delta H_i^\circ$, therefore $K_i = \dfrac{\Delta S_i^\circ}{R} - \dfrac{\Delta H_i^\circ}{RT}$, and from the data we obtain: $K_1 = 10.5$ atm; $K_2 = 2.052 \times 10^{-4}$ atm; $K_3 = 7.112 \times 10^{-5}$ atm, which allows us to obtain the following constants: $P_{H_2} < 1$; $\dfrac{K_3}{K_1} \ll 1$ and $\dfrac{K_2}{K_1^2} \# 2 \times 10^{-6} \ll 1$, giving the final approximation: $P_{H_2} + \dfrac{1}{K_1} P^2_{H_2} = P_{H_2} + P_{C_2H_6} \# 1$ the second-degree equation in P_{H_2} : $\dfrac{1}{K_1} P^2_{H_2} + P_{H_2} - 1 = 0$ whose positive root is P_{H_2} = 0.920 atm. The other values are as follows: $P_{CH_4} = 8.06 \times 10^{-2}$ atm, $P_{C_2H_6} = 1.45 \times 10^{-6}$ atm and $P_{C_2H_4} = 5.45 \times 10^{-7}$ atm (*).

2) From the expression for P, discuss its influence on each component: H_2: $P = \dfrac{K_2}{K_1^2} P^3_{H_2} + \dfrac{1}{K}\left(1 + \dfrac{K_3}{K_1}\right) P^2_{H_2} + P_{H_2}$. It is in the form: $P = P_{H_2} + a P^2_{H_2} + b P^3_{H_2}$ with a and b being positive; if P ↗, then P_{H_2} ↗. So $\dfrac{P}{P_{H_2}} = 1 + a P_{H_2} + b P^2_{H_2}$ therefore if P_{H_2} ↗, then $\dfrac{P}{P_{H_2}}$ ↗, i.e. $\dfrac{P_{H_2}}{P}$ ↘, which means that H_2 percentage decreases. CH_4: P # $P_{H_2} + P_{CH_4}$ so $\dfrac{P_{H_2}}{P} + \dfrac{P_{CH_4}}{P} \# 1$ and if $\dfrac{P_{H_2}}{P}$ ↘, then $\dfrac{P_{CH_4}}{P}$ ↗, therefore the CH_4 percentage increases. C_2H_6: $\dfrac{P_{C_2H_6}}{P} = K_2 \dfrac{P_{CH_4}}{P} . \dfrac{P_{CH_4}}{P} . \dfrac{P}{P_{H_2}}$, K_2 is constant and the three other terms increase with P, so $\dfrac{P_{C_2H_6}}{P}$

↗, which means the C_2H_6 percentage increases. C_2H_4: $\frac{P_{C_2H_4}}{P} = \frac{K_3}{K_1}\frac{P_{CH_4}}{P}$, the ratio of constants is constant and $\frac{P_{CH_4}}{P}$ ↗, therefore $\frac{P_{C_2H_4}}{P}$ ↗ and so the C_2H_4 percentage increases.

(*) The approximation is *a posteriori* justified.

EXERCISE 3.10.– Synthesis of methanol

1) The reaction is written as: $CO + 2H_2 \rightleftarrows CH_3OH$.

2) Therefore:

- n_1, n_2, n_3 are the number of respective moles of CO, H_2 and CH_3OH during the reaction, and therefore $n = n_1 + n_2 + n_3$;

- n_{10}, n_{20}, n_{30} are the same initial variables, and so $n_0 = n_{10} + n_{20} + n_{30}$.

By definition: $dn_1 = \upsilon_1 d\xi$; $dn_2 = \upsilon_2 d\xi$ and $dn_3 = \upsilon_3 d\xi$ with $\upsilon_1 = -1$, $\upsilon_2 = -2$ and $\upsilon_3 = 1$ giving $n = n_0 - 2\xi$; we can calculate the different partial pressures as a function of n_0 and ξ: $P_1 = \frac{n_1}{n}P = \frac{n_{10} - \xi}{n_0 - 2\xi}P$; $P_2 = \frac{n_2}{n}P = \frac{n_{20} - 2\xi}{n_0 - 2\xi}P$ and $P_3 = \frac{n_3}{n}P = \frac{n_{30} + \xi}{n_0 - 2\xi}P$, which allows us to calculate the affinity of the reaction, which is written as $A(T) = -\sum_i \upsilon_i g_i(T) = g_1(T) + 2g_2(T) - g_3(T)$. Knowing that for an ideal gas: $g_i(T) = g_i^o(T) + RT \ln P_i$, we obtain the following expression: $A(T) = \left[g_1^\circ(T) + 2g_2^\circ(T) - g_3^\circ(T)\right] - RT\ln\frac{P_3}{P_1 + P_2^2}$ so, using the partial pressures, $A(T,P) = A^\circ(T) + 2RT\ln P - RT\ln\frac{(n_{30}+\xi)(n_0-2\xi)^2}{(n_{10}-\xi)(n_{20}-2\xi)^2}$; the variation in G is directly deduced by defining the affinity at constant T and Ps: $dG = -Ad\xi$ and by integration: $G(\xi) - G(0) = -\int_0^\xi A d\xi$.

The numerical application gives $A(\xi)$: $A°(T)$ = –4,755 cal/mol and $A(T) = 2{,}584 - 10^3 \ln\left[\frac{\xi(300-2\xi)^2}{(100-\xi)^3}\right]$ is shown in Figure 3.5.

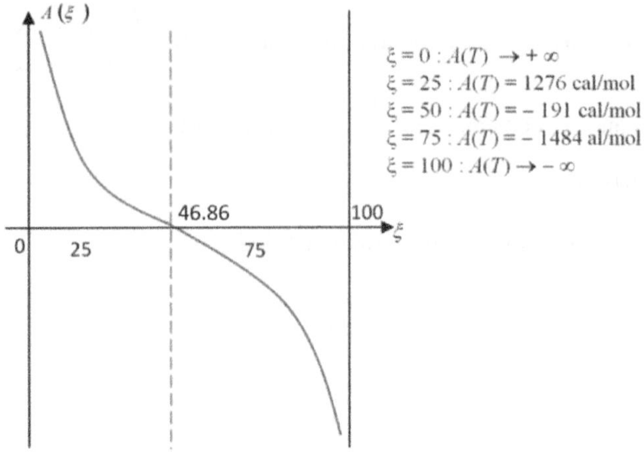

Figure 3.5. *Curve showing the variation in affinity as a function of ξ*

For $G(\xi)$: $G(\xi) - G(0) = 329{,}100 - 2{,}584\xi - 3 \times 10^5.\ln\dfrac{300-\xi}{100-\xi} + 10^3.\xi.\ln\dfrac{\xi(300-2\xi)^2}{(100-\xi)^3}$

as shown in Figure 3.6.

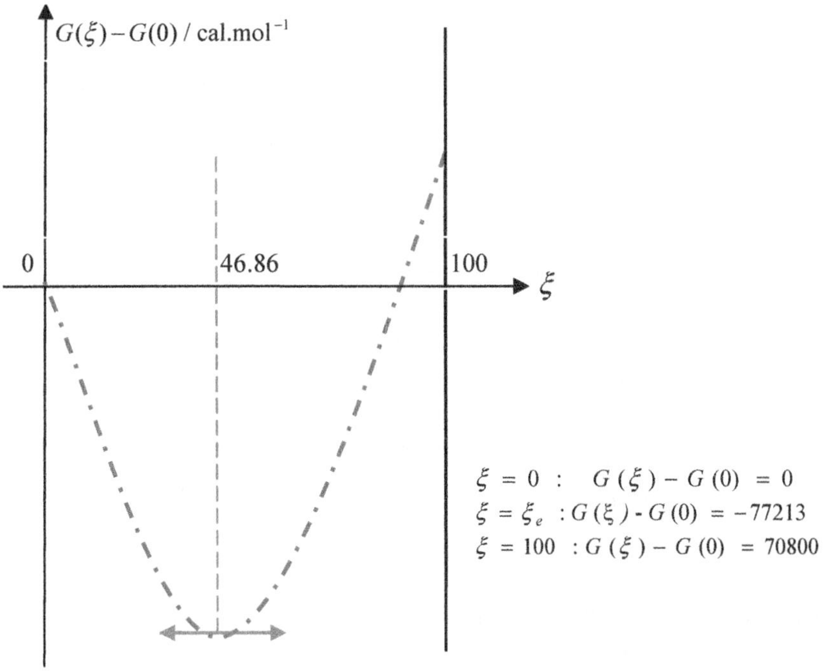

Figure 3.6. *Curve showing the variation in Gibbs free energy as a function of ξ*

3) The equilibrium conditions give $A(\xi_e)=0$, which implies: $\ln\dfrac{\xi_e(300-2\xi_e)^2}{(100-\xi_e)^3}=2.584$ and so $Z=\dfrac{\xi_e(300-2\xi_e)^2}{(100-\xi_e)^3}=e^{2.584}=13.29$. Its solution gives a root close to $\xi=50$; in fact: $\begin{cases}\xi=50 & \Rightarrow\; Z=16\\ \xi=45 & \Rightarrow\; Z=11.93\end{cases}$; by interpolation, we obtain the first approximation; and by extrapolation, the second approximation gives $\begin{cases}\xi_1=46.5 & \Rightarrow\; Z=13.1\\ \xi_2=46.86 & \Rightarrow\; Z=13.287\end{cases}$. The accuracy obtained is satisfactory, so we retain this value $\xi_e=46.86$ for the graph plot; ξ_e corresponds to the minimum of function $G(\xi)$, which is typical for a system at constant pressure.

EXERCISE 3.11.– Electromotive force of a battery cell

1) The battery cell has the following chemical equilibrium:

$$2Ag + Hg_2Cl_2 \rightleftarrows 2AgCl + 2Hg$$

This equilibrium gives rise to two reactions that occur at the electrodes: (1) $2Ag^+ + 2Cl^- \rightleftarrows 2AgCl + 2e^-$ and (2) $2Hg^+ + 2Cl^- \rightleftarrows Hg_2Cl_2 + 2e^-$.

We can see that the reaction involves two electrons with a difference of 1 mole.

2) By definition, the electromotive force (E) is given by relation $E=\dfrac{A(T)}{n.F}$, and the components are in their pure condensed state. It is possible to confuse $A(T)$ with $A°(T)=-\Delta G°(T)=-\Delta H°(T)+T\Delta S°(T)$ to calculate E, using thermodynamic data, knowing that:

$$\Delta H°(T)=2\Delta H_f°(AgCl)-\Delta H_f°(Hg_2Cl_2)$$

$$\Delta S°(T)=2S°(Hg)+2S°(AgCl)-S°(Hg_2Cl_2)-2S°(Ag)$$

We then obtain the following values: $\Delta H°(T) = 2.796$ kcal/mol and $\Delta S°(T) = 15.728$ cal/mol.K so $T\Delta S°(T) = 4.6893$ kcal/mol and $A(T) = 7.4853$ kcal/mol(*) giving $E = \dfrac{7,485.3 \times 4.184}{2 \times 96,500} = 0.1623$ V.

3) Starting with $E(T) = \dfrac{A(T)}{n.F} = -\dfrac{\Delta H°(T)}{n.F} + T\dfrac{\Delta S°(T)}{n.F}$, the variation in E with T is given by the partial derivative:

$$\left(\frac{\partial E}{\partial T}\right)_P = \frac{1}{n.F}\left[-\frac{\partial}{\partial T}(\Delta H°(T))_P + T\frac{\partial}{\partial T}(\Delta S°(T))_P + \Delta S°(T)\right]$$

$$= \frac{1}{n.F}\left[-\Delta C_p^\circ + T\frac{\Delta C_p^\circ}{T} + \Delta S°(T)\right] = \frac{1}{n.F}\Delta S°(T) = \frac{15.728}{2 \times 96\ 500}$$

so $\left(\dfrac{\partial E}{\partial T}\right)_P = 0.815 \times 10^{-4}$ V/K.

(*) $A(T) > 0$, which means that the reaction took place in the direction Ag → Hg, and so the cell can be described as follows: Ag/AgCl//$HgCl_2$/Hg.

EXERCISE 3.12.– Determining the electrode potential

1) Aqueous NaF is formed in two stages, represented by the following reactions: (1) $Na_{crist} + \frac{1}{2}F_{2gas} \rightleftarrows NaF_{aq}$ and (2) $NaF_{cr} \rightleftarrows NaF_{a}$; the sum of both gives the reaction: (3) $Na_{cr} + \frac{1}{2}F_{2gas} \rightleftarrows NaF_{aq}$ giving $\Delta G_3^\circ = \Delta G_1^\circ + \Delta G_2^\circ$; by calculating the values of ΔG_1° and ΔG_2°, we obtain:

$$\Delta G_1^\circ = \Delta H_1^\circ - T\Delta S_1^\circ = -136{,}330 - 298.15(13.1 - 24.3 - 12.2) = -129{,}353 \text{ cal/mol}$$

$$\Delta G_2^\circ = -RT\ln K_2 = \Delta G_2 - 2RT\ln a_\pm$$

Knowing that at equilibrium, for $m = \dfrac{42.2}{41.99} = 1.005$, then $\Delta G_2 = 0$ and $RT\ln K_2 = RT\ln a_\pm^2 \Leftrightarrow K_2 = a_\pm^2 = m_\pm^2.\gamma_\pm^2 = (1.005)^2 \times (0.60)^2 = 0.3636$ giving $\Delta G_2^\circ = 599.4$ cal/mol. This gives for the variation in G of the reaction (3): $\Delta G_3^\circ = -129{,}353 + 599.4 = -128{,}754$ cal/mol. So, by definition, we obtain:

$$\begin{cases} E_3^\circ = -\dfrac{\Delta G_3^\circ}{F} = \dfrac{128754}{23.0659\times 10^3} = 5.582\ \text{V} \\ E_3^\circ = V^\circ_{F_2/F^-} - V^\circ_{Na/Na^+} \\ V^\circ_{Na/Na^+} = -E^\circ_{Na/Na^+} = -2.714\ \text{V} \end{cases}$$

which gives a voltage of:

$$V^\circ_{F_2/F^-} = 5.582 - 2.714 = 2.868\ \text{V}$$

2) The same method allows us to write: $\Delta H^\circ_{f\text{NaF},aq} = \Delta H_3^\circ = \Delta H_1^\circ + \Delta H_2^\circ$ and $S^\circ_{\text{NaF},aq} = \Delta S_2^\circ + S^\circ_{\text{NaF},cr}$, with $\Delta S_2^\circ = \dfrac{\Delta H_2^\circ - \Delta G_2^\circ}{T} = \dfrac{190-599.4}{298.15} = -1.373$ cal/mol.K so $\Delta H^\circ_{f\ \text{NaF}_{aq}} = -136{,}330 + 190 = -136{,}140$ cal/mol and $S^\circ_{\text{NaF}_{aq}} = 13.10 - 1.373 =$ 11.727 cal/mol.K.

EXERCISE 3.13.– Dissociation of carbon dioxide

1) The reaction is written as: $2CO_2 \rightleftarrows 2CO + O_2$. At equilibrium, we obtain equilibrium constant K_p: $\Delta G_T^\circ = -RT\ln K_P = -RT\left(-\dfrac{135{,}280}{RT} + \dfrac{41.49}{R}\right) = 135{,}280 - 41.49T$.

2) Since $\Delta G_T^\circ = \Delta H_T^\circ - T\Delta S_T^\circ$, we obtain by identification $\Delta H_T^\circ = 135{,}280$ cal/mol and $\Delta S_T^\circ = 41.49$ cal/mol.K: note that these two values are independent of T, which means that the specific heats do not depend on T.

3) To determine ΔH°_{f298} and ΔS°_{298} for CO, imagine the following reaction cycle:

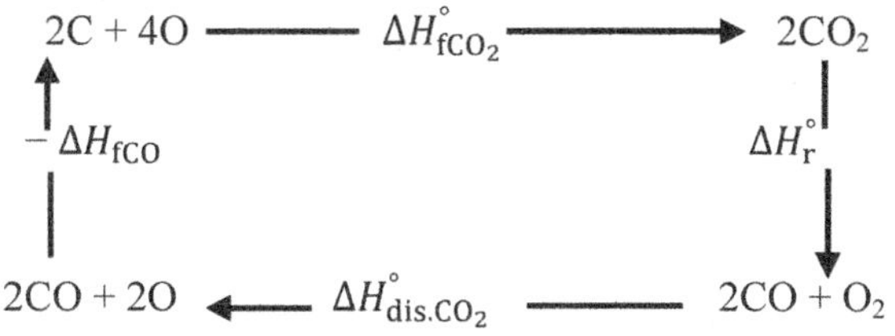

for which the variations in enthalpy and entropy are negligible.

Therefore, $\sum \Delta H = 0$, i.e. $2(-94) + 135.280 + 0 - 2\ \Delta H_{fCO} = 0$ so $2\Delta H_{fCO} = 52.72$ giving $\Delta H_{fCO} = 26.36$ kcal; similarly for entropy $\sum \Delta S = 0$, i.e. $2 \times 51 + 41.49 - 49 - 2\Delta S^{\circ}_{298(CO)} = 0$ so $2\Delta S^{\circ}_{298(CO)} = 94.48$ giving $\Delta S^{\circ}_{298(CO)} = 47.24$ cal/K.

EXERCISE 3.14.– Patent for benzene synthesis

1) This is prepared according to the following heterogeneous equilibrium: $C_2N_{2gas} + C_4H_{6gas} \rightleftarrows \alpha\text{-}C_6H_4N_{2sol} + H_{2gas}$ and reaction can only occur if $\Delta G^{\circ}_{298} < 0$; with $\Delta G^{\circ}_{298} = \Delta H^{\circ}_{298} - 298.\Delta S^{\circ}_{298}$, the calculations give:

$$\Delta H^{\circ}_{298} = (0 + 62{,}000) - (26{,}748 + 71{,}820) = -36{,}568 \text{ cal}$$

$$\Delta S^{\circ}_{298} = (31.21 + 77.09) - (57.64 + 66.42) = -15.76 \text{ cal/K}$$

hence $\Delta G^{\circ}_{298} = -36{,}568 + 298 \times 15.76 = -31{,}781.52$ cal.

2) At equilibrium, $\Delta G^{\circ}_{298} = -RT\ln K_P$, therefore $\ln K_P = -\frac{\Delta G^{\circ}_{298}}{RT} = -\frac{-31{,}781.52}{1.987 \times 298} = 53.826$ giving $K_P = 2.38 \times 10^{23}$.

EXERCISE 3.15.– Synthesis of benzene

1) The proposed reaction is written as: $6C + 3H_2 \xrightarrow{25°C} C_6H_6$.

2) To occur, the following conditions must be met $\Delta G^{\circ}_{298} < 0$ with $\Delta G^{\circ}_{298} = \Delta H^{\circ}_{298} - 298.\Delta S^{\circ}_{298}$.

As we do not know the standard enthalpy at 298 K for this reaction but instead that of combustion, at this temperature, for the elements involved, the reaction can be split into three intermediate steps:

1) $6C + 6O_2 \rightarrow 6CO_2$ with: $\Delta H_1 = -6 \times 94.05 = -564.30$ kcal.
2) $3H_2 + \frac{3}{2}O_2 \rightarrow 3H_2O$ with: $\Delta H_2 = -3 \times 68.32 = -204.96$ kcal.
3) $6CO_2 + 3H_2O \rightarrow C_6H_6 + \frac{15}{2}O_2$ with: $\Delta H_3 = 789.1$ kcal.

$6C + 3H_2 \rightarrow C_6H_6$ with $\Delta H^{\circ}_{298} = \sum \Delta H_i = 19.84$ kcal.

hence $\Delta G^{\circ}_{298} = 19.84 - 298\ (64.34 - 6 \times 1.36)10^{-3} = 30{,}991$ cal.

We find that $\Delta G^{\circ}_{298} > 0$: the reaction will never occur under these conditions.

EXERCISE 3.16.– Synthesis of iodine monobromide

1) The reaction corresponds at equilibrium to: $I_{2gas} + Br_{2gas} \rightleftarrows 2IBr_{gas}$. To verify the consistency of data, we must calculate for each component the value of $\Delta G^{\circ}_{298} = \Delta H^{\circ}_{298} - T.\Delta S^{\circ}_{298}$ between the condensed state and the gaseous state and compare it with the value given in the tables, i.e.:

- for Br_2: $\Delta G^{\circ}_{298} = 7.34 - 298.15(58.639 - 36.4) \times 10^{-3} = 0.709$ therefore acceptable with respect to 0.751;

- for I_2: $\Delta G^{\circ}_{298} = 14.876 - 298.15(62.28 - 27.9) \times 10^{-3} = 4.26$ is good with respect to 4.63;

- for IBr: in this case, we calculate ΔS°_{298} from $\Delta G^{\circ}_{\mathrm{f}}$ and $\Delta H^{\circ}_{\mathrm{f}}$ and then from S° values, we compare the calculations using the two methods as follows: $\Delta G^{\circ}_{\mathrm{f}298(IBr)} = 0.91 - 1/2(4.63 + 0.751) = -1{,}770.5$ cal/mol and $\Delta H^{\circ}_{\mathrm{f}298(IBr)} = 9.75 - 1/2(14.876 + 7.34) = -1{,}358$ cal/mol so $\Delta S^{\circ}_{298} = \frac{1770.5 - 1358}{298.15} = 1.3835$cal/mol.K. $\Delta S^{\circ} = S^{\circ}_{IBr} - \frac{58.639+62{,}28}{2} = S^{\circ}_{\mathrm{IBr}} - 60.595$ so $S^{\circ}_{\mathrm{IBr}} = 1.417 + 60.595 = 61.877$ cal/mol.K.

Therefore, the suitable value is $S^{\circ} = 61.80$ cal/mol.K.

2) The calculation of $\Delta H^{\circ}_{\mathrm{f}114(IBr)}$ is immediately possible since we have already calculated $\Delta H^{\circ}_{\mathrm{f}298(IBr)}$, and so: $\Delta H^{\circ}_{\mathrm{f}114(IBr)} = \Delta H^{\circ}_{\mathrm{f}298(IBr)} + \int_{298}^{387} \Delta C_P dT = -1{,}358 + (8.69 - \frac{8.81+8.6}{2})(387 - 298) = -1{,}358 - 1.3350 = -1{,}359.34$ cal/mol.

3) K_P is calculated from $\Delta G_{react.} = -RT\ln K_P$ with $\Delta G_{react.} = 2\,(-1{,}780.5) = -3{,}561$ cal/mol so $\ln K_P = \frac{3{,}561}{1.987 \times 298.15} = 6.009$ hence K_P # 407.

4) With the assumption that the composition is equimolar, we obtain:

	I_{2gas} +	Br_{2gas} ⇄	$2IBr_{gas}$
Initial	1	1	0
At equilibrium	$1 - x$	$1 - x$	$2x$

Giving the yield: $\rho = \frac{2x/2}{1/1} = x$; and $K_P = \frac{P_{IBr}^2}{P_{I2}P_{Br2}}$ with $P_{I_2} = P_{Br_2} = \frac{1-x}{2}$ and $P_{IBr} = P\frac{2x}{2} = P.x$ so $K_P = \frac{x^2}{\frac{(1-x)^2}{4}} = \frac{4x^2}{(1-x)^2} = \left(\frac{2x}{1-x}\right)^2$ hence $\frac{2x}{1-x} = \sqrt{K_p} = \sqrt{407}$ # 20 so $2x = 20 - 20x$, i.e. $x = \frac{20}{22} = 0.91$ that is to say a yield of $\rho = 91\%$.

EXERCISE 3.17.– Effect of pressure on chemical equilibrium

A stable equilibrium in state (A) characterized by (T, P_1, K_p) will evolve toward state (B) defined by (T, P_2, K_p). The variation in n_i as a function of P will be a function of the variation in P_i of each component and the number of moles of the reagents at t = 0 (initial state before reaction): $P_i = x_iP = \frac{n_i}{n}P$ and $n = \sum_i n_i$, so ξ_A and ξ_B are the chemical progress for each state, so the following table is established with $n_0 = n_{CO} = n_{CO_2}$:

		CO	Cl_2	$COCl_2$	Total
$t_0 = 0$	n_i	n_0	n_0	0	$2n_0$
$t_A > t_0$	n_i	$n_0(1-\xi_A)$	$n_0(1-\xi_A)$	$n_0\xi_A$	$n_0(2-\xi_A)$
	P_i	$\frac{1-\xi_A}{2-\xi_A}P_A$	$\frac{1-\xi_A}{2-\xi_A}P_A$	$\frac{\xi_A}{2-\xi_A}P_A$	P_A
$t_B > t_A$	nb moles	$n_0(1-\xi_B)$	$n_0(1-\xi_B)$	$n_0\xi_B$	$n_0(2-\xi_B)$
	P_i	$\frac{1-\xi_B}{2-\xi_B}P_B$	$\frac{1-\xi_B}{2-\xi_B}P_B$	$\frac{\xi_B}{2-\xi_B}P_B$	P_B

We can write the expression for K_P for the two states: $K_p = 15 = \frac{P_{COCl_2}^2}{P_{CO}P_{Cl_2}} = \frac{\xi_A(2-\xi_A)}{(1-\xi_A)^2P_A} = \frac{\xi_A(2-\xi_A)}{(1-\xi_A)^2}$, therefore $16\xi_A^2 + 32\xi_A + 15 = 0$; the second-degree equation whose positive root is ξ_A # 0.75 so $n = \sum_i n_i = n_0(2 - 0.75) = 1.25n_0$. $K_p = 15 = \frac{\xi_B(2-\xi_B)}{(1-\xi_B)^2P_B} = \frac{\xi_B(2-\xi_B)}{2(1-\xi_B)^2}$, so $31\xi_B^2 - 62\xi_B + 30 = 0$ whose positive root is $\xi_B = 0.82 \Rightarrow n = n_0(2 - 0.82) = 1.18n_0$.

CONCLUSION.– $\xi_B > \xi_A$ therefore ξ increases with P and consequently we obtain: n_{COCl_2} which increases; n_{CO} and n_{Cl_2} which decrease; and n which decreases.

EXERCISE 3.18.– Metal/refractory interaction

1) The reaction for the dissolution of alumina is written:

$Al_2O_{3\,(s)} \leftrightharpoons 2Al_{(dis.)} + 3O_{(dis.)}$

It can be considered as the weighted sum of the following two:

(1) $Al_2O_{3\,(s)} \leftrightharpoons 2Al_{(dis).} + \frac{3}{2}O_{2(g)}$

with an affinity written as:

$$A_1(T) = A_1^{\circ} - \frac{3}{2}\,4.57\,T\log P_{o_2} - 2\times 4.57\,T\log x_{Al}$$

(2) $\frac{1}{2}O_{2\,(s)} \leftrightharpoons O_{(dis)}$ with $A_2(T) = A_2^{\circ} - 4.57T\log x_O + \frac{1}{2}\,4.57T\log P_{O_2}$.

2) The affinity of the dissolution of $Al_2O_{3\,(s)}$ is therefore: $A(T) = A_1(T) + 3A_2(T) = (A_1^{\circ} + 3A_2^{\circ}) - 2\times 4.57T\log x_{Al} - 3\times 4.57T\ \log x_O$ and so: $A(T) = A^{\circ} - 2\times 4.57T\log x_{Al} - 3\times 4.57\,T\log x_O$.

We calculate A_1 using the relation $A_1^{\circ} = -\Delta H_1^{\circ} + T\Delta S_1^{\circ}$ with $-\Delta H_1^{\circ} = 402{,}120$ and $\Delta S_1^{\circ} = 2\times 20.80 + 3/2\times 62.22 - 57.56 = 77.37$ giving $A_1^{\circ} = -402{,}120 + 77.37\,T$ and $A_2^{\circ} = 27{,}900 - 6.08T$ hence $A = (-402{,}120 + 3\times 27{,}900) + (77.37 - 3\times 6.08)T$ whereby at $T = 1{,}873.15$ K: $A(T) = -207{,}660.64 - 17{,}120.59\ \log x_{Al} - 25{,}680.89\log x_O$. So at equilibrium $A(T) = 0$ and as we begin with pure Fe at 100%, we obtain $x_{Al} = \frac{2}{3}x_O$, i.e. $-207{,}660.64 - 17{,}120.59\log\frac{2}{3}x_O. -25{,}680.89\log x_O =$ $-207{,}660.64 - 17{,}120.59\log\frac{2}{3} - 17{,}120.59\log x_O - 25{,}680.89\log x_O = -207{,}660.64$ $- 17{,}120.59\log\frac{2}{3} - (17{,}120.59 + 25{,}680.89)\log x_O = 0$, therefore $\log x_O =$ -4.7818938674, which gives: $x_O = 1.652\times 10^{-5}$ and $x_{Al} = 1.101\times 10^{-5}$.

3) We verify that this value x_O gives a pressure of O_2 higher than the extremely low pressure of the liquid, which is impossible to eliminate even by keeping a strictly pure atmosphere.

EXERCISE 3.19.– Chemical variable, equilibrium constant and affinity

1) If ξ is the reaction progress, the table below shows the values at this equilibrium:

		SO_2	O_2	SO_3	Total
$t_0 = 0$		$n_1 = 100$	$n_2 = 25$	$n_3 = 0$	$n = 125$
$t > t_0$	n	$100 - 2\xi$	$25 - \xi$	2ξ	$125 - \xi$
	P_i	$\frac{100-2\xi}{125-\xi}P$	$\frac{25-\xi}{125-\xi}P$	$\frac{2\xi}{125-\xi}P$	P

Affinity is defined by $A(T) = \sum_i \nu_i g_i(T)$ which gives $A(T) = -\Delta G(T)$ and therefore at the standard state: $A^\circ(T) = -\Delta G^\circ(T)$ and $\Delta G^\circ(T) = \Delta H_r^\circ(298) - T\Delta S_r^\circ(298)$, hence $A^\circ(T) = T\Delta S_r^\circ(298) - \Delta H_r^\circ(298)$, which gives:

$$A^\circ(873) = 873\Delta S_r^\circ(298) - \Delta H_r^\circ(298)$$

We must now determine the reaction variables $\Delta H_r^\circ(298)$ and $\Delta S_r^\circ(298)$ from the table data, using Hess's law which expresses for any Z function: $\Delta Z_r^\circ(T) = \sum_i Z^\circ_{pro}(T) - \sum_i Z^\circ_{f\,reac}(T)$, hence at 298 K:

$$\Delta H_r^\circ(298) = 2\Delta H^\circ_{fSO_3}(298) - \left[2\Delta H^\circ_{fSO_2} + \Delta H^\circ_{fO_2}(298)\right]\Delta S_r^\circ(298)$$
$$= 2\Delta S^\circ_{fSO_3}(298) - \left[2\Delta S^\circ_{fSO_2} + \Delta S^\circ_{fO_2}(298)\right]$$

Calculate each term contained in these expressions as follows: $\Delta H^\circ_{fO_2}(298) = 0$ according to the simple body principle, $2\Delta H^\circ_{fSO_2}$: here the reaction is $S + O_2 \leftrightarrows SO_2$ with $\Delta H^\circ = \Delta G^\circ + T\Delta S^\circ$, therefore $\Delta H^\circ = \sum_i \nu_i g_i^\circ + T\sum_i \nu_i s_i^\circ$; hence:

$$\Delta H^\circ_{fSO_2}(298) = \left(g^\circ_{SO_2} - g^\circ_{O_2} - g^\circ_S\right) + T\left(s^\circ_{SO_2} - s^\circ_{O_2} - s^\circ_S\right) \# -70 \text{ kcal/mol}$$

$\Delta H^\circ_{fSO_3}(298)$: here the reaction is $S + \frac{3}{2}O_2 \rightleftharpoons SO_3$ so $\Delta H^\circ_{fSO_3} = \left(g^\circ_{SO_3} - \frac{3}{2}g^\circ_{O_2} - g^\circ_S\right) + T\left(s^\circ_{SO_3} - \frac{3}{2}s^\circ_{O_2} - s^\circ_S\right) = -8852 - 0 - 0 + 298(61.2 - \frac{3}{2}49 - 7.62)$ # 94.46 kcal/mol giving $\Delta H_r^\circ(298) = \Delta H^\circ_{fSO_3} - 2\Delta H^\circ_{fSO_2} = -46.99$ kcal/mol and $\Delta S_r^\circ(298) = 2\Delta S^\circ_{fSO_3} - \left[2S^\circ_{fSO_3}(298) + S^\circ_{O_2}(298)\right] = 2(61.2 - 59.4) - 49 = -45.4$ cal/K.mol, hence: $A^\circ(873) = -46.99 + 873(-45.4 \times 10^{-3}) = 7.36$ kcal/mol.

2) With $A^\circ(T) = RTln\mathrm{K_p}$, therefore $ln\mathrm{K_p} = \frac{A^\circ(T)}{RT} = \frac{7360}{1.987 \times 873}$ hence $\mathrm{K_p} = 67.52$.

3) Affinity expressed with regard to P_i is written as: $A(T) = A^\circ(T) - RTln\prod_i \left(\frac{P_i}{P_i^\circ}\right)^{\nu_i} = RTln\mathrm{K_p} - RTln\prod_i \left(\frac{P_i}{P_i^\circ}\right)^{\nu_i}$ with $P_i^\circ = 1$, therefore $\prod_i \left(\frac{P_i}{P_i^\circ}\right)^{\nu_i} =$

$\prod_i P_i^{\nu_i} = \frac{p_{SO_3}^2}{P_{SO_2}^2} = \frac{(125-\xi)\xi^2}{(25-\xi)(50-\xi)^2.P}$, hence $A(T) = RT\ln K_p \frac{(25-\xi)(50-\xi)^2}{(125-\xi)\xi^2}P$ with $P = 2$ giving $A(T) = RT\ln 2K_p \frac{(25-\xi)(50-\xi)^2}{(125-\xi)\xi^2}$.

4) At equilibrium, $A(T) = 0$, therefore $RT\ln 2K_p \frac{(25-\xi)(50-\xi)^2}{(125-\xi)\xi^2} = 0$, that is to say $2K_p \frac{(25-\xi)(50-\xi)^2}{(125-\xi)\xi^2} = 1$ so $(125 - \xi)\xi^2 = 2K_p (25 - \xi)(50 - \xi)^2$ and $(125 - \xi)\xi^2 = 135(25 - \xi)(50 - \xi)^2$; the third degree equation that is solved via successive iterations, knowing that $\xi \in [0; 25]$, gives the value ξ # 24.35.

PROBLEM 3.1.– Fuel cell

The chemical reaction is as follows: $H_2 + \frac{1}{2}O_2 \rightleftharpoons H_2O$.

1) From an electrochemical point of view, this reaction proceeds as follows: $2H^+ + O^{2-} \rightleftharpoons H_2O + 2e^-$, two electrons are therefore involved in the process; and by definition, $E = \frac{A}{nF}$, where n is the number of electrons involved and F is the Faraday constant, giving the relation $A = 2F\,E$.

2) At 1 atm and temperature T, its affinity is written as: $\overset{\circ}{A_T} = -\sum_i \upsilon_i \overset{\circ}{g}_{iT} = T\sum_i \upsilon_i \overset{\circ}{s}_{iT} - \sum_i \upsilon_i \overset{\circ}{h}_{iT}$ for relations $\overset{\circ}{h}_{iT} = \overset{\circ}{h}_{i298} + c_{pi}(T - 298.15)$ and $\overset{\circ}{s}_{iT} = \overset{\circ}{s}_{i298} + C_{pi}\ln\frac{T}{298.15}$. So at 1 atm at 100 °C, water evaporates. There is therefore a change in the physical state that we must take into account, as both temperature domains are dealt with separately:

a) 298.15 < T < 373.15 K: the C_{pi} and s_i of H_2O are those of the liquid therefore: $\overset{\circ}{g}_{iT} = \overset{\circ}{g}_{i298} + \left(C_{pi} - \overset{\circ}{s}_{i298}\right)(T - 298.15) - TC_{pi}\ln\frac{T}{298.15}$ and since $\overset{\circ}{g}_{i298} = \overset{\circ}{h}_{i298} - 298.15\,\overset{\circ}{s}_{i298}$ we obtain: $\overset{\circ}{A_T} = \overset{\circ}{A}_{298} - (T - 298.15)\sum_i \upsilon_i\left(C_{pi} - \overset{\circ}{s}_{i298}\right) + T\ln\frac{T}{298.15}\sum_i \upsilon_i C_{pi}$ giving:

$$A_T^\circ = A_{298}^\circ + 298.15\sum_i \upsilon_i\left(C_{pi} - s_{i298}^\circ\right) - T\left[\sum_i \upsilon_i\left(C_{pi} - s_{i298}^\circ\right) - \ln 298.15\sum_i \upsilon_i C_{pi}\right]$$
$$+T\ln T\sum_i \upsilon_i C_{pi}$$

So with all calculations completed, we obtain the expression for affinity:

$$A_T^\circ = 70{,}547 - 89.1T + 17.2T\log T/\text{cal}$$

and in Joules: $A_T^\circ = 295{,}168.6 - 372.794T + 71.965T\log T/\text{J}$.

So if $n = 2$ and $F = 96{,}485\text{C}$, then $nF = 192{,}970$ J, we obtain for E:

$$E_T = 1.5296 - 1.932 \times 10^{-3}T + 3.729 \times 10^{-4}T\log T/\text{V}$$

b) $373.15 < T < 473.15$: we must account for the vaporization with C_{pi} and s_i being those of the vapor, and we obtain the expression:

$$A_T^\circ = A_{373}^\circ + 373.15\sum_i \upsilon_i\left(C_{pi} - s_{i373}^\circ\right) - T\left[\sum_i \upsilon_i\left(C_{pi} - s_{i373}^\circ\right) - \ln 373.15\sum_i \upsilon_i C_{pi}\right]$$
$$+T\ln T\sum_i \upsilon_i C_{pi}$$

The calculation gives: $A_T^\circ = 53.319 + 16.2T - 5.8T\log T$ in calories and Joules: $A_T^\circ = 223{,}086.70 - 67.781T + 24.267T\log T$ in Joules with $nF = 192{,}970$ J. We obtain the expression for E:

$$E_T = 1.156 + 3.513 \times 10^{-4}T - 1.258 \times 10^{-4}T\log T/\text{V}$$

PROBLEM 3.2.– Dismutation of iron protoxide

This involves studying the reaction: $4\text{FeO} \rightleftarrows \text{Fe} + \text{Fe}_3\text{O}_4$ where all species are solids and do not form a solid solution, therefore its affinity will be the standard affinity: $A^\circ(T) = -\Delta H^\circ(T) + T\Delta S^\circ(T)$ so $A^\circ(10^3) = -\Delta H^\circ(10^3) + T\Delta S^\circ(10^3)$ with $\Delta H^\circ(10^3) = \Delta H^\circ_{\text{f(Fe}_3\text{O}_4)}(10^3) - 4\Delta H^\circ_{\text{f(FeO)}}(10^3) = -8{,}500$ cal/mol and $\Delta G^\circ(10^3) = \Delta G^\circ_{\text{f(Fe}_3\text{O}_4)}(10^3) - 4\Delta G^\circ_{\text{f(FeO)}}(10^3) = +1{,}300$ cal/mol. whereby: $\Delta S^\circ(10^3) = \dfrac{\Delta H^\circ(10^3) - \Delta G^\circ(10^3)}{10^3} = -9.8$ cal/mol.K. This results in the following equation $A^\circ(T) = 8{,}500 - 9.8\,T$, which gives Figure 3.6.

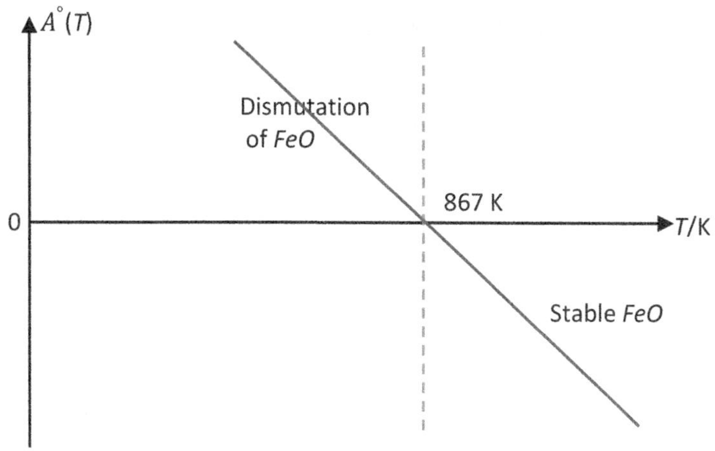

Figure 3.7. *Standard affinity as a function of T*

From this, we can see the three possible steps according to the sign of A° allowing the evolution of the reaction to be studied:

$T = 867$ K so $A°(T) = 0$: this is equilibrium;

$T < 867$ K so $A°(T) > 0$: FeO dismutes;

$T > 867$ K so $A°(T) < 0$: FeO is stable.

PROBLEM 3.3.– Gasification of carbon by water

1) The combustion equation is written as:

$$CO + \frac{1}{4}H_2 + \frac{3}{2}N_2 + O_2 + 4N_2 \rightarrow CO_2 + \frac{1}{4}H_2O + \frac{3}{8}O_2 + \frac{11}{2}N_2$$

Kirchoff's law expresses the variation in calorific capacity at constant V by: $\left(\frac{\partial Q_v}{\partial T}\right) = \sum n_j C_{vj} - \sum n_i C_{vi}$, where i and j are, respectively, the reaction reagents and products.

At constant V, $\left(\frac{\partial Q_v}{\partial T}\right) = \left(\frac{\partial U}{\partial T}\right)_v = \Delta C_v$, we must determine, under the reaction conditions, the physical state of the water formed: for this, we calculate P_{H_2O} in the mixture and compare it to its maximum value at 50 °C. If x is the number of moles of water vapor, we obtain the following relations: $\frac{1\text{ atm}}{P_{tot.}} = \frac{\frac{62}{8}}{\frac{55}{8}+x}$ and $\frac{P_{tot.}}{P_{H_2O}} = \frac{\frac{55}{8}}{x}$, therefore $P_{H_2O} = \frac{8x}{62} = 0.129x$/atm: compared to $P_{H_2O,max}$ (50 °C) = 9.2 cm Hg = 0.121 atm, which gives the maximum number of moles of water (as a vapor): $x = \frac{0.121}{0.129} = 0.94$ mol.

With 0.94 mol of water, the atmosphere is saturated with water vapor. Beyond this point, water will be in the liquid state; in the reaction considered, 0.25 mol of water is formed so 0.25 mol of water is therefore in the vapor state. The application of Kirchoff's law gives: $\int_0^{50} dQ_v = \int_0^{50} \sum n_j C_{vj} dT - \int_0^{50} \sum n_i C_{vi}\, dT$ i.e: $(Q_v)_{50} = (Q_v)_0 + \int_0^{50} \sum(\Delta n.\, C_v) dT$ with $(Q_v)_0 = \Delta H_{f\,CO_2} + ¼\, \Delta H_{f\,H_2O} + \Delta H_{v\,H_2O}$ and $(Q_v)_0 = -68{,}200 - \frac{68{,}400}{4} + \frac{606 \times 18}{4} = -82{,}573$ cal; $(\Delta n.\, C_v)_{diat.gas} = -\frac{15}{8}(4.8 + 0.0012\, T)$; $(\Delta n.\, C_v)_{CO_2} = 6.7 + 0.0052\, T$; $(\Delta n.\, C_v)_{H_2O} = \frac{1}{4}(5.9 + 0.0043)\, T$, so $\sum \Delta n.\, C_v = -36$ cal and hence $(Q_v)_{50} = -82{,}609$ cal.

The gases are deemed ideal and $(Q_P)_{50}$ is calculated using the equation: $(Q_P)_{50} - (Q_v)_{50} = \Delta n.RT$, therefore $(Q_P)_{50} = -82{,}609 - \left(\frac{57}{8} - \frac{62}{8}\right).\ 1.987 \times 323 = -82{,}609 - 401 \# -83{,}010$ cal.

2) The temperature of gases burned by the adiabatic combustion is given, at constant V and constant P, by writing $|Q\ | = 0$ in each case: $|Q_v| = 0 = \int_{50}^{T} \sum n_i C_{vi} dT = \int_{50}^{T} \sum \left[(6.7 + 0.0052T) + \frac{1}{4}(5.9 + 0.0043T) + \frac{47}{8}(4.8 + 0.0012T)\right] dT$ and so the equation: $6.675 \times 10^{-3}\, T^2 + 36.375T - 84.444 \times 10^3 = 0$ whose solution gives the positive root: $T_v = 1{,}757$ °C; $|Q_P| = |Q_v| + \int_{50}^{T} \Delta n.\, RdT = 6.675 \times 10^{-3}\, T^2 + 36.375T - 84.444 \times 10^3 + \int_{50}^{T} \frac{57}{8}\, 1.987dT$ so the second-degree equation in terms of T: $6.675 \times 10^{-3}\, T^2 + 50.625T - 85.561 \times 10^3 = 0$ whose solution (the positive root) is $T_P = 1{,}423$ °C.

PROBLEM 3.4.– Standard chemical potential of water

1) Let: $h^\circ_{lT} - h^\circ_{l298} = C^\circ_{pl}(\text{T} - 298)$ and:

$$h^\circ_{vT} - h^\circ_{l298} = C^\circ_{pv}(T - 373) + L^\circ_v + 75\, C^\circ_{pl}$$

$$s^\circ_{lT} - s^\circ_{l298} = C^\circ_{pl} \ln\left(\frac{T}{298}\right)$$

and:

$$\overset{\circ}{S}_{vT} - \overset{\circ}{S}_{l298} = C^{\circ}_{pv} \ln\left(\frac{T}{373}\right) + \frac{L^{\circ}_{v}}{373} + 0.225 C^{\circ}_{pl}$$

$$g^{\circ}_{lT} - g^{\circ}_{l298} = (h^{\circ}_{lT} - Ts^{\circ}_{lT}) - (h^{\circ}_{l298} - 298\ s^{\circ}_{l298}) = C^{\circ}_{pl}\left[T - 298 - T\ln\left(\frac{T}{298}\right)\right] - s^{\circ}_{l298}\ (T - 298)$$

$$g^{\circ}_{vT} - g^{\circ}_{l298} = (h^{\circ}_{vT} - Ts^{\circ}_{vT}) - (h^{\circ}_{l298} - 298\ s^{\circ}_{l298}) = L^{\circ}_{v}\left(1 - \frac{T}{373}\right) + C^{\circ}_{pv}\left(T - 373 - T\ln\frac{T}{373}\right)$$
$$+ C^{\circ}_{pl}(75 - 0.225\ T) - s^{\circ}_{l298}(T - 298)$$

2) The vaporization of water, considered as a chemical equilibrium, is written as: $H_2O_l \rightleftarrows H_2O_v$ giving the expression for its affinity:

$$A(T) = g_l(T\ ) - g_v(T\ ) = \left[g_l(T\ ) - g_l(298) - (g_v(T\ ) - g_l(298)\right]$$
$$= L^{\circ}_{v}\left(\frac{T}{373} - 1\right) + \left(C^{\circ}_{pl} - C^{\circ}_{pv}\right)\left(T - 373 - T\ln\frac{T}{373}\right)$$

So by disregarding the influence of ΔC°_{p} (which is the classic rough approximation), $A_T \# L^{\circ}_{v}\left(\frac{T}{373} - 1\right) \# A^{\circ}_{T} \# -\Delta H^{\circ}_{373} + T\Delta S^{\circ}_{373}$.

3) The variation in S of the universe is entropy created by any process, and by definition: $TdS_{created} = \mathrm{A}d\xi \Rightarrow dS_{created} = \frac{\mathrm{A}}{T}d\xi$, so for vaporization: $dS_{created} = L^{\circ}_{v}\left(\frac{1}{373} - \frac{1}{T}\right)d\xi$, which is integrated, giving $S_{created} = L^{\circ}_{v}\left(\frac{1}{373} - \frac{1}{T}\right)\xi$ and at 105 °C, we obtain: $S_{created} = 9{,}770\left(\frac{1}{373} - \frac{1}{378}\right)\frac{1{,}000}{18} = \Delta S_{univ} =$ 19.24 cal/mol.K.

Condensation is the inverse reaction of vaporization, therefore $A_T = -L^{\circ}_{v}\left(\frac{T}{373} - 1\right)$ hence $S_{created} = L^{\circ}_{v}\left(\frac{1}{T} - \frac{1}{373}\right)\xi$ and so at T = 95 °C = 368 K: $S_{created}$ = $9{,}770\left(\frac{1}{368} - \frac{1}{373}\right)\frac{1{,}000}{18}$, whereby: $S_{created} = \Delta S_{univ} =$ 19.75 cal/mol.K.

PROBLEM 3.5.– Partial pressure and equilibrium constant

1) This involves a chemical equilibrium where one of the bodies is in the solid state and hence not very sensitive to pressure. By definition, $K_p = \frac{\prod P_j^{\upsilon_j}}{\prod P_i^{\upsilon_i}}$. We complete the table below considering ξ as a chemical variable, whereby:

$$K_p = \left(\frac{2\xi}{1+\xi}\right)^2 \times \left(\frac{1+\xi}{1-\xi}\right)\frac{P^2}{P} = \frac{4\xi^2 P}{(1+\xi)(1-\xi)} = \frac{4\xi^2}{1-\xi^2}P$$

2) $K_p = 22.4$ and $P = 1$ so $\frac{4\xi^2}{1-\xi^2}P = 22.4$, i.e. $4\xi^2 = 22.4(1-\xi^2)$ and so $4\xi^2 + 22.4\xi^2 = 22.4$. Therefore, $\xi^2 = \frac{22.4}{26.4}$ and so $\xi = \sqrt{\frac{22.4}{26.4}}$ # 0.92, whereby: $P_{CO_2} = \frac{1-0.92}{1+0.92} = 0.04$ and $P_{CO} = \frac{2 \times 0.92}{1+0.92} = 0.95$.

		CO_2	CO	$2CO$	Total
$t_0 = 0$		n	Excess	0	N
$t > t_0$	nb moles	$n(1-\xi)$	/	2ξ	$n(1+\xi)$
	P_i	$\frac{1-\xi}{1+\xi}P$	/	$\frac{2\xi}{1+\xi}P$	P

3) The direction in which equilibrium will shift:

a) if P increases: direction 2;

b) if T increases: direction 1;

c) if a catalyst is used: direction 1;

d) if an inert gas is added: no effect;

e) if C is added: no effect;

f) is CO_2 is added: direction 1.

4) The relation between K_p and ΔH_T° is written as: $\ln K_p(T) = -\frac{\Delta H_T^\circ}{RT} + \text{Cte}$ therefore $\ln K_{P_{1120}} = -\frac{\Delta H_{1,120}^\circ}{1{,}120R} + C$ and $\ln K_{P_{1,150}} = -\frac{\Delta H_{1,150}^\circ}{1{,}150R} + C$, whereby: $\ln\frac{K_{P_{1,120}}}{K_{P_{1,150}}} = \frac{-\Delta H_T^\circ}{R}\left[\frac{1}{1{,}120} - \frac{1}{1{,}150}\right] = \frac{-\Delta H_T^\circ}{R}\left[\frac{1{,}120-1{,}150}{1{,}120 \times 1{,}150}\right] = -0.4$, thus $\frac{K_{P_{1,120}}}{K_{P_{1,150}}} = e^{-0.4}$ and so $K_{p1120} = \frac{K_{P_{1,120}}}{K_{P_{1,150}}}e^{-0.4} = 22.4e^{-0.4}$ # 13.84.

PROBLEM 3.6.– Corrosion of a pipe

1) Water, the main vector of oxygen, will react with chrome according to the reaction: $2Cr + 3H_2O \rightleftarrows Cr_2O_3 + 3H_2$. This reaction can be split into two linear combinations as follows: (a) $\left(2Cr+\frac{3}{2}O_2 \rightarrow Cr_2O_3\right)$ and (b) $3\left(H_2O \rightarrow \frac{1}{2}O_2 + H_2\right)$ so: $\Delta G^\circ = \Delta G^\circ_a + 3\Delta G^\circ_b = \Delta G^\circ_f\left(Cr_2O_3\right) - 3\times\Delta G^\circ_f + \left(H_2O\right) = -$ 178,600 − 3(−39,260) = −60,820 cal/mol; at equilibrium we observe $-\Delta G^\circ = A^\circ = RT\ ln\ K = RT\ ln\left(\frac{P_{H_2}}{P_{H_2O}}\right) = 3RT\ ln\frac{P_{H_2}}{P_{H_2O}}$ so $\Delta G^\circ = -3RT\ ln\frac{P_{H_2}}{P_{H_2O}}$, thus

$log\frac{P_{H_2O}}{P_{H_2}} = -\frac{60\,820}{3\times 4.576\times 1500} = -3.040$, whereby $\frac{P_{H_2O}}{P_{H_2}} \approx 10^{-3}$; so, the moisture content in hydrogen will not exceed the maximum value corresponding to $\frac{P_{H_2O}}{P_{H_2}} < 10^{-3}$, otherwise $A°(T)$ becomes positive and the pipe oxidizes.

2) Similar to water, CO_2 reacts with chrome according to the reaction: $2Cr + 3CO_2 \rightleftarrows Cr_2O_3 + 3CO$, which gives: $\Delta G^\circ = \Delta G^\circ_f\left(Cr_2O_3\right) + 3\Delta G^\circ_f\left(CO\right) - 3\Delta G^\circ_f\left(CO_2\right)$ and at equilibrium, as a function of P_i, we obtain: $\Delta G^\circ = -3RT\ ln\frac{P_{CO}}{P_{CO_2}} = -178{,}600 +$ 3(−58,370) − 3(−94,710) = − 69,580 cal/mol, whereby $\log\frac{P_{CO_2}}{P_{CO}} = \frac{-69{,}580}{3\times 4.576\times 1{,}500}$ $= -4.622$ so $\frac{P_{CO_2}}{P_{CO}} = 4.18\times 10^{-4}$. The moisture content in this case does not exceed the maximum value, corresponding to a pressure ratio $\frac{P_{CO_2}}{P_{CO}}$ 4.18×10^{-4}.

3) In this case, H_2O and CO_2 can react with Cr as we have just seen, but there will be an additional reaction for the transfer of oxygen between these two bodies as follows: $H_2O + CO_2 \rightleftarrows CO_2 + H_2$.

The combination of the two previous reactions shows that this relation is equivalent to the following:

$$\frac{1}{3}\left\{\left(2Cr+3H_2O \Leftrightarrow Cr_2O_3+3H_2\right)-\left(2Cr+3CO_2 \Leftrightarrow Cr_2O_3+3CO\right)\right\}$$

There are only two reactions that are linearly independent. Consequently, the system can be described by two progress variables. We can choose the first or the third reaction, by writing:

– ξ_1 and ξ_3, the respective progress;

– $K_1 = \frac{P_{H_2}}{P_{H_2O}}$ and $K_3 = \frac{P_{CO_2}}{P_{CO}} \times \frac{P_{H_2}}{P_{H_2O}}$, the equilibrium constants;

– n_1 and n_2 are the molar fractions of H_2O and CO_2 in 1 mole of gas.

These data are written in the table below:

	CO	H_2O	CO_2	H_2	Total
Initial state	$1-n_1-n_3$	n_1	n_2	0	1
Final state	$1-n_1-n_2-\xi_3$	$n_1-3\xi_1-\xi_3$	$n_2+\xi_3$	$3\xi_1+\xi_3$	1

With the total pressure being constant, the partial pressures are linked to molar fractions by $P_i = x_iP$, which gives the expressions: $K_1^{1/3} = \frac{3\xi_1+\xi_3}{n_1-3\xi_1+\xi_3}$ and $K_3 = \frac{(n_2+\xi_3)(3\xi_1+\xi_3)}{(1-n_1-n_2-\xi_3)(n_1-3\xi_1-\xi_3)}$; so the condition for no oxidation is that the first reaction does not see a shift to the right (formation of Cr_2O_3), i.e. $\xi_1 \leq 0$. If $\xi = 0$ in these two equations, we obtain $\xi_3 = \frac{n_1K_1^{1/3}}{1+K_1^{1/3}}$ and by removing ξ, it gives the limit relation between n_1 and n_2: $n_1+n_2 = \frac{K_3}{K_3+K_1^{1/3}}$. The calculation gives: $K_1^{1/3} = 912 \approx 10^3$ and $K_3 = 0.383$, i.e. $n_1+n_2 \cong 4.18 \times 10^{-4}$; therefore under these conditions, the minimum pipe supply level should be $n_1 + n_2 < 4.18 \times 10^{-4}$.

PROBLEM 3.7.– Synthesis of cyclohexane

1) This involves studying the reaction: $3H_{2gaz} + C_6H_{6liq} \rightleftarrows C_6H_{12liq}$ (1). Benzene and cyclohexane form an ideal solution. If x is the molar titer of C_6H_{12}, the law of mass action at equilibrium is written as: $K_1 = \frac{x}{1-x} P_{H_2}^{-3}$. K_1 is the equilibrium constant giving: $\ln K_1 = \ln K_{f(C_6H_{12})liq} - \ln K_{f(C_6H_6)liq}$. As the constants of formation are those of the gaseous state, we must consider the corresponding liquid/vapor equilibria, namely:

$$C_6H_{6_{liq}} \rightleftarrows C_6H_{6vap} (2) \text{ and } C_6H_{12_{liq}} \rightleftarrows C_6H_{12vap.} (3)$$

At equilibrium, we can write the following expressions for the equilibrium constants:

$$\ln K_2 = \ln K_{f(C_6H_6)gaz} - \ln K_{f(C_6H_6)liq} = \ln P^*_{C_6H_6}$$

$$\ln K_3 = \ln K_{f(C_6H_{12})gaz} - \ln K_{f(C_6H_{12})liq} = \ln P^*_{C_6H_{12}}$$

whereby $\ln K_1 = \ln K_{f(C_6H_{12})gaz} - \ln K_{f(C_6H_6)gaz} - \ln P^*_{C_6H_{12}} + \ln P^*_{C_6H6}$ and finally: $logK_1$ = 2.2999, which give: $K_1 \approx 200$ where $\frac{x}{1-x} = 200 P_{H_2}^3 = 2 \times 10^5$, i.e. $x \approx 1$ (the solution is basically pure cyclohexane).

2) Total pressure is the sum of partial pressures, i.e. $P = P_{H_2} + P_{C_6H_6} + P_{C_6H_{12}}$ with $P_{H_2} = 10; P_{C_6H_6} = 0$ and $P_{C_6H_{12}} = xP^S_{C_6H_{12}} = P^S_{C_5H_{12}} = 19.72$ atm, therefore $P = 29.72$ atm.

PROBLEM 3.8.– Synthesis of ethanol

1) a) The chemical reaction for equation: $C_2H_4 + H_2O \leftrightharpoons C_2H_5OH$.

b) K_p is defined by $RT\ln K_p = \Delta G_r^\circ$, therefore $\ln k_P = -\frac{\Delta G_r^\circ}{RT}$ and $\Delta G_{370}^\circ = \Delta H_{370}^\circ - \Delta S_{370}^\circ$. So the assumption that ΔH_r° is constant between 350 K and 500 K means that the effect of C_p is disregarded in this interval, which gives: $\Delta H_{370}^\circ = \Delta H_{420}^\circ - 370\Delta S_{370}^\circ$, whereby: $\Delta G_{370}^\circ = \Delta H_{420}^\circ - 370\Delta S_{420}^\circ$. We must therefore calculate ΔS_{420}° from $\Delta G_{420}^\circ = \Delta H_{420}^\circ - 420\Delta S_{420}^\circ$, where $\Delta S_{420}^\circ = \frac{\Delta G_{420}^\circ + \Delta H_{420}^\circ}{420} =$

$-\frac{9,875+40.17}{420}=-119.15$ J/K. We obtain: $\Delta G^\circ_{370}=40.17-370(-119.15)=3,915.5$ J so $\ln K_p=-\frac{3,915.5}{8,314\times 370}=-1.2728$, therefore $K_p=e^{-1.2728}=0.28$.

2) a) By definition, $dn_i=\gamma_i\, d\,\xi$, therefore $d\xi=\frac{dn_i}{\gamma_i}$. By applying this to each reagent, we obtain: $d\,\xi_1=\frac{dn_{C_2H_2}}{-1}=\frac{1.2}{-1}=-1.2$ and $d\xi_2=\frac{dn_{H_2O}}{-1}=\frac{0.133}{-1}=-0.133$ so $d\xi$ is different with respect to each body. Therefore, the stoichiometric conditions are not met: C_2H_4 is in excess, and it is H_2O that determines the rate of progress.

b) $dn_{C_2H_5OH}=\frac{d\xi}{1}=\frac{0.133}{1}=0.133$ and so $n_{C_2H_5OH}=0.133$. Since $n_{C_2H_5OH}=\frac{m_{C_2H_5OH}}{M_{C_2H_5OH}}$, $m_{C_2H_5OH}=0.133\ M_{C_2H_5OH}=0.133\times 46$ # 6.12 g.

3) The equilibrium shifts: a) in direction 1; b) in direction 2; c) in direction 1.

PROBLEM 3.9.– Preparation of sulfuric acid from gypsum

1) This concerns the reaction: (1) $CaSO_4 \rightleftarrows CaO+SO_2+\frac{1}{2}O_2$. If $P_{SO_2}+P_{O_2}=1\,\text{atm}$ and $P_{SO_2}=2P_{O_2}$, then $P_{SO_2}=\frac{2}{3}atm$ and $P_{O_2}=\frac{1}{3}$ atm. We can therefore calculate the affinity of the reaction:

$$A_1(T)=A_1^\circ(T)-RT\ln\left(P_{SO_2}\times P_{O_2}^{1/2}\right)=-\Delta H^\circ_{298}+T\Delta S^\circ_{298}-RT\ln\frac{2}{3\sqrt{3}}$$

Knowing that ΔH°_{298} and ΔS°_{298} are given by the following expressions:

$$\Delta H^\circ_{298}=\sum_i \upsilon_i \Delta H_{fi\ 298}=-151,800-70,900+342,400=119,700\ \text{cal/mol}$$

$$\Delta S^\circ_{298}=\Sigma\, \upsilon_i S^\circ_{298}$$

We can see that the standard entropy of SO_2 is missing. We calculate this using its formation reaction: $S + O_2 \rightleftarrows SO_2$, which gives: $\Delta S^\circ_{f298}=S^\circ_{SO_2}-S^\circ_S-S^\circ_{O_2}$ $=\frac{\Delta H^\circ_{f298}-\Delta G^\circ_{f298}}{298}=\frac{-70900+71700}{298}=2.684$ cal/mol.K; $S^\circ_{SO_2}=2.684+7.6$ so $+49.0=59.284\approx 59.3$ cal/mol.K, which gives: $\Delta S^\circ_{298}=9.5+59.3+24.5-25.5=$

67.8 cal/mol.K, and therefore an affinity of: $A_1(T) = -119\,700 + (67.8 + 4.576\,log\frac{3\sqrt{3}}{2})T = -119\,700 + 69.7$; so, at equilibrium, $A_1(T) = 0$, therefore T = 1,717.4 K = 1,444.2 °C.

2) Consider the equilibrium: (2) $SO_2 + \frac{1}{2}O_2 \rightleftarrows SO_3$ with $\sum_i P_i = 1$ atm; $P_{SO_2} = 2P_{O_2}$ and at equilibrium: $K_2(T) = \frac{1}{P^{1/2}_{O_2}}.\frac{P_{SO_3}}{P_{SO_2}}$. Knowing that $RT\ ln\ K_2(T) = \overset{\circ}{A}_1(T) = -\Delta H^\circ_2(T) + T\Delta S^\circ_2(T) \cong -\Delta H^\circ_2(298) + T\Delta S^\circ_2(298)$ with:

$$\Delta H^\circ_2(298) = -94{,}600 + 70{,}900 = -23{,}700 \text{ cal/mol}$$

$$\Delta S^\circ_2 = (298) = \frac{\Delta H^\circ_2(298) - \Delta G^\circ_2(298)}{298}$$

$$\Delta G^\circ_2(298) = -88{,}700 + 71{,}700 = -17{,}000 \text{ cal/mol}$$

We obtain the expression: $log\,K_2(T) = \frac{5{,}186}{T} - 4.92$ at 1,680 K, which gives $log\,P_{SO_2} = log\,K_1(T) + log\left(P_{SO_2} - P^{-1/2}_{O_2}\right) = -2$, i.e. $P_{SO_3} \cong 10^{-2}$ atm. Therefore, the hypothesis can be justified; we can disregard the SO_3 formation equilibrium.

3) This concerns the equilibrium: (3) $CaSO_4 + \frac{1}{2}S_iO_2 \rightleftarrows \frac{1}{2}(SiO_2.2CaO) + SO_2 + \frac{1}{2}O_2$; to calculate K₃($T$) we need the value of $\Delta H^\circ_2(298)$ but we are missing the $\Delta H^\circ_f(298)$ of the silicate, but we do have $\Delta H^\circ(298)$ for the following reaction: $S_iO_2 + 2CaO \rightleftarrows (SiO_2.2CaO)$. It can therefore be calculated as follows: $\Delta H^\circ(298) = \sum_i \upsilon\Delta H^\circ_{f(i)}(298) = \Delta H^\circ_{f(SiO_2.2CaO)} - \Delta H^\circ_{f(SiO_2)} - 2\Delta H^\circ_{f(CaO)}$ where: $\Delta H^\circ_{f(SiO_2.2CaO)} = (-30 - 216 - 303{,}6)10^3$ = 549.6 kcal/mol, which

gives the values of $\Delta H^{\circ}_{2(298)}$: $\Delta H^{\circ}_{2}(298) = \frac{-549.6}{2} - 70.9 + \frac{216}{2} + 342.4 =$ 104.7 kcal/mol. Hence, $\Delta S^{\circ}_{2}(298) = 67.85$ cal/mol.K.

As we assume no SO_3 is formed in any significant quantity, then $P_{SO_2} = \frac{2}{3}$, $P_{O_2} = \frac{1}{3}$ and $log\, K_2(T) = \frac{A^{\circ}_2(T)}{4.57T} = \frac{-22{,}900}{T} + 14.85$; and at equilibrium: $\log K_2(T) = \log P_{SO_2}.P^{1/2}_{O_2} = -0.42$, where $T = \frac{22{,}900}{14.85 + 0.42}$ # 1,500 K = 1,227 °C.

4) At T = 1,500 K, we obtain $log \frac{P_{SO_3}}{P_{SO_2}} = log\left(K_2 P^{1/2}_{O_2}\right) = -1.6$ hence $\frac{P_{SO_3}}{P_{SO_2}}$ # 2%.

CONCLUSION.– This verifies our hypothesis.

PROBLEM 3.10.– Conversion of methane

1) By definition $\Delta S^{\circ}_{T} = \sum \upsilon_i S^{\circ}_i(T)$, hence for the reaction: $CH_4 + H_2O \rightleftarrows CO + 3H_2$, we obtain the following value from the data provided: $\Delta S^{\circ}_{298} = -44.480 - 45.105 + 47.301 + 3 \times 31.21 = 51.346$ cal/K.

2) a) A standard thermodynamics approximation allows us to write: A°_{T} # $\Delta H^{\circ}_{298} + T\Delta S^{\circ}_{298}$ where: $A^{\circ}_{T} = -49{,}277 + 51.346T$.

b) By definition: $RT\, ln\, K_1 = A^{\circ}_{T}$, therefore at T = 10^3 K, we obtain $A^{\circ}_{1{,}000} = 2{,}069$ cal. so $K_1 = \exp\left(\frac{2{,}069}{1{,}987}\right) = 2.83$.

3) The following table summarizes the reaction between the initial and final state:

	CH4	H_2O	CO	H_2	Total
n (initial)	100	100n	0	0	100(n + 1)
n (final at ξ)	100 − ξ	100n − ξ	ξ	3ξ	100(n + 1) + 2ξ

By definition, $A_T = A_T^\circ - RT\,ln\,K_1(T) = A_T^\circ - RT\,ln\dfrac{P_{CO}.P_{H_2}^3}{P_{CH_4}.P_{H_2O}}$. This gives the expression for the affinity of the reaction at T:

$$A_T = -49\,277 + 51.346T - RT\,ln\frac{27P^2\xi^4}{(100n-\xi)(100-\xi)\left[100(n+1)+2\xi\right]^2}$$

4) a) At equilibrium, $A_T = 0$, whereby:

$$ln\,K_1(T) = ln\frac{27P^2\xi^4}{(100n-\xi)(100-\xi)\left[100(n+1)+2\xi\right]^2} = \frac{A_T^\circ}{RT} = \frac{51.346-\dfrac{49\,277}{T}}{1.9872}$$

i.e: $\dfrac{27P^2\xi^4}{(100n-\xi)(100-\xi)\left[100(n+1)+2\xi\right]^2} = K_1(T) = \exp\dfrac{51.346-\dfrac{49\,277}{T}}{1.9872}$. Solving the equation for T with $P = 1$, $n = 3$ and $\xi = 99$ gives:

$$T_1 = \frac{49\,277}{51.346-1.9872\,ln\left[\dfrac{27\times(99)^4}{201\times(598)^2}\right]} = 1{,}115 \text{ K } \# \; 841.85\text{ °C}$$

b) Incremental values of 1,200 K are used to obtain the values of enthalpy and standard entropy as follows:

$\Delta H_{1200}^\circ = \Delta H_{298}^\circ + \sum \upsilon_i (h_{1200}^\circ - h_{298}^\circ)_i = 49{,}277 + 3\times 6{,}404 + 6{,}798 - 8{,}248 - 12{,}732 =$ 54,307 cal

$\Delta S_{1200}^\circ = \Delta S_{298}^\circ + \sum_i \upsilon_i (s_{1200}^\circ - s_{298}^\circ)_i = 51.346 + 3 \times 9.83 + 10.28 - 12.54 - (62.424 - 44.490) = 60.842$ cal

where $T_1' = \dfrac{54{,}307}{60.842-1.9872\ln\left[\dfrac{27\times(99)^4}{201\times(598)^2}\right]} = 1{,}011 \text{ K } \# \; 738\text{ °C}.$

c) The equation for the law of mass action, of (4) above, shows that at fixed T and ξ, P and n are linked by the proportionality of P at $\left[100(n+1)+2\xi\right]\sqrt{(100n-\xi)}$, whereby replacing T_1' and n by their respective values, we obtain:

$$P=\frac{(200+2\xi)\sqrt{100-\xi}}{(400+2\xi)\sqrt{300-\xi}}$$

therefore, for a 99% conversion: $P = 0.047$ atm.

6) At $P = 1$ atm and $n = 1$, at equilibrium we would obtain a progress of 99% at temperature $T_1''=1\ 307\,\text{K}$ # 1,033.85 °C.

CONCLUSION.– It is a good idea to work with an excess of water vapor ($n = 3$).

PROBLEM 3.11.– Cyclization of hexane

1) Applied to each component, the VK-C rule gives, for the Gibbs free energy between $300K \le T \le 600K$, the following calculations:

- n-hexane: $\Delta G_f^\circ \Leftrightarrow 4(CH_2)+2(CH_3)+RTln\sigma_1$. The calculation of each term gives the following results:

$$\begin{cases}4(CH_2)= 4\times(-5\,193+24.3T) =-20\,772+97.20T\\2(CH_3)= 2\times(-10\,943+22.15T) =-21\,886+44.30T\\RTln\sigma_1 =1.9872Tln2=1.38T\end{cases}$$

where: $\Delta G_f^\circ = -42658+142.88T$ and at T = 298.15 K, we obtain: $\Delta G_f^\circ = -\ 58.3$ cal/mol;

- cyclohexane: $\Delta G_f^\circ \Leftrightarrow 6(CH_2)+\text{cycle }6+RT\ln\sigma_2$. The calculation of each term gives: $\begin{cases}6(\text{CH}_2)=-\ 31{,}158+145.80\,T\\ \text{cycle6}=-\ 1{,}128\ -16.35\,T\\ RTln\sigma_2=1.9872\,T\ \ln 6=3.56\,T\end{cases}$ where $\Delta G_f^\circ=-\ 32{,}286+133.01T$ and at $T = 298.15$ K: $\Delta G_f^\circ = 7{,}371$ cal/mol.

2) With regard to the group contribution to the group, the cyclization reaction is written as: $2(CH_3)+\text{symmetry}\,\sigma_1 = 2(CH_2)+1(H_2)+\text{cycle }6+\text{symmetry}\,\sigma_2$ for which we obtain: $\Delta G^\circ \Leftrightarrow 2(CH_2)-2(CH_3)+\text{cycle }6+RTln\frac{\sigma_2}{\sigma_1}$. The calculation of each term gives:

$$\begin{cases} 2(CH_2) = 2(-5{,}830+25.44\,T) = -11{,}660+50.88T \\ -2(CH_3) = -\,2(-12{,}310+24.36\,T) = 24{,}620-48.72T \\ \text{cycle } 6 = -1{,}930-15.04T \\ RTln\dfrac{\sigma_2}{\sigma_1} = 1.9872\,T\,ln3 = 2.16T \end{cases}$$

so: $\Delta G^\circ = 11\ 030 - 10.70T$ therefore at 1,000 K: $\Delta G^\circ = 330$ cal/mol; and for the same reaction, we obtain: $RT\,ln\,K = -\Delta G^\circ$ with K = 0.847.

At equilibrium, under 1 atm: $P_{H_2} = P_{C_6H_{12}} = P$ and $P_{C_6H_{14}} = (1-2P)$, where $K = \dfrac{P^2}{1-2P}$ therefore $P^2 - 2KP - K = 0$, the second-degree equation whose solutions proposed and by the tables as follows:

– VK-C:

$$\Delta' = (0.847)^2 + 0.847 = 1.5725 \Rightarrow \sqrt{\Delta'} = 1.254 : \begin{cases} P = 1.254 - 0.847 = 0.40 \text{ atm} \\ 1-2P = 1-0.80 = 0.20 \text{ atm} \end{cases}$$

– tables:

$$\Delta' = (0.7)^2 + 0.7 = 1.19 \Rightarrow \sqrt{\Delta'} = 1.09 : \begin{cases} P = 1.09 - 0.7 = 0.39 \text{ atm} \\ 1-2P = 1-0.78 = 0.22 \text{ atm} \end{cases}$$

3) At 298.15 K, for the cyclization reaction: $\Delta G^\circ = \Delta G^\circ_{fC_6H_{12}} - \Delta G^\circ_{fC_6H_{14}} = 10{,}372 - 9.87T = \Delta H^\circ - T\Delta S^\circ$, where ΔS° = 9.87 cal/mol.K; and according to the tables, for this reaction: $\Delta S^\circ = S^\circ_{C_6H_{12}} + S^\circ_{H_2} - S^\circ_{C_6H_{14}} =$ 71.27+31.20 – 92.44 = 10.03 cal/mol.K.

The values are of the same order of magnitude. However, when we calculate these values from just the entropies of the two bodies, i.e. $\Delta S^\circ = S^\circ_{C_6H_{14}} - S^\circ_{C_6H_{12}}$, we obtain:

– VK-C: $\Delta S^\circ = 31.20 - 9.87 = 21.33$ cal./mol.K ;

– tables: $\Delta S^\circ = 92.44 - 71.27 = 21.17$ cal./mol.K .

This difference essentially comes from the cyclization whose entropic term would be approximately 16.35 cal/mol.K according to VK-C.

4) With no decrease in entropy due to the cyclization, the entropy of the reaction (9.87 + 16.35 = 26.22) should correspond to the formation of 1 mole of hydrogen, i.e. approximately 30 cal/mol.K.

PROBLEM 3.12.– Dissociation of water

1) a) The reaction for the dissociation of water is written as: $2H_2O_{vap.} \rightleftarrows 2H_{2\ gas} + O_{2\ gas}$ with a particular relation: $P_{H_2} = 2P_{O_2}$.

b) The variance is given by the following rule: v = c + 2 – ϕ, where c is the number if components and ϕ is the number of phases; the initial assumption that we only have water vapor implies that c = 3 – 2 = 1 and, with ϕ = 1, we obtain: v = 1 + 2 – 1 = 2.

2) a) As a function of the dissociation rate α, we determine the reaction progress to equilibrium given in the table below:

	H_2O	H_2	O_2	Total
Initial	2	0	0	2
At equilibrium	2(1– α)	2 α	A	2 + α
Partial pressures	$\frac{2(1-\alpha)}{2+\alpha}P$	$\frac{2\alpha}{2+\alpha}P$	$\frac{\alpha}{2+\alpha}P$	P

The equilibrium constant is defined as: $K_P = \frac{P_{H2}^2 . P_{O2}}{P_{H2O}^2} = \frac{4\alpha^2 P^2}{(2+\alpha)^2} \frac{\alpha P}{2+\alpha} \frac{(2+\alpha)^2}{4(1-\alpha)^2 P^2} = \frac{\alpha^3 P}{(1-\alpha)^2 (2+\alpha)}$.

b) The constant K_P is a function of T and it is expressed as a function of P; knowing these two variables allows us to determine α and as a result all the partial pressures, hence v = 2.

c) K_p is related to ΔH_T° and ΔS_T° by the relation: $log\ K_P = \frac{-\Delta H_T^\circ}{2.3.RT} + \frac{\Delta S_T^\circ}{2.3.R}$, with $\Delta H_T^\circ = -2\ (-68.3 + 9.8) = 117.00$ kcal/mol and $\Delta S_T^\circ = 62.4 + 49 - 2(16.7 + \frac{9{,}800}{373.15})$ $= 25.47$ cal/mol.K, where $logK_P = \frac{-117{,}000}{4.57\times 2{,}273.15} + \frac{25.47}{4.57} = -5.689$, therefore $K_P = 2.05.10^{-6}$ atm.

3) Taking into account the assumption that $\alpha \ll 1$, then a $(1-\alpha) \approx 1$ and $(2+\alpha) \approx 2$, which allows us to write: $K_P = \frac{\alpha^3 P}{2}$ and for $P = 1$ atm, we obtain: $K_P = \frac{\alpha^3}{2}$ hence $\alpha = \sqrt[3]{2K_P} = \sqrt[3]{4.1\ \times\ 10^{-6}} = 1.6 \times 10^{-2}$.

4) From the expression for K_P, we find that: $P = \frac{(1-\alpha)^2(2+\alpha)}{\alpha^3}K_P$. We know the value of K_P at 2,000 °C, so for $\alpha = 0.1$, we just replace this value. We obtain: $P = \frac{0.9^2\times 2.10}{10^{-3}} \times 2.05 \times\ 10^{-6} = 3.49 \times 10^{-3}$ atm $= 2.65$ torr.

PROBLEM 3.13.– Synthesis of nitrogen monoxide

1) This involves studying the equilibrium: $N_2 + O_2 \rightleftarrows 2NO$, where K_p is related to the variation in Gibbs energy via the relation: $\Delta G_T^\circ = -RTlnK_P$ with: $\Delta G_T^\circ = \Delta H_T^\circ - T\Delta S_T^\circ$.

a) Assuming $\Delta C_P = 0$, then $\Delta H_T^\circ = 2\Delta H_{f298}^\circ = 2 \times 21{,}600 = 43{,}200$ cal/mol and $\Delta S_T^\circ = 2S_{T(NO)}^\circ - (S_{T(N_2)}^\circ + S_{T(O_2)}^\circ) = 2\ \times\ 50.34\ -\ (49.03\ +\ 45.77) =$ 5.88 cal/mol.K, so for K_p: $lnK_P = -\frac{\Delta G_T^\circ}{RT} = -\frac{\Delta H_T^\circ}{RT} + \frac{\Delta S_T^\circ}{R} = -\frac{43{,}200}{1.987T} + \frac{5.88}{1.987} = -\frac{21{,}741.32}{T} +$ 2,959.

b) Taking into account C_P but with ΔC_P = Cte, in this case we find that $\Delta H_T^\circ = \Delta H_{298}^\circ + \int_{298}^{T}\Delta C_P dT$ where $\int_{298}^{T}\Delta C_P dT = [2C_{P(NO)} - (C_{P(N_2)} + C_{P(O_2)})]\Delta T$ $= -0.5(T - 298)$, therefore $\Delta H_T^\circ = 43{,}200 - 0.5T$ and $\Delta S_T^\circ = \Delta S_{298}^\circ + \int_{298}^{T}\Delta C_P \frac{dT}{T} =$ $5.88 - 0.5ln\frac{T}{298}$ so $\Delta S_T^\circ = 8.73 - 0.5lnT$, which gives: $\Delta G_T^\circ = 43{,}200 - 0.5T - 8.73T -$ $0.5ln\frac{T}{298} = 43{,}349 - 9.23T + 0.5TlnT$, where $lnK_P = -\frac{\Delta G_T^\circ}{RT} = -\frac{43{,}349}{1.987T} - \frac{9{,}23T}{1.987T} +$ $\frac{0{,}5T}{1.987T}lnT = -\frac{21{,}816{,}3}{T} - 0.2516lnT + 4.645$. The values obtained are shown in the following table:

	$\Delta C_P = 0$		$\Delta C_P = Cte$	
T/K	lnK$_P$	K$_P$	lnK$_P$	K$_P$
10^3	−18.91	6.136×10^{-9}	−18.782	6.965×10^{-9}
2×10^3	−8.176	2.813×10^{-4}	−7.912	3.664×10^{-4}
3×10^3	−4.642	9.64×10^{-3}	−4.288	1.37×10^{-2}
4×10^3	−2.896	5.53×10^{-2}	−2.476	8.41×10^{-2}

2) To calculate the yield of NO, we must determine the variation in the composition for this reaction:

	N_2	+	O_2	⇄	2NO
At the start	1		1		0
At equilibrium	0 – x ; 1 – x ; 2x	–	$x1$	–	$x2x$

for which $K_P = K_x.P^{\Delta v}$. But at 1 atm pressure, $K_P = K_x = \frac{(2x/2)^2}{\left(1-\frac{x}{2}\right)\left(1-\frac{x}{2}\right)} = \frac{4x^2}{(1-x)^2}$, therefore $\sqrt{K_P} = \frac{2x}{1-x}$ giving the value $x = \frac{\sqrt{K_P}}{2+\sqrt{K_P}}$.

The yield is defined by $\rho = \frac{2x/2}{1/1} = x$, therefore $\rho = \frac{\sqrt{K_P}}{2+\sqrt{K_P}}$. We obtain: $\rho_{a3,000} = \frac{\sqrt{0,0137}}{2+\sqrt{0,0137}} = 5.54\%$; $\rho_{a4,000} = 12.66\%$; $\rho_{b3,000} = 4.68\%$ and $\rho_{b4,000} = 10.52\%$.

3) Using air, equilibrium is displaced but the reaction that remains is the same, so the equilibrium constant is also the same.

4) In this case, given the composition of air, which is $4N_2$ for 1 O_2, the reaction is as follows:

	$(3\ N_2)$	+	N_2	+	O_2	⇄	2 NO	+	$(3N_2)$
At start	3		1		1		0		0
At equilibrium	0		$1 - x$		$1 - x$		$2x$		3

So, the expression for $K_P = \frac{(2x/5)^2}{\left(\frac{1-x}{5}\right)\left(\frac{1-x}{5}\right)} = \frac{4x^2}{(1-x)(4-x)} = \frac{4x^2}{x^2-5x+4}$ # x^2, therefore $x = \sqrt{K_P}$; since $\rho = x$, $\rho \approx \sqrt{K_P}$, which gives the following two values for the proposed T: $\rho_{3,000} = \sqrt{0.00964} = 9.82\%$ and $\rho_{4,000} = \sqrt{0.0553} = 23.52\%$.

5) With respect to O_2, we obtain $\rho = x$; with respect to N_2, we obtain: $\rho = \frac{2x/2}{4/1} = \frac{x}{4}$ and so $\rho_{/O_2} = 4\ \rho_{/N_2}$.

PROBLEM 3.14.– Synthesis of alumina

1) a) The chemical equilibrium has the general formula: $\sum_i \upsilon_i\ A_i \rightleftharpoons \sum_j \upsilon_j\ B_j$, where υ_i and υ_j are the stoichiometric coefficients, A_i are the reagents and B_j are the reaction products; υ_i are negative and υ_j are positive ($\upsilon_i < 0$ and $\upsilon_j > 0$). The reaction is therefore written as: $2Al + \frac{3}{2}O_2 \rightleftarrows Al_2O_3$.

b) By definition, the progress ξ of a chemical reaction: $dn_i = \upsilon_i d\xi$ where: $dn_{Al} = -2d\xi$; $dn_{O_2} = -\frac{3}{2}d\xi$ and $dn_{Al_2O_3} = d\xi$.

2) We know that with dn_{Al} we can deduce $d\xi$ by: $-0.2 = -2d\xi$ so $d\xi = 0.1$ and $dn_{Al_2O_3} = 0.1$.

3) a) Any chemical reaction is accompanied by heat, i.e. received if it is endothermic, or released if it is exothermic, so the exact equilibrium equation is: $\sum_i \upsilon_i\ A_i \rightleftharpoons \sum_j \upsilon_j\ B_j + Q$; heat is released so the reaction is exothermic and the corresponding heat is the heat of reaction at constant V: Q_V.

The variation in U of the system is given by: $dU = \delta Q + \delta W = \delta Q - PdV = \delta Q$, since $dV = 0$ therefore $\Delta U = Q_v = 39.86$ kcal.

b) We burn 5.4 g of aluminum in an excess of O_2, therefore $dn_{Al} = \frac{5,4}{26,98}$ mol which gives Q_V, whereby $q_v = \frac{Q_V}{0.1} = -398.6$ cal/mol.

PROBLEM 3.15.– Combining hydrogen and oxygen

1) We write the variation in the reaction composition, between the start of the reaction and equilibrium, i.e.:

$2H_2O$	$\rightleftarrows$	$2H_2$	+	O_2
2		0		0
$2-2x$		$2x$		x

so $n = 2 - 2x + 2x + x = 2 + x$ and by definition $P = \sum P_i$ and $P_i = Px_i$, therefore $P = P\,(x_{H_2O} + x_{O_2} + x_{H_2})$ and $P_{H_2O} = P.x_{H_2O} = P\frac{2-2x}{2+x}$, where $x = \frac{2(P - P_{H_2O})}{2P + P_{H_2O}}$.

2) The partial P of H_2 and O_2 are given by: $P_{H_2} = P\frac{2x}{2+x}$ and $P_{O_2} = P\frac{x}{2+x}$ where $\frac{P_{H_2}}{P_{O_2}} = \frac{2x}{x} = 2$, therefore $P_{H_2} = 2P_{O_2}$.

3) The equilibrium constant is written as: $K_P = \frac{(P_{H_2})^2 P_{O_2}}{(P_{H_2O})^2} = \frac{4(P_{O_2})^3}{(P_{H_2O})^2}$ where:

a) as a function of P and x: $K_P = \frac{4\left(P\frac{x}{2+x}\right)^2}{\left(P\frac{2-2x}{2+x}\right)^2} = P\frac{x^3}{(1-x)^2(2+x)}$;

b) as a function of P and P_{H_2}: $K_P = \frac{4(P_{O_2})^3}{(P_{H_2O})^2}$; and by replacing x with its expression, we obtain: $K_p = \frac{4\left(\frac{P - P_{H_2O}}{3}\right)^3}{(P_{H_2O})^2} = \frac{4}{27}\frac{(P - P_{H_2O})^3}{(P_{H_2O})^2}$.

4) The numerical application for $P = 0.540$ and $P_{H_2O} = 0.5$ gives for K_p and x the following values: $K_P = \frac{4}{27}\frac{(0.54 - 0.5)^3}{(0.5)^2} = 37.93 \times 10^{-6}$ and $x = \frac{2(0.54 - 0.5)}{2 \times 0.54 + 0.5} = \frac{0.08}{1.58} = 0.0506$.

5) a) From the two reactions, the second is multiplied by a factor of 2 and then subtracted member by member, so that their constants are divisible(*). We obtain:

$2CO_2 \rightleftarrows 2CO + O_2$ with $K_{P1} = 1.4 \times 10^{-3}$

$2CO_2 + 2H_2 \rightleftarrows 2CO + 2H_2O$ with $K'_{P2} = K^2_{P2} = (6.4)^2 = 40.96$

$2H_2O \rightleftarrows 2H_2 + O_2$

where $K_P = \frac{K_{P1}}{K'_{P2}} = \frac{1{,}4.10^{-3}}{40{,}96} = 34.18 \times 10^{-6}$.

b) In this case, we assume that $\Delta C_P = 0$. The equilibrium constant is given by $\Delta G^\circ_T = \Delta G^\circ_{298} = -RT\ln K_P$ with $\Delta G^\circ_{298} = \Delta H^\circ_{298} - T\Delta S^\circ_{298}$. Here, $\Delta H^\circ_{298} = -\Delta H^\circ_{f298}$ and so $\Delta H^\circ_{298} = -\ 2(-57{,}800) = 115{,}600$ cal/mol and $\Delta S^\circ_{298} = 2 \times 31.2 + 49 - 2 \times 45.1 = 21.2$ cal/mol.K, where $RT\ln K_P = 115{,}600 - 2{,}500 \times 21.2 = 62{,}600$ cal/mol and so for the equilibrium constant, we obtain: $\ln K_P = -\frac{62{,}600}{1.987 \times 2{,}500} = -\ 12.602$. Hence, $K_p = 3.37 \times 10^{-6}$.

c) In this case, $\Delta C_P = \text{Cte} \neq 0$. The following calculations are required: $\Delta H^{\circ}_T = \Delta H^{\circ}_{298} + \int_{298}^{2,500} \Delta C_P dT$ and $\Delta S^{\circ}_T = \Delta S^{\circ}_{298} + \int_{298}^{2,500} \Delta C_P \frac{dT}{T}$. This gives: $\Delta C_P =$ $16.6 + 8.6 - 21.6 = 3.6$, so $\Delta H^{\circ}_T = 115,600 + 3.6(2,500 - 298) = 123,527$ cal/mol and $\Delta S^{\circ}_T = 21.2 + 3.6 ln\frac{2500}{298} = 28.857$ cal/mol.K. Therefore, $RTln\text{K}_P = 123,527 -$ $28.857 \times 2,500 = 51,352$ and so: $ln\text{K}_P = -\frac{51,352}{1.987 \times 2,500} = -10.27$ and $\text{K}_P = 34.64 \times 10^{-6}$.

CONCLUSION.– By disregarding the terms pertaining to the calorific capacities, we find the constant is 10 times smaller than in the other cases.

6) The enthalpy of the combustion of water vapor is given by the relation: $\Delta H^{\circ}_{c(298)} = -\Delta H^{\circ}_{f298} = \int_{298}^{T} \Delta C_P dT$, i.e. $57,800 = \int_{298}^{T} (7.5 + 0.002.TdT = [7.5.T + 10^{-3}T^2]_{298}^{T}$, which gives the equation: $10^{-3}T^2 + 7.5.T - 60,124 = 0$ whose positive root is the solution, i.e. T # 4,850 K.

(*) The expressions of ΔG°_T being in lnKp, when we add up the reactions, the Kp are multiplied and when we subtract them, the Kp are divided, as they are involved in their logarithm.

PROBLEM 3.16.– Manufacture of ferromanganese

1) a) The reduction of MnO by graphite is represented by the reaction: $C_{(s)} + MnO_{(liq)} \rightarrow Mn_{(pure\ liq)} + CO_{(g)}$.

b) The affinity of this reaction is written as: $A(T) = A^{\circ} - RTln\ P_{CO}$ with $A^{\circ} = -\Delta G^{\circ} = T\Delta S - \Delta H^{\circ} = 39.13T - 65.600$, with the expression for $A(T)$: $A(T) = 39.13T - 65,600 - RTln\ P_{CO}$; since the reaction occurs at 1 atm $P_{co} = 1$ and so $ln\ P_{CO} = 0$ and $A(T) = 39.13T - 65,600$. At 1,500 K, we obtain $A(T) = 58,695 - 65,600 = -6,905$ cal/mol.

c) $A(T)$ is negative, the reaction is impossible: MnO is not reduced in these conditions.

2) a) The reduction is always the same: $C_{(s)} + MnO_{(liq)} + Fonte_{(liq)} \rightarrow$ $(Mn, Fe)_{solution.} + CO_{(g)}$, but Mn forms a liquid solution with iron.

b) For this reaction, the affinity is written as: $A(T) = A^{\circ} - RTln\ x_{Mn} P_{CO}$. Or since $P_{CO} = 1$ atm, therefore $A(T) = A^{\circ} - RT\ ln\ x_{Mn}$.

3) a) At equilibrium, $A(T) = 0$ hence $ln\ x_{Mn} = \frac{A^{\circ}}{RT} = \frac{6,905}{2,980.5} = -2.3167$ therefore $x_{Mn} = 0.098$.

b) This value is the upper limit that the titer can reach under these conditions of T and P. This is also the same value that is obtained with the law of mass action.

4

Mixtures or Solutions

4.1. Exercises

EXERCISE 4.1.– Partial molar state function

A total molar state function z of a binary mixture is represented by the following relation: $z = x_1 x_2 f(x_1)$ with $f(x_1) = A + Bx_1 + Cx_1^2$.

Calculate the partial molar state functions z_1 and z_2.

Answers: $z_1 = x_2^2\left(A + 2Bx_1 + 3Cx_1^2\right)$ and $z_2 = x_1^2\left(A - B + 2(B - C)x_1 + 3Cx_1^2\right)$.

EXERCISE 4.2.– Partial molar volume and molarity

The volume (in cm^3) of a solution of sodium chloride in 1 kg of water at constant temperature is given by the expression: $V = 1001.38 + 16.6253m +$, where m is the molarity of the salt.

Determine the expressions giving the partial molar volumes of two bodies.

Data: M_{H_2O} = 18.015 g/mol. *Answers*: v_1: see corrections and $v_2 = 18.039 - 0.0157\, m^{\frac{3}{2}} - 0.0202m^2$.

EXERCISE 4.3.– Enthalpy of dissolution water/sulfuric acid

When we mix n mol of $H_2O(1)$ at 1 mol of $H_2SO_4(2)$, an amount of heat is released: $Q = \frac{17.860\, n}{n+1.798}$ cal.

1) Calculate as a function of n, the partial molar enthalpy of the dissolution of water in a given mixture.

2) Calculate as a function of n, the partial molar enthalpy of dissolution of the acid in this same mixture.

3) Calculate their respective values.

Answers: (1) $h_1 = -\frac{ab}{(b+n)^2}$; (2) $h_2 = -\frac{an^2}{(b+n)^2}$; (3) $h_1 = -4{,}101.81$ cal/mol; $h_2 = -2{,}281.32$ cal/mol.

EXERCISE 4.4.– Partial molar enthalpy of mixing

The total molar enthalpy of mixing for a binary mixture (1)/(2) is represented by: $h^m = x_1x_2(A_1x_1 + A_2x_2)$ with A_1 and A_2 being constants.

1) Establish the general expression of h_1^m.

2) Deduce the expression for the partial molar enthalpies of mixing for the two constituents.

Answers: (1) see corrections; (2) $h_1^m = x_2^2[A_2 + 2x_1(A_1 - A_2)]$ and $h_2^m = x_1^2$ $[A_1 + 2x_2(A_2 - A_1)]$.

EXERCISE 4.5.– Ideal mixture of alkaline fluorides

Experimentally determining activities of potassium fluoride (KF) and lithium fluoride (LiF) in the mixtures at a temperature of 1,148 K showed that for an equimolar mixture: $a_{KF} = a_{LiF} = 0.295$.

Calorimetry measurements show that the heat released, when 1 mol of KF is mixed with 1 mol of LiF, is 2,400 cal.

1) Calculate the molar entropy of mixing in this solution.

2) Show that this mixture is an ideal solution.

Answers: (1) s^m # 1.380 cal/K.mol; (2) $s^{mid} = 1.377$ # 1.38 cal/K.mol, therefore s^m # s^{mid}, the solution is ideal.

EXERCISE 4.6.– Chloroform/acetone mixture

The total molar enthalpy of mixing for an ideal solution made with chloroform (1) and acetone (2), expressed in J/mol at constant T and P, is given by the relation: $h^m = x_1x_2[7647+2161(x_1-x_2)]-1745.05\ (x_1-x_2)^2$

Calculate h_2^m for a solution made with 1 mol of acetone in 1 kg of chloroform.

Data: $M_2 = 119$ g/mol.

Answer: $h_2^m = 17.463$kJ/mol.

EXERCISE 4.7.– Total and partial molar volumes of mixing

The change in total molar volume of the methanol(1)/water(2) binary, as a function of the molar fraction in the alcohol x_1, is given in the following table:

x_1	0	0.2	0.4	0.6	0.8	0.9	1
v (cm^3.mol^{-1})	17.5	20.4	23.3	27.3	32.7	36.3	42.5
v^m (cm^3.mol^{-1})							

1) Calculate the values for the total molar volume of mixing to complete the previous table.

2) Determine graphically the partial molar volumes of mixing of each constituent for fractions 0.3 and 0.7 in methanol.

3) Deduce the values for the partial molar volumes of each constituent for these same molar fractions.

Answers: (1) see corrections; (2) $x_1 = 0.3$: $v_1^m = -9.95$ cm^3/mol and $v_2^m = -0.55$ cm^3/mol; $x_1 = 0.7$: $v_1^m = -4.45$ cm^3/mol and $v_2^m = -6.65$ cm^3/mol; (3) $x_1 = 0.3$: $v_1 = 32.5$ cm^3/mol and $v_2 = 16.9$ cm^3/mol; $x_1 = 0.7$: $v_1 = 38.05$ cm^3/mol and $v_2 = 10.85$ cm^3/mol.

EXERCISE 4.8.– Excess volume of the water/methanol mixture

One mole of water is added to a large volume of a water(1)/methanol(2) solution at 15 °C (with composition $x_2 = 0.4$) and we find that the total volume of the solution increases by 17.35 cm^3, whereas the addition of 1 mol of methanol to the same volume increases the solution volume by 39 cm^3.

1) Calculate the total molar volume of this solution.

2) Calculate the total excess molar volume of the solution.

Data: Methanol: M = 12 g/mol and $\rho_{15\,°C}$ = 0.8 g.cm^{-3}; water: M = 18 g/mol and $\rho_{15\,°C}$ = 1 g.cm^{-3}.

Answers: (1) v = 26.01 cm^3/mol; (2) v^{ex} = −0.79 cm^3/mol.

EXERCISE 4.9.– Excess molar enthalpy, entropy and Gibbs free energy

The total excess molar Gibbs free energy of the methanol(1)/ethanol(2) binary at constant pressure is represented by the relation: $g^{ex} = RT\text{B}\, x_1x_2$, where B is a function of temperature according to the below table.

T (°C)	35	40	45
B	17.5	0.458	0.439

1) Calculate s^{ex} and h^{ex} of the titer mixture x_1 = 0.3 at 45 °C.

2) Deduce the activity coefficients γ_1 and γ_2.

Answers: (1) s_1^{ex} = 0.35 cal/K.mol and h_1^{ex} = 168.94 cal/mol; (2) γ_1 = 1.24 and γ_2 = 1.04.

EXERCISE 4.10.– Activity coefficients at infinite dilution

The total excess molar Gibbs free energy of the carbon(1)/acetone(2) mixture is given at constant pressure and at T = 35 °C by the expression: $\frac{g^{ex}}{RT} = \frac{ABx_1x_2}{Ax_1+Bx_2}$ with A = 1.5. B = 3, independent of T.

1) Establish the expression for the activity coefficient of each component.

2) Deduce the values of these coefficients at infinite dilution γ_1^{∞} and γ_2^{∞} at 90 °C.

3) Calculate the Henry constant for each species present at the molar fraction scale, assuming that the evaporation is an ideal gas.

Data: At 90 °C: P_1^S = 1 bar and P_2^s = 0.95 bar.

Answers: (1) $\ln g_1 = A\left(\dfrac{Bx_2}{Ax_1 + Bx_2}\right)^2$ and $\ln g_2 = B\left(\dfrac{Ax_1}{Ax_1 + Bx_2}\right)^2$; (2) $\gamma_1^{\infty} = 4.4817$ and $\gamma_2^{\infty} = 20.085$; (3) $H_1 = 4.4817$ and $H_2 = 19.08$.

EXERCISE 4.11.– Dissolution of oxygen in water

The dissolution of O_2 in H_2O at atmospheric pressure and at ambient temperature of 20 °C imposes a partial pressure of O_2 equal to 190 mm Hg. In these conditions, the Henry constant of O_2 is 3.3×10^7 mm Hg.

Determine the solubility of O_2 in water.

Answer: $m = 3.2 \times 10^{-4}$ mol/L.

EXERCISE 4.12.– Modified Roozeboom method

In 1959, H.C. Van Ness and R.V. Mrazeck proposed an improvement to the Roozeboom method by practicing the construction of the tangent, not on the total molar state function $z(x_1, x_2)$ but on the state function $f(x_1, x_2) = \dfrac{z(x_1, x_2)}{x_1 x_2}$, with z being a state function of mixing or excess.

1) Give the expressions for the partial molar state functions f_1 and f_2 as a function of f_1 and f_2, and their partial derivatives with respect to the molar titers.

2) Express z_i as a function of f and its partial derivatives; deduce the relation between z_i and total and partial f.

The experimental results in the table below give the mixing volume of the cyclohexane(1)/cyclopentane(2) binary at 288.15 K.

x_1	0.115	0.20	0.28	0.36	0.457	0.572	0.746
v^m 10^3.cm^3.mol^{-o}	17.2	28.9	38.5	46.6	53.8	57.4	50.1
v^m/x_1x_2 .cm^3.mol^{-o}	0.169	0.1806	0.191	0.2023	0.2168	0.2345	0.2644

3) Calculate, applying this method, the partial molar volume of mixing of cyclohexane in the solution $x_1 = 0.4$.

Answers: (1) see corrections; (2) $z_1 = x_2^2(2f - f_2)$ and $z_2 = x_1^2(2f - f_1)$; (3) $v_1^m = 0.0961$ cm^3/mol.

EXERCISE 4.13.– Dissolution of hydrogen in iron

At atmospheric pressure, hydrogen dissolves in molten iron at 1,600 °C, at 25 p.p.m.

What pressure is required for H_2 so that only 5 p.p.m of H_2 is dissolved at the same temperature?

Answer: $P_{H_2} = 0.04$ atm.

EXERCISE 4.14.– Dissolution of alumina in silica

At 1,900 °C, alumina(1) dissolves in molten silica(2) and forms an ideal solution with the liquid solution containing 50% moles of alumina in equilibrium with pure alumina. Several reference choices for the activities can be selected.

Give the different reference states of interest and calculate the values of the corresponding activities for alumina and silica in a 10% (in moles) solution of alumina.

Data $T_f(Al_2O_3) = 2{,}100$ °C and $T_f(S_iO_2) = 1{,}700$ °C.

Answers: $a_1 = 0.10$ $a_2 = 0.90$ and $a_1 = 0.20$.

EXERCISE 4.15.– Composition of water gas

Water gas is obtained by dissolution in water at atmospheric pressure and a temperature of 20 °C, with a composition of: 45% H_2, 45% CO and 10% N_2 in volume.

Determine the mass composition of 1 m^3 of water gas.

Data: The Henry constants in mm Hg under these conditions are as follows: for H_2: 6.83×10^4; for CO: 5.36×10^4 and for N_2: 8.04×10^4.

Answers: $m_{H_2} = 0.732$ g; $m_{CO} = 13.06$ g; $m_{N_2} = 1.93$ g.

EXERCISE 4.16.– Activities of iron and carbon in steel

Studying the equilibrium of austenite with mixtures of methane and hydrogen give results that accurately represent the activity of carbon at any temperature using the following relation: $\ln a_C = \ln \frac{x_c}{x_{Fe}} + 6.6 \frac{x_c}{x_{Fe}}$, where a_c is the activity of carbon, and x_c and x_{Fe} are the atomic titers of carbon and iron, respectively.

Establish the relation that allows us to calculate the activity of iron in this steel.

Answer: $\ln a_{Fe} = -\frac{x_c}{x_{Fe}} - 3.3\left(\frac{x_c}{x_{Fe}}\right)^2$.

EXERCISE 4.17.– Athermal solution of benzene/butyl sebacate

Brönsted and Colmant determined at 18 °C the vapor tension of benzene on top of the benzene(1)/butyl sebacate(2) solution. With the latter being a non-volatile liquid, they found that from these experimental results the activity coefficient of benzene was accurately represented by relation: $\log \gamma_1 = -1.816\, x_2^2 + 2.473\, x_2^{5/2} - 0.950\, x_2^3$ and that the dissolution of butyl sebacate in benzene occurs at any titer or temperature, without any thermal effect.

1) Show that these results allow us to determine the activity coefficient of benzene in solution at any temperature.

2) Give the expression of $\log \gamma_1$ at $T = 55$ °C.

Answers: (1) $g_1^{ex} = -Ts_1^{ex}$; (2) $\log\gamma_1 = -\frac{s_1^{ex}}{4.57}$.

EXERCISE 4.18.– Binary solution of benzene/hexane

Measurements of the activity coefficient of benzene in benzene(1)/hexane(2) solutions at 70 °C, as a function of the composition, can be seen in the table below:

x_1	0.1	0.2	0.3	0.4	0.5	0.6	0.7	0.8	0.9
γ_1	1.1667	1.1468	1.1235	1.0986	1.0738	1.0506	1.0303	1.0143	1.004

Here, we represent the excess Gibbs free energy by a Margules development with two terms $\frac{g^{ex}}{RT} = x_1 x_2[B_0 + B_1(x_2 - x_1)]$.

1) Calculate the values of B_0 and B_1 at 70 °C.

2) Calculate the activity coefficient limits at this temperature.

Answers: (1) $B_0 = 0.226$ and $B_1 = -0.059$; (2) $\gamma_1^{\infty} = 1.1818$ and $\gamma_2^{\infty} = 1.3298$.

EXERCISE 4.19.– Binary solution of methyl acetate/cyclohexane

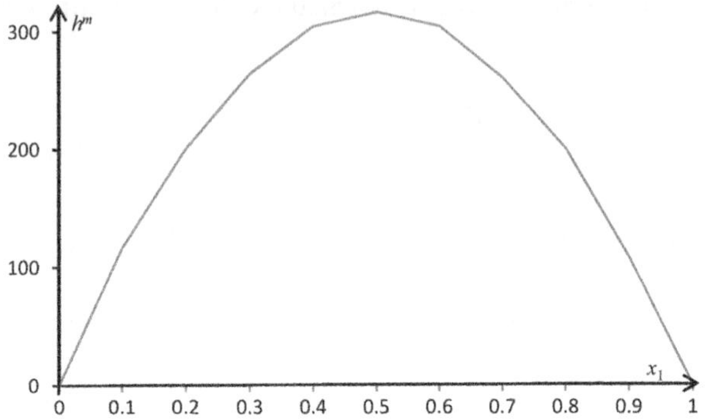

Figure 4.1. *Molar enthalpy of a mixture as a function of x_1*

The experimental measurements for the total molar enthalpy of the liquid methylacetate (1)/cyclohexane(2) binary mixture at 303.15 K are represented in Figure 4.1, and the values of the product Ts^{ex} are represented in Figure 4.2.

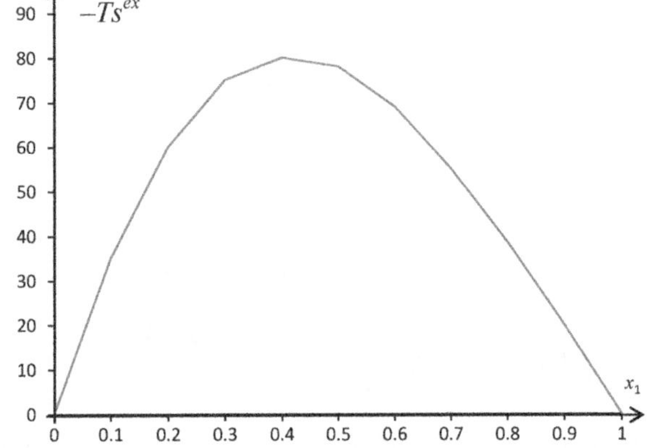

Figure 4.2. *$Ts^{ex} = f(x_1)$ curve*

1) Calculate for the solution $x_1 = 0.3$ the partial molar enthalpy of mixing and the partial state function of methyl acetate at 303.15 K.

2) Deduce the activity of acetate in this same solution at a temperature of 323.0 K (specify the approximations used).

Answers: (1) $h_1^m = 674\ \mathrm{J/mol}$ and $Ts_1^{ex} = 138\ \mathrm{J/mol}$; (2) $a_1 = 0.365$

EXERCISE 4.20.– Ternary solution of benzene/cyclohexane/hexane

According to Redlich and Kister, we can predict the thermodynamic properties of a ternary solution from the properties of the three binary constituents and from measurements carried out on the ternary solution.

To verify this method, we consider the solution at 70 °C, formed of a mixture of benzene(1)/cyclohexane(2)/hexane(3), whose excess molar Gibbs free energy of binary mixtures is represented as follows:

– binary(1)/(2): $\dfrac{g_{1/2}^{ex}}{RT} = \mathrm{A}(T)x_1x_2$;

– binary(1)/(3): $\dfrac{g_{1/3}^{ex}}{RT} = x_1x_3[\mathrm{B}_0 + \mathrm{B}_1(x_3 - x_1)]$;

– binary(2)/(3): behaves as an ideal solution in any composition.

1) Determine the expression of the activity coefficient of benzene at trace state in the ternary mixture.

2) Deduce the value of this coefficient in a solution composed of 2 mol of cyclohexane and 2 mol of hexane.

Data: At 70 °C. $\mathrm{A}(T) = 0.349$; B_0 0.226 and $\mathrm{B}_1 = -0.059$.

Answers: (1) $\ln \gamma_1^{\infty} = \mathrm{A}(T)x_2 + x_3(\mathrm{B}_0 + \mathrm{B}_1x_3)$; 2) $\gamma_1^{\infty} = 1.3137$

EXERCISE 4.21.– Nitridation of iron by ammonia

When a gaseous mixture of ammonia and hydrogen is passed over iron, with certain precautions, there is a strong dissolution of nitrogen in the iron, accompanied by a very low release of nitrogen gas.

The experiment of a gas phase containing 10% mol of NH_3 at 700 °C and at atmospheric pressure give a solution, containing 4.5% by weight, of nitrogen in iron.

Determine the percentage by weight of nitrogen dissolved under the same conditions in the presence of a phase containing 20% mol of NH_3.

Answer: $x_N \approx 0.107$ # 11%.

EXERCISE 4.22.– Influence of oxygen on the melting point of silver

The solubility of oxygen in liquid silver was determined experimentally at 962 °C for 100 g of silver. We obtained the following results:

P_{O2} (torr)	128	488	760	1,203
O_2 dissolved TPN (cm^3)	93.5	180.3	222.0	284.1

1) Show that the dissolution of oxygen in silver verifies Sievert's law, with regard to low concentrations.

2) Deduce the error from the previous results that can arise during the calibration of a couple determining the solidification point of silver in air.

Data: $T_f(Ag_{\text{pure}}) = 960.8$ °C; $h_f(Ag_{\text{pure}}) = 2{,}690$ cal/mol.

Answers: (1) $P = 1.68 \times 10^6$ x; 2) $\Delta T = -11$ °C.

EXERCISE 4.23.– Hydrolysis of chlorine

Chlorine is hydrolyzed in an aqueous solution according to the reaction: $Cl_2 + H_2O \rightleftarrows HOCl + H^+ + Cl^-$. Assuming the activity coefficients are equal to 1 and disregarding the dissociation of water and hypochlorous acid:

1) calculate the equilibrium constant of this reaction at 298 K;

2) express this constant as a function of the molarity and molar fraction of dissociated chlorine.

Note that 1,000 g of pure water is saturated with gaseous chlorine at 298 K and under atmospheric pressure;

3) theoretically determine the number of moles of non-hydrolyzed chlorine;

4) deduce the total number of moles of dissolved chlorine.

Data:

	Cl_{2gaz}	$Cl_{2sol.aq}$	HOCl	Cl^-	H^+	H_2O_{liq}
ΔG°_{298} (cal.mol^{-1})	0	1 650	−1,911	−3,135	0	−5,669

Answers: (1) $K_P = 4.37 \times 10^{-4}$; (2) $K_P = m^2 \cdot \frac{x^3}{1-x}$; (3) $m_{Cl} = 6.16 \times 10^{-2}$; 4) $m = 0.0916$.

4.2. Problems

PROBLEM 4.1.– Nitridation of iron with ammonia

Nitrogen dissolves in iron in its atomic state "N". Passing pure nitrogen over iron under atmospheric pressure, we find that, at equilibrium, iron contains 0.003% of N_2 by weight. If we pass a mixture of "H_2/NH_3" over iron containing 10% of NH_3 under the same conditions, we obtain a solution containing 0.45% by weight of N_2.

1) Calculate the amount of dissolved nitrogen in both cases and explain the disparity that is recorded or observed.

2) Calculate the amount of nitrogen that can be dissolved using a mixture containing 20% of NH_3.

Data: At 700 °C, $\Delta G^\circ_f(NH_3) = 13.9$ kcal/mol.

Answers: (1) $x_N = 1.2 \times 10^{-4}$; x'_N 0.0425; (2) $x_N = 0.472\%$.

PROBLEM 4.2.– Molar volumes of acetone/chloroform mixtures

Acetone(1) and chloroform(2) form a mixture with a variation in volume. To explain this phenomenon, the binary mixture(1)/(2) can be considered as an ideal ternary solution between non-associated (1) and (2) and the complex(1/2) formed following equilibrium: (1)+(2) ⇄ (1/2) whose constant is $K_p = 0.7$ at 328 K.

1) Give the expression of the total molar volume of the binary mixture of n_1 mol of (1) and n_2 mol of (2) as a function of molar volumes of pure substances.

2) Give the expression for the molar volume of acetone in the binary.

Answers: (1) $v=\left(\dfrac{n_1-\xi}{n_1+n_2}\right)v_1^P+\left(\dfrac{n_2-\xi}{n_1+n_2}\right)v_2^P+\left(\dfrac{\xi}{n_1+n_2}\right)v_{1/2}^P$;

$$(2)\ v_1=v_1^P+v_1=v_1^P+\left(v_{12}^P-v_1^P-v_2^P\right)\frac{(1+K_P)\xi-K_P n_2}{(1+K_P)\left[2\xi-(n_1+n_2)\right]}.$$

PROBLEM 4.3.– Proprieties of carbonic anhydride

Consider carbonic anhydride at 100 °C and 100 atm. We propose to determine the following thermodynamic properties:

1) v (molar volume);

2) $h-h^*$ (difference to ideal gas);

3) f (fugacity).

We use each of the following methods: (a) Redlich–Kwong state equation; (b) Houguen–Watson–Ragatz tables; (c) Lee–Kesler tables; (d) generalized diagrams.

Data: $P_C = 72.8$ atm; $T_C = 304.2$ K; $z_C = 0.274$ and $\omega = 0.225$.

The experiment gives: $v = 229.5$ cm^3/mol and $h-h^* = -760$ cal/mol.

Answers: (1) $v_a = 222.5$; $v_b = 225.0$: $v_c = 228.2$ and $v_d = 223$; (2) $(h-h^*)_a = -709$; $(h-h^*)_b = -724$; $(h-h^*)_c = -779$; (3) $f_a = 76.7$; $f_b = 79.2$; $f_c = 79.4$ and $f_d = 77$.

PROBLEM 4.4.– Methane/propane mixture

The gaseous mixture of methane(1) and propane(2) has the following state equation: $\dfrac{Pv}{RT}=1+\dfrac{B}{v}$, where B expressed as a function of molar titers is written as $B=B_{11}x_1^2+B_{12}x_1x_2+B_{22}x_2^2$. At 344 K, the coefficients B_{ij} have the following values in cm^3/mol: $B_{11} = -27$; $B_{12} = -93.5$ and $B_{22} = -290$.

1) a) Calculate the total molar volume of the equimolar mixture and the titer mixture under a pressure of 15 atm.

b) Compare the previous result to that you would obtain if $B_{12} = \frac{(B_{11} + B_{22})}{2}$. Then with that obtained from the application of the Lewis rule.

2) Establish the expression of the difference to an ideal gas with a Gibbs free energy of a gaseous mixture of n_1 mol of methane and n_2 mol of propane.

3) Express the fugacity of each gas as a function of the mixture's total molar volume.

4) Compare the fugacity values (f_i) obtained under the conditions and the assumptions from question 1).

5) a) Reuse the calculations from the previous questions, except numerical applications with the equation of state $B_{12} = \frac{(B_{11} + B_{22})}{2}$.

b) Find the relation between B'_{11}, B'_{22} and B'_{12} such that the equimolar mixture satisfies the Lewis rule.

Answers: (1) (a) $v(0.5) = 1{,}747$ cm^3 and $v(0.1) = 1{,}583.2$ cm^3; (b) $v(0.5) =$ 1,708 cm^3 and $v(0.1) = 1{,}565.8$ cm^3; $v(0.5) = 1{,}690.1$ cm^3 and $v(0.1) = 1{,}557$ cm^3; (2) $F - F^* = \frac{RT}{V}\left(B_{11}n_1^2 + 2B_{12}n_1n_2 + B_{22}n_2^2\right)$; (3) $f_1 = x_1 \frac{RT}{v} \exp\left[\frac{2}{v}\left(B_{11}x_1 + B_{12}x_2\right)\right]$; $f_2 = x_2 \frac{RT}{v} \exp\left[\frac{2}{v}\left(B_{22}x_2 + B_{12}x_1\right)\right]$; (4) see corrections; (5) (a) $f_1 = x_1 P \exp\left[P(2B'_{11} + 2B'_{12}x_2 - B')\right]$; (b) $B'_{12} = \frac{(B'_{11} + B'_{22})}{2}$.

PROBLEM 4.5.– Partial and mixing total molar volumes

The experimental results of the methane(1)/propane(2) mixture obtained at $P = 13.6$ atm and $T = 344$ K are given in the table below:

x_1	0	0.1	0.2	0.3	0.4	0.5
v cm^3/mol	?	1,791.1	1,835.4	1,872.8	1,910.3	1,941.0
x_1		0.6	0.7	0.8	0.9	1
v cm^3/mol		1,966.5	1,991.5	2,016.4	2,035.2	?

1) Determine graphically:

a) the molar volumes of methane and propane;

b) the partial molar volumes for the 80% mixture of methane.

2) Calculate the total molar volume of mixing for each composition in the table.

3) Deduce the partial molar volumes of mixing of the binary with 80% methane.

Answers: (1) (a) v_1^{pure} = 2,047 cm^3 and v_1^{pure} = 1,748 cm^3; (b) v_1= 18.40 cm^3 and v_2= 20.6cm^3; (2) see corrections; (3) v_1^m = 13 cm^3 and v_2^m = 92 cm^3/mol.

PROBLEM 4.6.– Methane/propane mixture under pressure

Experimentally measuring the molar volume (v) of an equimolar mixture of methane and propane under a pressure of 13.6 atm and at a temperature of 344 K gave v = 1,938 cm^3/mol.

Determine the values of v obtained for the following methods:

1) virial equation at P limited to the second term;

2) the Lewis rule from pure products;

3) the corresponding states: Kay's rule; Gunn and Prausnitz rule;

4) Rédlich–Kwong equation.

Data: Virial coefficients in cm^3/mol: B_{11} = –31; B_{22} = –330 and B_{12} = –93.5.

	v^{pure} (cm^3.mol^{-1})	Pc (atm)	Tc (K)	Tb (K)	Ω	Zc
CH_4	2,044	45.4	190.6	111.7	0.008	0.288
C_3H_8	1,745	41.9	369.8	231.1	0.152	0.281

Answers: (1) v = 1,946 cm^3/mol; (2) v = 1,895 cm^3/mol; (3) v = 1,959 cm^3/mol and v = 1,957 cm^3/mol; (4) v = 1,944 cm^3/mol.

PROBLEM 4.7.– Water/methanol mixture

The total volume of a solution obtained by addition of n_1 mol of methanol to 10^3 cm^3 water is given by the expression: $V = 10^3 + 35n_1 + 0.5n_1^2$

1) Determine the partial molar volumes of both components as a function of n_1.

2) Calculate the total volume of mixing as a function of n_1.

3) Deduce:

a) the total molar volume of mixing;

b) the partial molar volumes of mixing.

4) From the partial molar volumes of mixing, find the total partial molar volumes.

Data: $v_{H_2O}^{pure} = 18$ cm^3/mol; $v_{CH_3OH}^{pure} = 35.5$ cm^3/mol.

Answers: (1) $v_1 = n_1 + 35; v_2 = 18 - 9 \times 10^{-3} n_1^2$; (2) $V^m = 0.5\, n_1 (n_1 - 1)$; (3) (a) $v^m = \dfrac{9n_1(n_1-1)}{18n_1 + 10^3}$; (b) $v_1^m = n_1 - 0.5$; $v_2^m = -9 \times 10^{-3} n_1^2$; (4) $v_1 = 9 \times 10^{-3} n_1^2 + 18$ and $v_2 = n_1 + 35$.

PROBLEM 4.8.– Activity of copper and zinc in alloys

The results from the calorimetric study of liquid Cu/Zn alloys at 1,300 K and the measurement of vapor pressure on top of this binary are given in the table below:

x_{Zn}	0	0.1	0.2	0.3	0.4	0.5
$-h^m$(cal.mol^{-o})	0	650	1,180	1,510	1,700	1,790
γ_{Zn}	0	0.136	0.199	0.279	0.377	0.489
x_{Zn}		0.6	0.7	0.8	0.9	1
$-h^m$(cal.mol^{-o})		1,720	1,460	1,150	640	0
γ_{Zn}		0.607	0.837	0.922	0.981	0

1) Calculate graphically the amount of heat released when 0.1 mol of Cu is added to a liquid alloy of 10 mol of titer $x_{Zn} = 0.7$ at 1,300 K.

2) Compare the result with that obtained from the analytical method with h^m as the development term of type: $h^m = Ax(1 - x)$.

3) Show that these experimental results can be represented by the equation:

$$\log \gamma_{Zn} = B(1 - x_{Zn})^2.$$

4) From the previous result, deduce the expression of γ_{Cu} and calculate activity of copper in an alloy of titer $x_{Cu} = 0.5$.

5) Show that the previous experimental results could predict the activity of C_u in the liquid alloy at any temperature, and calculate its value for $x_{Cu} = 0.5$ at 1,600 K.

Answers: (1) $Q = 352$ cal; (2) $Q = 350.9$ cal; (3) B = −0.863; (4) $a_{Cu} = 0.5854$; (5) see corrections; $a_{Cu} = 0.336$.

PROBLEM 4.9.– Dichloromethane/acetone solution degreasing

The total molar enthalpy of mixing binary solutions CH_2Cl_2 *(1)/*C_3H_6O*(2)* have allowed their representation by relation $h^m = x_1 x_2[A_0 + A_1(2x_1 - 1) + A_2(2x_1 - 1)^2]$. With the three parameters being independent of the composition but varying with temperature at 25 °C, these measurements gave the following values:

$A_0 = 3{,}736.7$ J/mol; $A_1 = 387.76$ J/mol; $A_2 = 264.11$ J/mol and at 0 °C, the measurement for an equimolar solution gave the value $h^m = -962\text{J}/\text{mol}$.

1) Establish the relations allowing the calculation of partial molar enthalpies of mixing for both components as a function of x_1.

2) Calculate the heat released when 3 g of acetone is added to 1 kg of dichloromethane.

Ten mole of each component is adiabatically mixed at an initial temperature of $T = 25$ °C and we see that the calorific capacities of the pure bodies and solutions formed are independent of temperature.

3) Calculate the final temperature of the mixture.

A degreasing solution must be prepared with equal masses of these products. To create a continuous mixture, an adiabatic reactor is used in which pure products enter, respectively, at $T_1 = 15$ °C and $T_2 = 5$ °C with mass flow rates of 1 kg/s.

4) Calculate the output temperature of the mixture (state the approximations).

Data: At 25 °C. $C_{P1} = 101.4$ J/mol.K and $C_{P2} = 126$ J/mol.K.

Answers: (1) $h^m = x_2^2\left[A_0 - A_1 + A_2 + (4A_1 - 8A_2)x_1 + (12A_2 x_1^2)\right]$ and $h_2^m = x_1^2$ $\left[A_0 - 3A_1 + 5A_2 + (4A_1 - 16A_2)x_1 + 12A_2 x_1^2\right]$; (2) $Q \# 200$ J; (3) $T_f = 33.21$°C, (4) $T_f = 16.10$ °C.

PROBLEM 4.10.– Adiabatic mixer: cyclohexane/methyl acetate

In an adiabatic mixer, one part pure cyclohexane is added at 30 °C with a flow rate of 10 kg/min to one part methyl acetate(1)/cyclohexane(2) mixture at 50 °C with a molar titer of $x_2 = 0.1$ with a flow rate of 5 kg/min.

The enthalpy of mixing a binary (1)/(2) mixture is represented by the following: $h^m = x_1 x_2 \left[A_0 + A_1(x_1 - x_2) + A_2(x_1 - x_2)^2 \right]$, where the coefficients A_i are independent of T, in a wide range close to 300 K and has the following values: $A_0 = 7{,}067$; $A_1 = -446$ and $A_2 = 1{,}322$.

1) What can be said about the calorific capacity of mixing (1)/(2)?

2) Express the molar calorific capacity of the binary mixture as a function of x_1.

3) Calculate the molar titer and the temperature of the solution on exiting the mixer, considered to be perfectly homogenized.

Data:

	$C_3H_6O_2$	C_6H_{12}
M (g.mol^{-1})	74	84
C_p (cal.mol^{-1}.K^{-1})	16.0	12.1

Answers: (1) null; (2) $C_P = x_1 C_{P1}^{pure} + x_2 C_{P2}^{pure}$; (3) $x_1 = 0.323$; $x_2 = 0.677$ and $T = 13.84$ °C.

PROBLEM 4.11.– Benzene/cyclohexane binary solution

The total excess molar Gibbs free energy is represented by the relation $\frac{g^{ex}}{RT} = A(T) x_1 x_2$ and the experimental study gave the following results:

T (°C)	35	40	45	50
$A(T)$	0.479	0.458	0.439	0.420

1) Establish, from this relation, the analytical expressions that allow the dimensionless functions $\frac{h^m}{RT}$ and $\frac{S^{ex}}{R}$ to be described as a function of T and x.

2) Calculate the heat released when 0.8 mol of benzene is mixed with 1.2 mol of cyclohexane at 40 °C and 50 °C.

3) Calculate the activity coefficients of both components at 40 °C of a solution where $x_1 = 0.40$.

4) Establish the relations that allow the activity coefficient γ_1^{∞} and γ_2^{∞} to be calculated at any temperature.

Answers: (1) ; $\dfrac{h^m}{RT} = x_1 x_2 \dfrac{dA}{d(\ln T)}; \dfrac{S^{ex}}{R} = -x_1 x_2 \left(A - \dfrac{1}{T}\dfrac{dA}{d(1/T)} \right)$; (2) $Q = -1563.5\text{J}$;
(3) $\gamma_1 = 1.1795$; $\gamma_2 = 1.0761$; (4) $\gamma_1^{\infty} = \gamma_2^{\infty} = \exp\left(\dfrac{391.776}{T} - 0.793 \right)$.

PROBLEM 4.12.– n-Heptane/aniline two-phase mixtures

Heptane(1) and aniline(2) are not miscible in any proportions at normal temperature. At a temperature of 39 °C, they form two phases (a) and (b) in equilibrium whose compositions are as follows: $x_{1a} = 0.883$ and $x_{2b} = 0.918$.

Experimental measurements helped establish for g^{ex} the following Margules development: $g^{ex} = x_1 x_2 \; (A + Bx_1)$.

1) Show that these data allow us to calculate the coefficients A and B.

2) a) Calculate the activity coefficient of heptane in each phase.

b) Deduce the corresponding activities and verify them.

3) Calculate the vapor tension of this mixture.

Data: $P_1^{\circ} = 89$ mm Hg and P_2° is negligible at this temperature.

Answers: (1) A = 1,804.470 and B = −194.736; (2) (a) $\gamma_{1a} = 1.0328$ and $\gamma_{1b} = 11.095$; (b) $a_{1a} = a_{1b}$ # 0.91; (3) P # 81 mm Hg.

PROBLEM 4.13.– Activities of zinc/cadmium liquid alloys

The calorimetric measurements of liquid mixtures Zn(1)/Cd(2) were carried out in such a way that at any temperature and titer zone of interest, the partial enthalpy of mixing of zinc in solution from pure liquid zinc is given by the following relation:

$$h_{Zn}^{m} = (1 - x_{Zn})^2(1{,}200 + 1{,}900\, x_{Zn})$$

1) Show how the activity of zinc is obtained along the deposition curve and calculate its value for the titer 0.3 at $T = 273$ °C.

2) Show how the activity of zinc is obtained at any other solution temperature if the enthalpies and entropies of excess do not depend on temperature. Calculate a_{Zn} for $x_{Zn} = 0.3$ and $T = 527$ °C.

3) Show how you would solve the problem in the absence of calorimetric data, assuming it is a regular solution. Calculate a_{Zn} for a titer of 0.3 and $T = 527$ °C.

Data: Enthalpy of fusion of pure *Zn*: $L_f = 1{,}765$ cal/mol.

Deposition diagram of pure Zn:

x_{Zn}	0.26	0.3	0.4	0.5	0.6	0.7	0.8	0.9	1.0
T (°C)	266	273	292	309	322	334	352	380	420

Answers: (1) $a_{Zn} = 0.707$; (2) $\ln a_{Zn} = \frac{L}{R}\left(\frac{1}{T_1} - \frac{1}{T_0}\right) - \frac{h^{ex}}{R}\left(\frac{1}{T_1} - \frac{1}{T}\right)$; $a_{Zn} = 0.549$; (3) $a_{Zn} = 0.593$.

PROBLEM 4.14.– Properties of a ternary mixture

Consider the ternary solution of methylcyclohexane(1)/n-heptane(2)/aniline(3) at 39 °C with the following information for each of the three binary mixtures:

– binary (1)/(2): it is an ideal solution;

– binary (2)/(3): they are not miscible in any way at 39 °C, both phases are in equilibrium for molar titers:

phase(a): $x_{2a} = 0.883$; $x_{3a} = 0.117$;

phase(b): $x_{2b} = 0.082$; $x_{3b} = 0.918$.

And the data allow calculation of the coefficients A and B of the Margules development for the excess Gibbs free energy: $g^{ex} = x_2 x_3[A + B(x_2 - x_3)]$;

– binary (3)/(1): these are not miscible in any proportions; the two-phase domain is symmetrical and at 39 °C both phases are in equilibrium for the titers:

phase(a): $x_{3a} = 0.346$; $x_{1a} = 0.654$;

phase(b): $x_{3b} = 0.654$; $x_{1b} = 0.346$.

Knowing that the Redlich–Kister method predicts the ternary properties from the three binary mixtures using the total excess molar Gibbs free energy as the calculation per the following analytical expression: $g^{ex} = g^{ex}_{(1,2)} + g^{ex}_{(2,3)} + g^{ex}_{(3,1)} + D\, x_1 x_2 x_3$ (where D is the only unknown ternary term).

1) Express the partial excess Gibbs free energy of methylcyclohexane as a function of D.

2) Express the activity coefficient γ_1 as a function of D at point M ($x_1 = 0.0517$, $x_2 = 0.8305$).

3) Determine D knowing that the partial pressure of methylcyclohexane at point M is 4.91 torr and its vapor pressure in its pure state at 39 °C is 89 mm Hg.

4) Calculate the distribution coefficient of methylcyclohexane between both phases rich in aniline and heptane for low concentrations of methylcyclohexane.

Answers: (1) $g_1^{ex} = Dx_2x_3(1 - 2x_1) + 1{,}282.21\, x_3\,(x_2 + x_3) - x_2x_3\,[1{,}707.64 + 194.(x_2 - x_3)]$; (2) $\log \gamma_1 = 10^{-4}\,D - 0.0965$; (3) D = 576.62 cal/mol; (4) $k_d = \dfrac{x_{1b}}{x_{1a}} = 0.187$.

PROBLEM 4.15.– Studying the tin/lead binary by the addition of mercury

At 600 K, tin(1) and lead(2) form a liquid solution with low vapor tension. We propose to determine the activities of these two metals in their solution via the temperature and vapor tension P of the mixture obtained by adding to a solution of known titers of (1) and (2) with variable amounts of mercury (3) whose vapor tension at 600 K is 430 mm Hg.

The results obtained are shown in the table below:

x_{Hg}	P (mm Hg)		
	Pb^{pure}	$\dfrac{x_{Sn}}{x_{Pb}} = 1$	Sn^{pure}
0.1	128.0	117.5	152.5
0.2	204.0	190.0	234.0
0.4	279.5	269.0	302.0

1) a) Give expression of the activity coefficient of Hg under these conditions.

b) Verify that the expression obtained from these results can be represented by the Margules development: $\ln \gamma_{Hg} = a x_{Pb}^2 + b x_{Pb} x_{Sn} + c x_{Sn}^2$

c) Determine the values of a, b and c.

2) Assume that this result can be interpolated in any ternary domain, express the variable g^{ex} as a function of x_{Pb}, x_{Sn} and x_{Hg}.

3) Calculate the activities of tin and lead in the binary using this expression.

Answers: (1) $\gamma_3 = \dfrac{P}{430\,x_3}$ (b) see corrections; (c) a = 0.586; b = 0.890; c = 0.680; (2) $\dfrac{g^{ex}}{RT} = a x_{Pb} x_{Hg} + (a - b + c) x_{Pb} x_{Sn} + c x_{Sn} x_{Hg}$; (3) $\ln \gamma_{sn} = (a - b + c)$ x_{Sn}^2 and $\ln \gamma_{Pb} = (a - b + c) x_{Pb}^2$.

PROBLEM 4.16.– Properties of magnesium/bismuth/lead alloys

We aim to study thermodynamic properties of the ternary mixture Mg(1)/Bi(2)/Pb(3) at $T = 900$ K by the progressive addition of solid Mg in an initial binary solution Bi(2)/Pb(3), formed of 0.2 mol of Bi and 1.2 mol of Pb liquid at the same temperature.

For a certain number of additions, the amount of heat released Q is measured, corresponding to the amount Δn_1 of Mg added as well as the molar titer x_1 of the ternary solution at the moment of addition, as shown in the following table. The total molar enthalpy of mixing of the binary liquid Bi(2)/Pb(3) is given at 900 K by the relation $h^m = B x_1 x_2$ with $B = -4{,}310$ J/mol.

1) Calculate the amount of heat Q that must be released for the binary solution to remain at 900 K.

2) Determine the values of h^m in the solution (reference: pure liquid Mg), which can be deduced from these experiments.

x_1	0.0	0.1	0.2	0.3
Δn_1	0.0010	0.0014	0.0012	0.0016
Q (J)	15.732	14.795	7.314	3.648
x_1	0.4	0.45	0.5	/
Δn_1	0.0012	0.0012	0.0015	
Q (J)	–1.092	–2.741	–5.272	

3) a) Represent graphically the values of h_1^m as a function of the number n_1 of moles of Mg added.

b) Show that this plot can help determine a graphical integration h^m of the ternary mixture in the composition domain where the measurements were taken (the trapezoidal rule is used mainly for these integrations[1]).

c) Calculate the value of h^m for $n_1 = 0.4$.

4) Find the result obtained using the Darken method.

Data: Enthalpy of fusion of pure Mg at 900 K $L_{f1} = 8954$ J/mol and $T_{f1} =$ 923 K.

Answers: (1) $Q = 2{,}068.8\,\text{J}$; (2) $h_1^m = -\dfrac{Q}{\Delta n_1} - L_{f1}$; (3) (a and b) see corrections; (c) $h^m = -6.314\,\text{J}$; (4) $h^m = -6{,}296\,\text{J}$.

4.3. Tests

TEST 4.1.– Enthalpy of mixing of water/sulfuric acid

Mixing n_1 moles of $H_2O(1)$ and n_2 moles of $H_2SO_4(2)$ releases an amount of heat Q. We attempt to represent the experimental results using the following expression: $Q = H^m = \frac{an_1n_2(n_1+bn_2+cn_1n_2)}{(Q_1+Q_2)^2}$, where a, b and c are constants dependent on temperature only, which we must adjust.

Is the chosen expression correct as it stands? Justify this answer.

TEST 4.2.– Volume of a cyclohexane/carbon tetrachloride mixture

At 20 °C, the addition of 100 g of cyclohexane to 10 g of the binary solution $C_6H_{12}(1)/CCl_4(2)$ of a given titer x_2 increases the volume of the solution (in cm^3) data by: $\delta V = 128.50 + 0.71x_2^2$.

1 Note: This involves sharing the integration interval in n intervals equal to $h = \frac{b-a}{n}$ and in this case: $\int_a^b f(x)dx = h\left[\frac{f(a)}{2} + f(a+h) + f(a+2h) + \cdots + f(b-h) + \frac{f(b)}{2}\right]$.

1) Determine the partial molar volume of C_6H_{12} in the solution as a function of x_2.

2) Determine the partial molar volume of CCl_4 in the solution.

3) Determine the total molar volume of the solution.

TEST 4.3.– Enthalpy of a water/sulfuric acid mixture

When n moles of $H_2O(1)$ are mixed with 1 mol of $H_2SO_4(2)$, an amount of heat $Q = \frac{17.860\,n}{n+1.798}$ cal is released.

1) Determine the partial molar enthalpy of mixing (h_2^m) of H_2SO_4.

2) Determine the partial molar enthalpy of mixing (h_1^m) of H_2O.

3) How by knowing only the expression of (h_1^m) can one establish the expression of Q?

TEST 4.4.– Activity of calcium in molten chloride

A small amount of calcium metal dissolves in molten chloride. At 1,200 K, the binary solution $CaCl_2(1)/Ca(2)$ is in equilibrium with pure liquid Ca metal for the titer $x_2 = 0.112$.

However, determining the activity of $CaCl_2$ in any domain this solution exists, gives: $a_1 = 1 - 2x_2$.

1) Express the activity of calcium as a function of x_2 specifying the activity reference point chosen.

2) Calculate a_2 for $x_2 = 0.047$.

TEST 4.5.– Partial molar state functions of a reacting system

The synthesis of methanol at a constant temperature of 500 K and constant pressure of 20 atm occurs as per the following reaction: $CO+2H_2 \rightleftarrows CH_3OH$.

From the initial values: $n_{CO} = 1{,}000$, $n_{H_2} = 2{,}000$ and $n_{CH_3OH} = 0$ and assuming the mixture is ideal, equilibrium is reached with a progress of $\xi_{éq} = 411.72$. Next, we add 1 mol of pure CO to the system at 500 K and at 20 atm. There are three stages for this transformation I, II and III:(I): the mixture is not allowed to form; (II): the mixture is formed without reaction with the introduced CO; (III): CO is allowed to react until equilibrium is once again reached.

1) Determine the variation in volume and the variation in enthalpy:

a) when we move from stage (I) to stage (II);

b) when we move from stage(I) to stage (III).

2) Deduce the specific partial molar volume and enthalpy of mixing for CO in the gaseous reaction mixture, assuming the reaction is in constant equilibrium.

Data:

At 500 K	ΔH_f° (kcal.mol^{-o})	s° (cal.mol^{-o}.K^{-o})
CO	–26.3	50.93
H_2	/	34.808
CH_3OH	– 49.7	63.58

TEST 4.6.– Wilson development

Wilson proposed the following general development for the excess Gibbs free energy of a solution of n constituents: $\frac{g^{ex}}{RT} = -\sum_i^n x_i ln(1 - \sum_j^n x_j \mathrm{A}_{ji})$ with j≠1.

1) a) Establish the relation: $h = \frac{\partial(^g/_T)}{\partial(^1/_T)}$.

b) Application to the pentanol(1)/n-hexane(2) binary with the following data: $\mathrm{A}_{12} = 1 - \frac{v_2}{v_1}\frac{\mathrm{a}_{12}}{RT}$, $\mathrm{A}_{21} = 1 - \frac{v_1}{v_2}\frac{\mathrm{a}_{21}}{RT}$, $v_1 = 109.2$ cm^3/mol, $v_2 = 132.5$ cm^3/mol, $\mathrm{a}_{12} = 1718.3$ cal/mol and $\mathrm{a}_{21} = 166.6$ cal/mol.

c) Establish the expression for the molar enthalpy of a binary mixture and calculate its value for $x_2 = 0.2$.

2) a) Application to the ternary mixture benzene(1)/hexane(2)/cyclohexane(3) at 343.15 K and the data: $\mathrm{A}_{12} = \mathrm{A}_{21} = 0.175$; $\mathrm{A}_{23} = \mathrm{A}_{32} = 0$; $\mathrm{A}_{31} = -0.366$ and $\mathrm{A}_{13} = 0.478$.

b) Calculate a_1 in solution with the titers of $x_1 = 0.1$ and $x_2 = 0.5$.

4.4. Detailed corrections

EXERCISE 4.1.– Partial molar state functions

If a total molar state function z is given by the relation: $z = x_1 x_2 \mathrm{f}(x_1)$, where $\mathrm{f}(x_1) = \mathrm{A} + \mathrm{B}x_1 + \mathrm{C}x_1^2$, the derivative partial molar state function can be calculated from the following relations: $z_1 = z + x_2\left(\frac{\partial z}{\partial x_1} - \frac{\partial z}{\partial x_2}\right)$ and

$$z_2 = z + x_2\left(\frac{\partial z}{\partial x_2} - \frac{\partial z}{\partial x_1}\right)$$

where $\begin{cases} \frac{\partial z}{\partial x_1} = x_2 \mathrm{f}(x_1) + x_1 x_2 \mathrm{f}'(x_1) \\ \frac{\partial z}{\partial x_2} = x_1 \mathrm{f}(x_1) \end{cases}$ which gives: $\begin{cases} z_1 = x_2^2\left[\mathrm{f}(x_1) + x_1 \mathrm{f}'(x_1)\right] \\ z_2 = x_1^2\left[\mathrm{f}(x_1) + x_1 \mathrm{f}'(x_1) - \mathrm{f}'(x_1)\right] \end{cases}$

therefore $\begin{cases} z_1 = x_2^2\left(\mathrm{A} + 2\mathrm{B}x_1 + 3\mathrm{C}x_1^2\right) \\ z_2 = x_1^2\left[\mathrm{A} - \mathrm{B} + 2(\mathrm{B} - \mathrm{C})x_1\, 3\mathrm{C}x_1^2\right] \end{cases}$.

EXERCISE 4.2.– Partial molar volume and molarity

By definition, the molarity (m) is the number of moles of a body (1) in 1 kg of another (2), i.e. $m = \frac{n_1}{1\mathrm{kg}_{(2)}}$; here, the solution proposed is n_1 moles of NaCl in 1 kg of H_2O, therefore $m = n_1$, whereby writing V in the form: $V = A + Bm + Cm^{\frac{3}{2}} + Dm^2$, we obtain for $n_O = \frac{2V_{O_2}}{22414}$ the expression: $v_1 = \left(\frac{\partial V}{\partial n_1}\right)_{T,P,n_2} = \left(\frac{\partial V}{\partial m}\right)_{T,n_2} =$ $\left(\frac{\partial\left(A + Bm + Cm^{\frac{3}{2}} + Dm^2\right)}{\partial m}\right)$, which gives: $v_1 = B + \frac{3}{2}Cm^{\frac{1}{2}} + 2Dm =$ $16.6253 + 2.6157m^{\frac{1}{2}} + 0.2388m$; we know that $V = n_1 v_1 + n_2 v_2$, therefore:

$$v_2 = \frac{V - n_1 v_1}{n_2} = \frac{V - m v_1}{n_2} = \frac{A + Bm + Cm^{3/2} + Dm^2 - Bm - \frac{3}{2}Cm^{3/2} - 2Dm^2}{n_2}$$

and $v_2 = \dfrac{A + (B-B)m + \left(C - \frac{3}{2}C\right)m^{\frac{3}{2}} + (D-2D)m^2}{n_2} = \dfrac{A}{n_2} - \dfrac{C}{2}\dfrac{m^{\frac{3}{2}}}{n_2} - \dfrac{Dm^2}{n_2}$; since $m_2 = 10^3$ g and $M_2 = 18.015$ g/mol, therefore $n_2 = \dfrac{10^3}{18.015} \# 55.51$ and so $v_2 =$

$$v_2 = \frac{1001.38}{55.51} - \frac{1.7438}{2 \times 55.51} m^{\frac{3}{2}} - \frac{1.1194}{55.51} m^2 = 18.039 - 0.0157\, m^{\frac{3}{2}} - 0.0202\, m^2 .$$

EXERCISE 4.3.– Enthalpy of dissolution water/sulfuric acid

By definition, the enthalpy of reaction at constant P is $Q_p = \Delta H$ with $\Delta H = \sum n_i h_i$ with $n_1 = n$ and $n_2 = 1$, $\Delta H = nh_1 + h_2$.

1) By definition the partial molar state function of a constituent (i) during a given transformation is: $h_i = \left(\dfrac{\partial \Delta H}{\partial n_i}\right)_{T,P,n_j}$ and as dissolution occurs at constant T, P and n_2,

we obtain: $h_1 = \left(\dfrac{\partial \Delta H}{\partial n}\right)_{T,P,n_2} = \left(\dfrac{\partial\left(-\dfrac{an}{b+n}\right)}{\partial n}\right)_{T,P,n_2} = -\dfrac{ab}{(b+n)^2}$.

2) From the expression of ΔH, we find that: $h_2 = \Delta H - nh_1$ and so: $h_2 = -\dfrac{an}{(b+n)} + \dfrac{abn}{(b+n)} = -\dfrac{an^2}{(b+n)^2}$, where for 1 mol of (1), i.e. $n = 1$,a = 17860 and b = 1.798, then: $\Delta H = -6{,}380$ cal. We obtain: $h_1 = -4{,}101.81$ cal/mol and $h_2 = -2{,}281.32$ cal/mol.

EXERCISE 4.4.– Partial molar enthalpy and enthalpy of mixing

1) This involves determining the partial state function knowing the total state function. The Roozeboom general rule $h_2^m = h^m + (1 - x_2)\dfrac{dh^m}{dx_2}$ applied to a binary

gives: $h_i^m = h^m + (1 - x_i)\left[\left(\dfrac{\partial h^m}{\partial x_i}\right)_{xj} - \left(\dfrac{\partial h^m}{\partial x_j}\right)_{xj}\right]$.

2) From the expression $h^m = x_1x_2\left(A_1x_1 + A_2x_2\right)$, we obtain the relations:

$$\begin{cases} \dfrac{\partial h^m}{\partial x_1} = x_2\left(A_1x_1 + A_2x_2\right) + A_1x_1x_2 \\ \dfrac{\partial h^m}{\partial x_2} = x_1\left(A_1x_1 + A_2x_2\right) + A_2x_1x_2 \end{cases}$$

. This results in the partial state functions: $h_1^m = x_2^2\left[A_2 + 2x_1\left(A_1 + A_2\right)\right]$ and $h_2^m = x_1^2\left[A_1 + 2x_2\left(A_2 + A_1\right)\right]$.

EXERCISE 4.5.– Ideal mixture of alkaline fluorides

1) By definition, $g = h - Ts$ and so $g^m = h^m - Ts^m$. We know that g^m and the activity a are linked by relation: $g^m = RT\ln a$, therefore $s^m = \dfrac{h^m - g^m}{T} = \dfrac{h^m}{T} - R\ln a$. The mixture is made without any chemical reaction, so the heat released is the total enthalpy of mixing: $H^m = -2400$ cal and corresponds to $n = 2$ mol, which means $h^m = \dfrac{H^m}{2}$ and so $s^m = -\dfrac{2{,}400}{2\times 1{,}148} - 1987\ln 0.295 = 1.380$ cal/K.mol.

2) $s^{mid} = -R\ln x = -1.987\ln 0.5 = 1.377$ # 1.38 cal/K.mol.

This shows that s^m # s^{mid}, and hence the mixture, is practically an ideal solution.

EXERCISE 4.6.– Chloroform/acetone mixture

Determining the partial molar state function when the total molar state function obtained from the Roozeboom analytical rule is known: $h_2^m = h^m + \left(1 - x_2\right)\dfrac{dh^m}{dx_2}$.

Here, $dh^m = \left(\dfrac{\partial h^m}{\partial x_1}\right)_{TPn_2} dx_1 + \left(\dfrac{\partial h^m}{\partial x_2}\right)_{TPn_1} dx_2$ with $dx_1 = -dx_2$, this becomes $dh^m = \left(\dfrac{\partial h^m}{\partial x_2}\right)_{TPn_1} - \left(\dfrac{\partial h^m}{\partial x_1}\right)_{TPn_2} dx_2$, which gives: $h_2^m = h^m + x_1\left[\left(\dfrac{\partial h^m}{\partial x_2}\right)_{TPx_1} - \left(\dfrac{\partial h^m}{\partial x_1}\right)_{TPx_2}\right]$; we calculate the two partial derivatives from $h^m = h^m = x_1x_2\left[a + b\left(x_1 - x_2\right) - c\left(x_1 - x_2\right)^2\right]$ where:

$$\left(\frac{\partial h^m}{\partial x_1}\right)_{x_2} = x_2\left[a + b(x_1 - x_2) - c(x_1 - x_2)^2\right] + x_1 x_2\left[b - 2c(x_1 - x_2)\right]$$

$$\left(\frac{\partial h^m}{\partial x_2}\right)_{x_1} = x_1\left[a + b(x_1 - x_2) - c(x_1 - x_2)^2\right] + x_1 x_2\left[bx_2 - 2c(x_1 - x_2)\right]$$

$x_i = \frac{n_i}{n}$; we can calculate them since $n_2 = \frac{1000}{119} = 8.40$ and $n_1 = 1$, therefore $n = 9.40$ and $x_1 = \frac{1}{9.4} = 0.106$; $x_2 = \frac{8.4}{9.4} = 0.894$ therefore $x_1 - x_2 = -0.788$ and $x_1 x_2 = 0.095$, where the calculation of each term is given as:

$$\left(\frac{\partial h^m}{\partial x_1}\right)_{x_2} = 0.894[7{,}607 + 2{,}161 \times 0.788 - 1{,}745(0.788)^2]$$
$$+0.095[2161 - 2 \times 1{,}745 \times 0.788] = 7{,}637.201 \text{ J/mol}$$

$$\left(\frac{\partial h^m}{\partial x_2}\right)_{x_1} = 0.106[7{,}607 + 2{,}161 \times 0.788 - 1{,}745(0.788)^2]$$
$$-0.095[2{,}161 \times 0.894 - 2 \times 1{,}745 \times 0.788] = 9{,}044.547 \text{ J/mol}$$

$$h^m = 0.095[7{,}607 + 2{,}161 \times 0.788 - 1{,}745(0.788)^2] = 781.50 \text{ J/mol}$$

where $h_2^m = 781.50 + 9.044.547 + 7.637.201 = 17.463$ kJ/mol.

EXERCISE 4.7.– Total and partial molar volumes of mixing

1) To complete the table, it is necessary to calculate the total molar volume of mixing given by the relation: $v^m = v - \sum_i x_i v_i^{pur}$ with $i = 2.v_1^{pur} = 42.5$ cm^3/mol and $v_2^{pur} = 17.5$ cm^3/mol, which gives the relation: $v^m = v - (42.5x_1 + 17.5x_2)$. Knowing that $x_2 = 1 - x_1$, we can calculate the corresponding value for each composition , as shown below:

x_1	0	0.2	0.4	0.6	0.8	0.9	1
v (cm^3.mol^{-1})	17.5	20.4	23.3	27.3	32.7	36.3	42.5
$\sum x_i v_i^{pur}$ (cm^3.mol^{-1})	0	−22.5	−27.5	−32.5	−37.5	−40.0	0
v^m (cm^3.mol^{-1})	0	−2.1	−4.2	−5.2	−4.8	−3.7	0

2) On the graph we plot $v^m = f(x_1)$, shown in Figure 4.3, as we know that the tangent at any point on the curve intersects the axes of the pure bodies at 2 points corresponding to the value of v_i^m. We obtain for the compositions 0.3 and 0.7 the values of the second table:

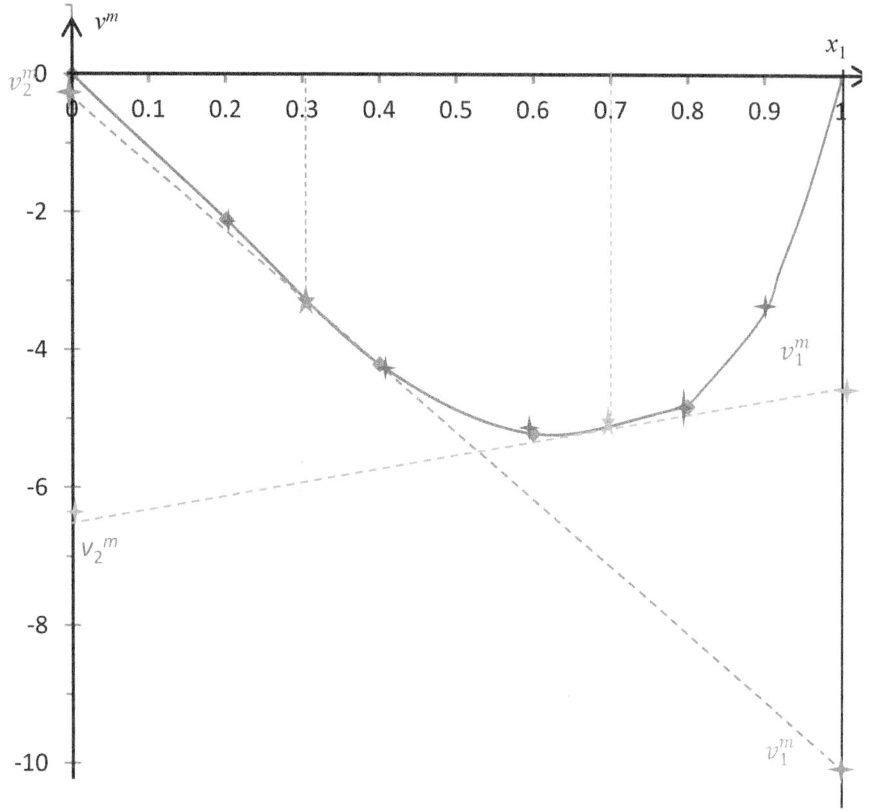

Figure 4.3. *$v^m = f(x_1)$ curve*

x_1	v_1^m (cm^3.mol^{-1})	v_2^m (cm^3.mol^{-1})
0.3	−9.95	−0.6
0.7	−4.45	−6.65

3) By definition, the partial molar volume is: $v_i^m = v_i - v_i^{pur}$. So: $v_i = v_i^{pur} + v_i^m$ where the values for the two compositions (0.3 and 0.7) are:

x_1	v_1 (cm^3.mol^{-1})	v_2 (cm^3.mol^{-1})
0.3	32.5	16.9
0.7	38.05	10.85

EXERCISE 4.8.– Excess volume of water/methanol mixture

1) The volume of the solution is significant with respect to 1 mol of water or alcohol. We can consider that the volumes added correspond to the partial molar volumes, where: $v = x_1v_1 + x_2v_2$, with $x_1 = 0.6$; $v_1 = 17.35$ cm^3.mol^{-1}; $v_2 = 39$ cm^3.mol^{-1} and so: $v = 0.6 \times 17.35 + 0.4 \times 39 = 26.01$ cm^3.mol^{-1}.

2) The total excess molar volume of the solution is the variation in volume due to the non-ideality of the solution. It corresponds to the difference in volume between that of the real solution obtained above and that of the ideal solution given by the relation: $v^{id} = x_1 v_1^{pur} + x_2 v_2^{pur} = 0.6\,\frac{18}{1} + 0.4\frac{32}{0.8} = 26.08$ cm^3.mol^{-1}, where: $v^{ex} = v - v^{id} = 26.01 - 26.80 = -0.79$ cm^3.mol^{-1}.

There is a decrease in volume when water and methanol are mixed.

EXERCISE 4.9.– Molar excess enthalpy, entropy and Gibbs free energy

1) We know that $s = -\left(\frac{\partial g}{\partial T}\right)_P$ and $s^{ex} = -\left(\frac{\partial g^{ex}}{\partial T}\right)_P$ and so $s^{ex} = -\frac{\partial\left(RTBx_1x_2\right)}{\partial T} =$ $-x_1x_2R\left(B + T\frac{dB}{dT}\right)$, where $\frac{dB}{dT}$ is the slope of the curve B = f(T) given in Figure 4.4 and the value: $\frac{dB}{dT} = = \frac{0.479 - 0.439}{308.15 - 318.15} = -4 \times 10^{-3}$. So for $x_1 = 0.3$ at 45 °C, we obtain: $s^{ex} = -0.21 \times 1.987(0.439 - 318.15 \times 0.004) = 0.348$ cal/mol.K.

By definition, $g = h - Ts$, therefore $h = g + Ts$ and $h^{ex} = g^{ex} + Ts^{ex}$ and so: $h^{ex} = RTBx_1x_2 - RTBx_1x_2 - x_1x_2RT^2\frac{dB}{dT} = -x_1x_2RT^2\frac{dB}{dT}$, which gives: $h^{ex} = 0.21 \times 1.987 \times 318.15^2 \times 0.004 = 168.94$ cal/mol.

2) The activity coefficient of a body in solution is defined by: $g_i^{ex} = RTln\gamma_i$, i.e.

$ln\gamma_i = \frac{g_i^{ex}}{RT}$.

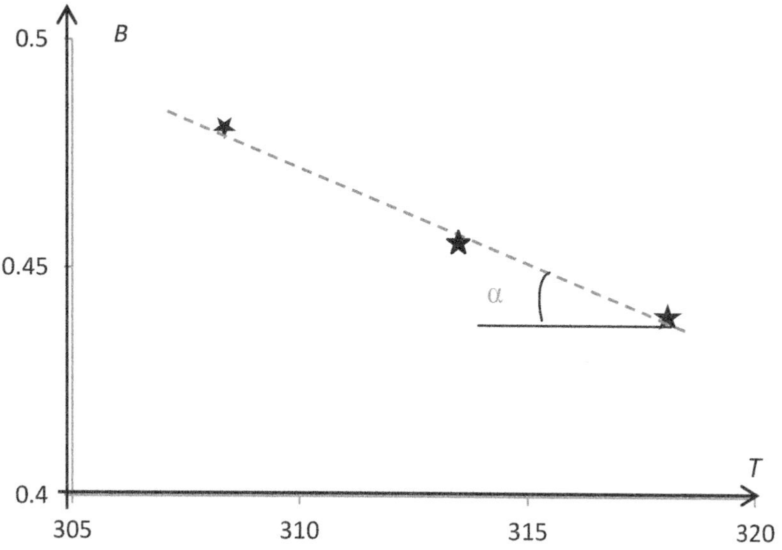

Figure 4.4. *B = f(T) curve*

We need to determine the $v_i^{pur} = v_i^* + B_{ij} = \frac{RT}{P} + B_{ij}$ using relation:

$$g_i^{ex} = g^{ex} + x_j\left(\frac{\partial g^{ex}}{\partial x_i} - \frac{\partial g^{ex}}{\partial x_j}\right)$$

Where for each component:

$$g_1^{ex} = RTBx_1x_2 + x_2^2RTB - x_1x_2RTB = RTBx_2^2$$

Hence $ln\gamma_1 = Bx_2^2 = 0.439\times0.7^2 = 0.2151$ # 0.22 and $\gamma_1 = 1.24$.

$$g_2^{ex} = RTBx_1x_2 + x_1^2RTB - x_1x_2RTB = RTBx_2^2$$

hence $ln\gamma_2 = Bx_1^2 = 0.439\times0.3^2 = 0.03951$ # 0.04, therefore $\gamma_2 = 1.04$.

EXERCISE 4.10.– Activity coefficients at infinite dilution

1) The activity coefficient γ_i of a constituent is defined with respect to its excess variable by $g_i^{ex} = RTln\gamma_i$, i.e. $ln\gamma_i = \frac{g_i^{ex}}{RT}$. We must now establish the expressions of g_1^{ex} and g_2^{ex}, which are defined by the relation: $g_i^{ex} = g^{ex} + (1-x_i)\left[\left(\frac{\partial g^{ex}}{\partial x_i}\right)_{TPx_j} - \left(\frac{\partial g^{ex}}{\partial x_j}\right)_{TPx_i}\right]$, whereby from the expression of g^{ex}:

$\frac{g^{ex}}{RT} = \frac{ABx_1x_2}{Ax_1 + Bx_2}$, we obtain:

$$\left(\frac{\partial\left(g^{ex} / RT\right)}{\partial x_1}\right) = \frac{ABx_2\left(Ax_1 + Bx_2\right) - A^2Bx_1x_2}{\left(Ax_1 + Bx_2\right)^2} = \frac{AB^2x_2^2}{\left(Ax_1 + Bx_2\right)^2}$$

$$\left(\frac{\partial g^{ex} / RT}{\partial x_2}\right) = \frac{ABx_1\left(Ax_1 + Bx_2\right) - AB^2x_1x_2}{\left(Ax_1 + Bx_2\right)^2} = \frac{A^2Bx_1^2}{\left(Ax_1 + Bx_2\right)^2}$$

where:

$$\frac{g_1^{ex}}{RT} = \frac{ABx_1x_2}{\left(Ax_1 + Bx_2\right)} + x_2\left[\frac{AB^2x_2^2 - A^2Bx_1^2}{\left(Ax_1 + Bx_2\right)^2}\right] = \frac{AB^2x_2^2\left(x_1 + x_2\right)}{\left(Ax_1 + Bx_2\right)^2} = A\left(\frac{Bx_2}{Ax_1 + Bx_2}\right)^2$$

$$\frac{g_2^{ex}}{RT} = \frac{ABx_1x_2}{\left(Ax_1 + Bx_2\right)} + x_1\left[\frac{A^2Bx_1^2 - AB^2x_2^2}{\left(Ax_1 + Bx_2\right)^2}\right] = \frac{A^2Bx_1^2\left(x_1 + x_2\right)}{\left(Ax_1 + Bx_2\right)^2} = B\left(\frac{Ax_1}{Ax_1 + Bx_2}\right)^2$$

Which gives the two relations: $ln\gamma_1 = A\left(\frac{Bx_2}{Ax_1 + Bx_2}\right)^2$ and $ln\gamma_2 = B\left(\frac{Ax_1}{Ax_1 + Bx_2}\right)^2$.

2) Infinite dilution means that the composition of one of the constituents is close to zero ($x_i \to 0$). This therefore implies that:

$$ln\gamma_1^{\infty} = \lim_{x_1 \to 0} A\left(\frac{Bx_2}{Ax_1 + Bx_2}\right)^2 = A = 1.15 \text{ hence } \gamma_1^{\infty} = 4.4817$$

$$ln\gamma_2^\infty = \lim_{x_2 \to 0} B\left(\frac{Ax_1}{Ax_1 + Bx_2}\right)^2 = B = 3 \text{ hence } \gamma_2^\infty = 20.085$$

3) Henry's law was established for diluted solutions to describe the relationship between the titer in the solution and the fugacity in the vapor and gives $H_i(T,P) = \lim_{x_i \to 0} \frac{f_i}{x_i}$ with $f_i = x_i \gamma_i f_i^{ref}$; so, assuming that the vapor is an ideal gas, then $f_i = P_i$ and $f_i^{ref} = P_i^s$, so at infinite dilution $H_i(T,P) = \lim_{x_i \to 0} \gamma_i^\infty P_i^s$ and the calculation gives: $H_i(T,P) = 4.482$ and $H_2(T,P) = 19.08$.

EXERCISE 4.11.– Dissolution of oxygen in water

By definition, the solubility (s) is the maximum amount of a body that can be dissolved in a given volume of another body, which corresponds to the formation of a solution saturated in solute (1) in a solvent (2): this is the number of moles of a constituent in the solvent volume. In this case, $s_{O_2} = \frac{n_{O_2}}{1l_{H_2O}}$, which means we need to determine n_{O_2}.

Henry's law is written as: $P_{O_2} = H\,x_{O_2}$ with $x_{O_2} = \frac{n_{O_2}}{n}$, therefore $\frac{P_{O_2}}{H} = \frac{n_{O_2}}{n} = \frac{n_{O_2}}{n_{O_2} + n_{H_2O}}$ and so $n_{O_2} = \left(n_{O_2} + n_{H_2O}\right)\frac{P_{O_2}}{H} \Rightarrow n_{O_2}\left(1 - \frac{P_{O_2}}{H}\right) = n_{H_2O}\frac{P_{O_2}}{H}$

$\Rightarrow n_{O_2} = n_{H_2O}\frac{P_{O_2}}{H}\frac{H}{\left(H - P_{O_2}\right)}$ giving $n_{O_2} = n_{H_2O}\frac{P_{O_2}}{\left(H - P_{O_2}\right)}$. With $n_{H_2O} = \frac{10^3}{18} = 55.56$, we obtain: $n_{O_2} = 55.56\frac{190}{\left(3.3.10^7 - 190\right)} = 3.2 \times 10^{-4}$ mol and therefore a solubility of $s_{O_2} = 3.2 \times 10^{-4}$ mol/L.

EXERCISE 4.12.– Modified Roozeboom method

1) This concerns expressing a partial molar state function using a total molar state function for a binary: $f_1 = f + x_2\left(\frac{\partial f}{\partial x_1} - \frac{\partial f}{\partial x_2}\right)$ and $f_2 = f + x_1\left(\frac{\partial f}{\partial x_1} - \frac{\partial f}{\partial x_2}\right)$.

2) Applying to function z, we obtain: $z_1 = z + x_2\left(\frac{\partial z}{\partial x_1} - \frac{\partial z}{\partial x_2}\right)$ and $z_2 = z + x_1\left(\frac{\partial z}{\partial x_2} - \frac{\partial z}{\partial x_1}\right)$. So $z = x_1 x_2 f$ and therefore $\begin{cases} \frac{\partial z}{\partial x_1} = x_2 f + x_1 x_2 \frac{\partial f}{\partial x_1} \\ \frac{\partial z}{\partial x_2} = x_1 f + x_1 x_2 \frac{\partial f}{\partial x_2} \end{cases}$, giving the expression of z_i: $z_1 = x_1 x_2 f + x_2^2\left(f + x_1 \frac{\partial f}{\partial x_1}\right) - x_1 x_2 f - x_2^2\left(x_1 \frac{\partial f}{\partial x_2}\right) = x_2^2\left[f + x_1\left(\frac{\partial f}{\partial x_1} - \frac{\partial f}{\partial x_2}\right)\right]$

$= x_2^2(2f - f_2)$

Since $x_1\left(\frac{\partial f}{\partial x_1} - \frac{\partial f}{\partial x_2}\right) = f - f_2$, similarly for constituent(2), $z_2 = x_1^2(2f - f_1)$.

4) In this case, we replace the state function z with v^m, the total molar volume of mixing allows us to obtain v_1^m using one of the following methods:

– Roozeboom method: with this method, we plot the tangent to the $v^m(x_1)$ curve as shown in Figure 4.5;

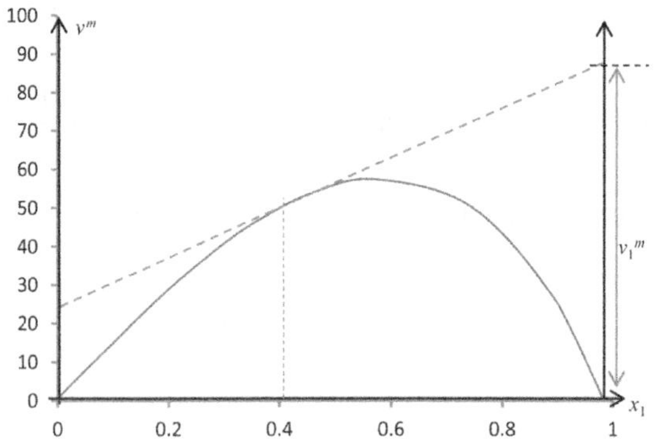

Figure 4.5. *vm = f(x₁) curve*

– modified Roozeboom method: with this method, the curve $\frac{v^m}{x_1 x_2} = f(x_1)$ is plotted, as shown in Figure 4.6.

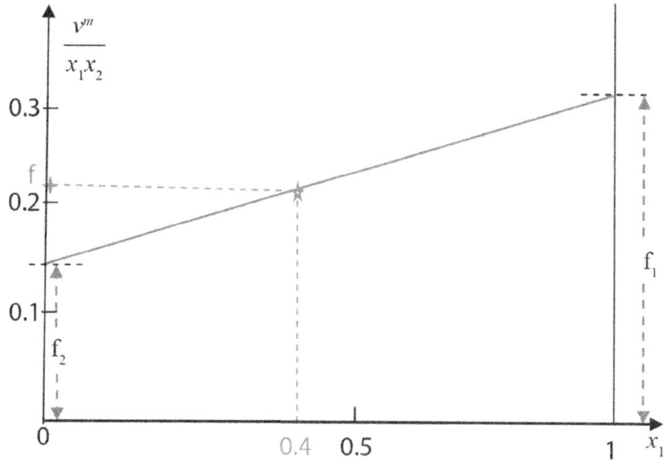

Figure 4.6. $\frac{v^m}{x_1 x_2} = f(x_1)$ *curve*

The intersection of this curve with axes of the pure bodies gives the values of f_i: $f_1 = 0.305$ and $f_2 = 0.149$; as the variables correspond to $x_1 = 0.4$, we obtain the value $f = 0.208$, whereby the partial molar volume of mixing of each constituent is:

$$v_1^m = x_2^2(2f - f_2) = 0.6^2\,(2\times 0.208 - 0.149) = 0.0961$$

$$v_2^m = x_1^2(2f - f_1) = 0.4^2\,(2\times 0.208 - 0.305) = 0.0178$$

EXERCISE 4.13.– Dissolution of hydrogen in iron

Note that $T_f(Al_2O_3) = 2{,}100$ °C dissolves in a monoatomic form: $H_{2gas} \rightleftarrows 2H_{dis}$. So at equilibrium, $K(T) = \frac{a_H^2}{P_{H_2}}$. Given the low concentrations, applying Henry's law gives: $a_H = k.x_H$, that is to say $a_H^2 = k^2 x_H^2 = K(T)P_{H_2}$. We can therefore write: $x_H = \alpha(T).\sqrt{P_{H_2}}$, i.e. Sievert's law, or even in the form $P_{H_2} = \beta(T)x_H^2$. So at constant T, if we change the pressure of hydrogen, we obtain $P'_{H_2} = \beta(T)x_H'^2$ and so

$P'_{H_2} = P_{H_2}\left(\frac{x'_H}{x_H}\right)^2$; if the molar titers are equal to the mass titers, then the result is

$P_{H_2} = 1$, we have the value: $P'_{H_2} = \left(\frac{5}{25}\right)^2 = \frac{1}{25} = 0.04$ atm.

EXERCISE 4.14.– Dissolution of alumina in silica

At 1,900 °C, the solution is saturated Al_2O_3 for $x_1 = 0.5$ and the solution is ideal. So for any reference: $a_1 = k_1(T)x_1$. Several reference states are possible:

– Al_2O_3 pure supercooled liquid ($T_{f1} > T$): $a_1 = x_1 = 0.10$.

– SiO_2 pure liquid ($T_{f2} < T$): $a_2 = x_2 = 0.9$.

The stable state of pure Al_2O_3 is solid at 1,900 °C; therefore, we can take pure solid alumina as the reference state. In this case, $a_1 = kx_1$ and at saturation $a_1 = 1$ and so $k = \frac{1}{x_1^s}$ where: $a_1 = \frac{x_1}{x_1^s} = \frac{0.1}{0.5} = 0.20$.

– The pure solid reference SiO_2 is not of interest.

EXERCISE 4.15.– Composition of water gas

Henry's law is written as: $P_i = H.x_i$ and Dalton's law is: $P_i = y_iP$; we know P and y_i, where:

$$x_{H_2} = \frac{y_{H_2}}{H_{H_2}} = \frac{0.45}{6.83}10^{-4} = 6.59 \times 10^{-6}$$

$$x_{CO} = \frac{y_{CO}}{H_{CO}} = \frac{0.45}{5.36}10^{-4} = 8.40 \times 10^{-6}$$

$$x_{N_2} = \frac{y_{N_2}}{H_{N_2}} = \frac{0.1}{8.04}10^{-4} = 1.24 \times 10^{-6}$$

Now $x_i = \frac{n_i}{\sum n_i} = \frac{m_i/M_i}{\sum_i n_{i-1} + n_{H_2O}}$, where $\sum_i n_{i-1}$ relates to dissolved gaseous bodies in water. It is negligible with respect to n_{H_2O} and so $x_i = \frac{m_i/M_i}{\sum_i n_{i-1} + n_{H_2O}}$ and $m_i = x_iM_ix_{H_2O}$. We obtain:

$$m_{H_2} = \frac{6.59\times10^{-6}\times28\times10^{6}}{18} = 0.732 \text{ g}$$

$$m_{CO} = \frac{8.40\times10^{-6}\times28\times10^{6}}{18} = 13.06 \text{ g}$$

$$m_{N_2} = \frac{1.24\times10^{-6}\times28\times10^{6}}{18} = 1.93 \text{ g}$$

EXERCISE 4.16.– Activity of iron and carbon in steel

For a binary, knowing the activity of a constituent as a function of the titer allows us to calculate the activity of the other constituent using the Gibbs–Duhem relation: $x_c dg_c + x_{Fe} dg_{Fe} = 0$, with $dg_c = d(ln a_c)$ and $dg_{Fe} = d(ln a_{Fe})$ which equates to: $x_c d(ln a_c) + x_{Fe} d(ln a_{Fe}) = 0$, giving the expression: $d(ln a_{Fe})$

$$d(\ln a_{Fe}) = -\frac{x_c}{x_{Fe}} d(\ln a_c) = -\frac{x_c}{x_{Fe}} d\left[\ln\frac{x_c}{x_{Fe}} + 6.6\frac{x_c}{x_{Fe}}\right]$$, which is written in the form: $d(\ln a_{Fe}) = -\, d\left(\frac{x_c}{x_{Fe}}\right) - 6.6\frac{x_c}{x_{Fe}} d\left(\frac{x_c}{x_{Fe}}\right)$ and so:

$$\ln a_{Fe} = \int_0^{x_c} d(\ln a_{Fe}) = \int_0^{x_c/x_{Fe}} \left[-d\left(\frac{x_c}{x_{Fe}}\right) - 6.6\frac{x_c}{x_{Fe}} d\left(\frac{x_c}{x_{Fe}}\right)\right],$$ which gives:

$$\ln a_{Fe} = -\frac{x_c}{x_{Fe}} - 3.3\left(\frac{x_c}{x_{Fe}}\right)^2.$$

EXERCISE 4.17.– A thermal benzene/butyl sebacate solution

1) By definition, we have $g_1^{ex} = RT\, ln\, \gamma_1 = h_1^m - Ts_1^{ex}$. The solution formed by this dissolution is athermal, and therefore $h_1^m = 0$ and: $g_1^{ex} = -Ts_1^{ex}$. This gives $ln\, \gamma_1 = -\frac{s_1^{ex}}{R}$, which is independent of *T*, therefore the expression proposed by the authors is valid at any temperature.

2) At $T = 55$ °C, we obtain the relation: $\log \gamma_1 = -\frac{s_1^{ex}}{4.57}$

EXERCISE 4.18.– Binary solution of benzene/hexane

1) From the proposed equation, $g^{ex} = x_1 x_2[\mathrm{B}_0 + \mathrm{B}_1(x_2 - x_1)]RT$, we can determine one of the terms of $\frac{g_i^{ex}}{RT}$ using the Roozeboom relation:

$$\frac{g_1^{ex}}{RT} = \frac{g^{ex}}{RT} + x_2\left[\frac{\partial\left(\frac{g^{ex}}{RT}\right)}{\partial x_1} - \frac{\partial\left(\frac{g^{ex}}{RT}\right)}{\partial x_2}\right] = ln\gamma_1$$

Then, we can determine the terms $\begin{cases} \frac{\partial(g^{ex}/RT)}{\partial x_1} = x_2(\mathrm{B}_0 + \mathrm{B}_1(x_2 - x_1) - \mathrm{B}_1 x_1.x_2 \\ \frac{\partial(g^{ex}/RT)}{\partial x_2} = x_1(\mathrm{B}_0 + \mathrm{B}_1(x_2 - x_1) + \mathrm{B}_1 x_1.x_2 \end{cases}$

where: $ln\gamma_1 = x_2^2(\mathrm{B}_0 + \mathrm{B}_1(x_2 - x_1) - 2x_1\mathrm{B}_1$ and so $\frac{ln\gamma_1}{(1-x_1)^2} = \mathrm{B}_0 + \mathrm{B}_1(1-4x_1)$. Therefore, the Darken function $y = \frac{ln\gamma_1}{(1-x_1)^2}$ is linear with respect to x_1. The results are presented in the table below:

x_1	0.1	0.2	0.3	0.4	0.5
y	0.190	0.214	0.238	0.261	0.285
x_1	0.6	0.7	0.8	0.9	/
y	0.309	0.332	0.356	0.379	/

This allows us to plot $y = \mathrm{f}(x_1)$ as shown in Figure 4.7; with two particular values, we determine both constant, as follows:

$$\begin{cases} x_1 = 0 \Rightarrow y_0 = \mathrm{B}_0 + \mathrm{B}_1 = 0.167 \\ x_1 = 0.25 \Rightarrow y_{1/4} = \mathrm{B}_0 = 0.226 \end{cases} \text{hence} \begin{cases} \mathrm{B}_0 = 0.226 \\ \mathrm{B}_1 = 0.059 \end{cases}$$

b) The activity limit corresponds to the infinite dilution of the constituent concerned. Therefore, for(1): $x_1 \approx 0$, we obtain $x_2 \approx 1$ and $ln\gamma_1^{\infty} = \mathrm{B}_0 + \mathrm{B}_1 = 0.167$

and so $\gamma_1^{\infty} = 1.1818$; for (2): relation $x_1 g_1^{ex} + x_2 g_2^{ex} = g^{ex}$, we deduce: $ln\,\gamma_2 = \left[\frac{(g^{ex}/RT) - x_1\,ln\,\gamma_1}{x_2}\right]$, i.e. by replacing them with the corresponding values: $ln\,\gamma_2 = x_1\left[(B_0 + B_1(x_2 - x_1)\right] - x_1.x_2\left[B_0 + B_1(1-4x_1)\right]$, which allows us to obtain: $x_2 \approx 0$ and $x_1 \approx 1$, therefore $ln\gamma_2^{\infty} = B_0 - B_1 = 0.285$ and hence $\gamma_2^{\infty} = 1.3298$.

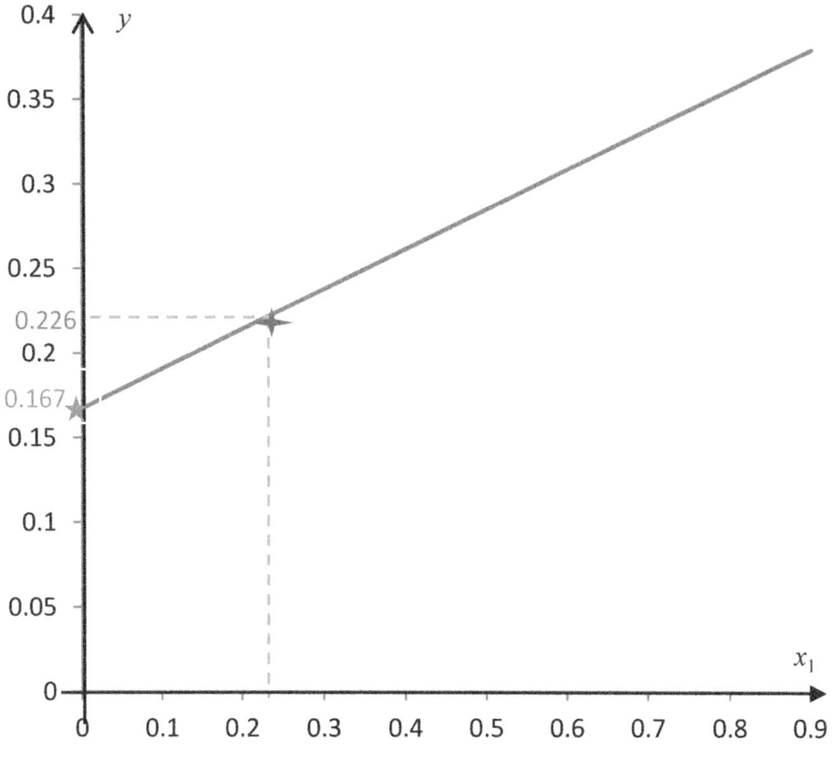

Figure 4.7. *y = f(x₁) curve*

EXERCISE 4.19.– Binary solution of methyl acetate/cyclohexane

1) From the graph shown in Figure 4.8, using the Roozeboom method, we find: $h_1^m = 674$J./mol and from the graph in Figure 4.9, we find: $Ts_1^{ex} = 138$ J/mol.

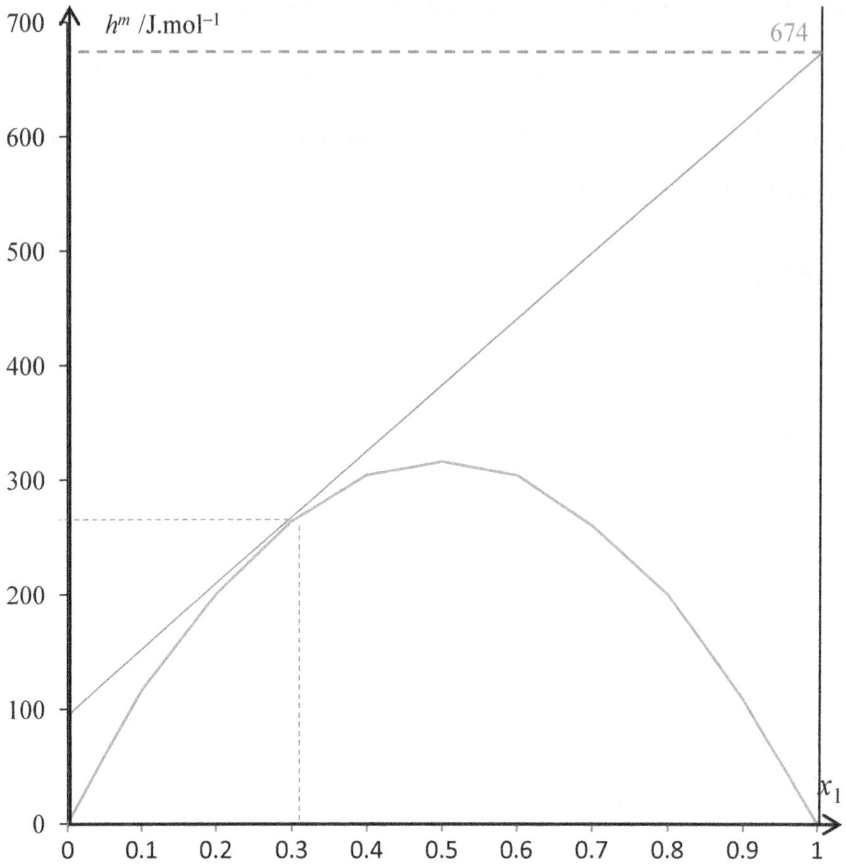

Figure 4.8. *$h^m = f(x_1)$ curve*

2) With no indications on the variations in h^m and s^{ex} with T, we initially assume that they are independent of T (which means that there is no influence of C_p^m terms on these two variables). We can therefore write the two expressions:

$$\begin{cases} h_1^m(323.0\text{ K}) = h_1^m(303.15\text{ K}) = 674 \\ s_1^{ex}(323.0\text{ K}) = s_1^{ex}(303.15\text{ K}) = \dfrac{138}{303.15} \end{cases}$$

From the expression $ln\gamma_1 = ln\left(\frac{a_1}{x_1}\right) = ln\left(\frac{h_1^m - Ts_1^{ex}}{RT}\right)$, we obtain:

$$a_1 = 0.3.exp\left[\left(\frac{674}{323} - \frac{138}{303.15}\right)\frac{1}{8.3144}\right] = 0.365.$$

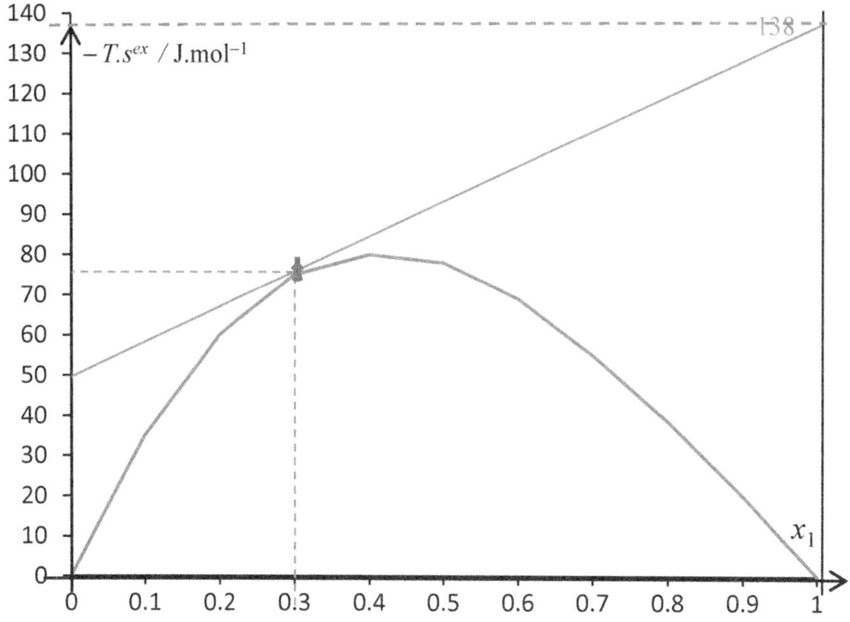

Figure 4.9. *[$-Ts^{ex} = f(x_1)$] curve*

EXERCISE 4.20.– Ternary solution of benzene/cyclohexane/hexane

1) The Redlich–Kwong hypothesis can be represented as the relation $\frac{g^{ex}}{RT} = \frac{g^{ex}_{1/2}}{RT} + \frac{g^{ex}_{1/3}}{RT} + \frac{g^{ex}_{2/3}}{RT}$. Since $g^{ex}_{2/3}$, then the binary (2/3) is ideal $\frac{g^{ex}}{RT} = A(T)x_1x_2 + x_1x_3\left[B_0 + B_1(x_3 - x_1)\right]$, where the expression of the activity coefficient is: $ln\,\gamma_1 = \frac{g_1^{ex}}{RT} = \frac{g^{ex}}{\mathrm{RT}} + (1 - x_1)\frac{\partial\left(\frac{g^{ex}}{RT}\right)}{\partial x_1} - x_2\frac{\partial\left(\frac{g^{ex}}{RT}\right)}{\partial x_2} - x_3\frac{\partial\left(\frac{g^{ex}}{RT}\right)}{\partial x_3}$; calculating each term gives the three following expressions:

$\frac{\partial(g^{ex}/RT)}{\partial x_1} = A(T)x_2 + x_3\left[B_0 + B_1(x_3 - x_1)\right] - x_3x_2B_1$; $\frac{\partial(g^{ex}/RT)}{\partial x_2} = A(T)x_1$ and $\frac{\partial(g^{ex}/RT)}{\partial x_3} = x_1\left[B_0 + B_1(x_3 - x_1)\right] + B_1x_1x_3$. The multiplication is carried out using the respective factors as follows:

$$\begin{cases} \frac{\partial(g^{ex}/RT)}{\partial x_1} = A(T)x_2 + x_3\left[B_0 + B_1(x_3 - x_1)\right] - x_3x_2B_1 \\ \frac{\partial(g^{ex}/RT)}{\partial x_2} = A(T)x_1 \\ \frac{\partial(g^{ex}/RT)}{\partial x_3} = x_1\left[B_0 + B_1(x_3 - x_1)\right] + B_1x_1x_3 \end{cases} \begin{matrix} (1 - x_1) \\ (-x_2) \\ (-x_3) \end{matrix}$$

And when $x_1 \approx 0$, we obtain $\ln\gamma_1^\infty = A(T)x_2 + x_3(B_0 + B_1x_3)$.

2) The mixture is equimolar, and $x_2 = x_3 = 0.5$. It results in: $\ln\gamma_1^\infty = 0.349 \times 0.5 + 0.50(226 - 0.059 \times 0.5) = 0.2528 \Rightarrow \gamma_1^\infty = 1.3137$.

EXERCISE 4.21.– Nitridation of iron by ammonia

Assuming that N_2 dissolves in atomic form in iron, we can represent this dissolution by: $2NH_3 \rightarrow 2N + 3H_2$ whose equilibrium constant is $K = a_N^2 \cdot \frac{P_{H_2}^3}{P_{NH_3}^2}$ with $a_N = \gamma_N x_N$ and hence $K = \gamma_N^2 . x_N^2 \cdot \frac{P_{H_2}^3}{P_{NH_3}^2}$; close to $x_N = 0$, γ_N does not depend on x_N (law of diluted solutions). Here, $K' = \sqrt{\frac{K}{\gamma_N^2}} = x_N \frac{P_{H_2}^{3/2}}{P_{NH_3}}$. K' is independent of titers and pressure. Applying this relation to both experiments, assuming that the relationship between the molar titers and mass titers are constant, allows us to write $K'_{10\%} = K'_{20\%}$ $0.045\frac{(0.9)^{3/2}}{0.1} = x_N \frac{(0.8)^{3/2}}{0.2}$, which gives us the titer $x_N = 0.107$, let $x_N \approx 11\%$.

NOTE.– For high accuracy, we should calculate the molar titer corresponding to the mass titer of 0.045. We find that it is 0.159; we then determine the new molar titer, using the relation above: we find that $x_N = 0.379$ to which a mass titer of 0.133 ≈ 13.3% corresponds.

EXERCISE 4.22.– Influence of oxygen on the melting point of silver

1) For the different parameters, we adopt the following notation:

– n_O: number of moles of O in the solution;

– $n_{O_2} = \frac{1}{2} n_O$ of moles of O_2 in the solution;

– x_O: titer of atomic oxygen in the solution;

– n_{Ag} : number of moles of Ag;

– V_{O_2} : volume at $T.P.n$ of dissolved oxygen.

Here: $n_{Ag} = \frac{100}{107.88} = 0.92696$ mol and $n_{O_2} = \frac{V_{O_2}}{22414}$ therefore $n_O = \frac{2V_{O_2}}{22414}$ and $x_O = \frac{n_O}{n_O + n_{Ag}}$; Sievert's law is written: we establish the table with the following values:

V_{O_2}	93.5	180.3	222.0	284.1
n_O	0.00834	0.01609	0.01981	0.02535
x_O	0.008917	0.01706	0.02092	0.02662
P_{O_2}	128	488	760	1203
$\frac{P_{O_2}}{x_O^2} 10^{-6}$	1.61	1.68	1.74	1.70

It gives us the relation: $P_{O_2} = 1.68 \times 10^6 x_O^2$, where Sievert's law is verified for k = 1.68 × 10[6] (k depends only on T).

2) The dissolution of O_2 is written as: $\frac{1}{2} O_2 \rightleftarrows O$. Henry's law is applied to O: $\frac{1}{2} g_{O_2} = g_O$. As for the compound, we have $g_{O_2} = g_{O_2}^o + RT\ln P_{O_2} =$

$g^o_{O_2} + RT\,ln k + RT\,ln x_O$, which gives: $g_O = \frac{1}{2}(g^o_{O_2} + RT\,ln k) + RT\,ln x_O$, which can be written in the form $g_O = g^1_O + RT \ln x_O$, where g^1_O is only a function of T and represents the reference chemical potential of species O: the solution is therefore ideal for small x_O.

In the diluted solution where solidification is beginning: $g_{Ag(s)} = g_{Ag(liq)}$, i.e. $g^1_{Ag(s)} = g^1_{Ag(liq)} + RTln\gamma_{Ag} xAg$, and by taking the pure liquid as the reference state of Ag at the same temperature, we obtain $\gamma_{Ag} \approx 1$.

Here, $RT\,ln x_{Ag} = g^1_{Ag(s)} - g^1_{Ag(liq)} = -\Delta h_f + T\Delta s_f$ so by disregarding the effect of specific heat, we obtain: $RT\,ln x_{Ag} = -\Delta h_f \left(1 - \frac{T}{T_f}\right)$. So, $x_{Ag} = 1 - x_O$ and we can therefore write: $ln\, x_{Ag} = ln(1 - x_O) \approx -x_O \approx -\frac{\Delta h_f}{RT}\left(1 - \frac{T}{T_f}\right)$, where the T of solidification will be given by: $\frac{1}{T_f} = \left(\frac{1}{T_f} + \frac{Rx_O}{\Delta h_f}\right)$ and so for the titer $x_O = 0.0099$ (corresponding to $P_{O_2} \approx 160\ mmHg$): $T = 1222.9$ K = 949.8 °C, with a decrease in melting point $\Delta T = -11$ °C.

EXERCISE 4.23.– Hydrolysis of chlorine

1) At equilibrium, the hydrolysis reaction corresponds to:

$Cl_2 + H_2O \rightleftarrows ClOH + H^+ + Cl^-$

And by definition, the chemical potential of the constituent is written as: $g_i = \overset{\circ}{g_i} + RTln(m_i)$, where m_i is the molarity of each constituent in 1,000 g of water, giving:

$$g_{Cl_2} = \overset{\circ}{g}_{Cl_2} + RTln.\, m_{Cl_2} . g_{H_2O} = \overset{\circ}{g}_{H_2O} + RTln.\, m_{H_2O}$$

$$g_{ClOH} = \overset{\circ}{g}_{ClOH} + RTln.\, m_{ClOH} . g_{H^+} = \overset{\circ}{g}_{H^+} + RTln.\, m_{H^+}$$

$$g_{Cl^-} = \overset{\circ}{g}_{Cl^-} + RTln.\, m_{Cl^-}\text{: with } \Sigma_i\, g_i = \Delta G \text{ and } \Delta G^\circ = \Sigma_i\, \overset{\circ}{g_i}$$

which gives $\Delta G = \Delta G^{\circ} +$ RTlnK $= \Delta G^{\circ} +$ RT$ln\frac{m_{Cl2}.m_{Cl^-}.m_{H^+}}{m_{Cl2}.x_{H2O}}$, therefore $\frac{m_{Cl2}.m_{Cl^-}.m_{H^+}}{m_{Cl2}}$ = K, since we have x_{H_2O} # 1; the calculation gives: $\Delta G^{\circ} = -19.11 - 31.35 + 0 - 1.65 + 56.69 = 4.58$. At equilibrium, we always have $\Delta G = 0$, and so we obtain: $RTln$K = −4.58 and therefore lnK $= -\frac{4.58}{1.987 \times 298} = -7.735$, which gives K = 4.37×10^{-4}.

2) The variation in reaction composition from the initial moment to equilibrium is written as:

Cl_2	+	H_2O	⇄	ClOH	+	H^+	+	Cl^-
m_I		1		0		0		0
m (1−x)		1		mx		mx		mx

And by referring to the expression of K, we obtain the equation in its final form:

$$K = \frac{mx.mx.mx}{m(1-x)} = \frac{m^3x^3}{m\,(1-x)} = m^2\left(\frac{x^3}{1-x}\right)$$

3) Theoretically, the number of moles of non-hydrolyzed Cl_2 that remain in solution at equilibrium obeys the equality potentials of the species in the vapor phase and in the solution, i.e. $g_{Cl_2gas} = g_{Cl_2dis}$, therefore $\overset{\circ}{g}_{Cl_2gas} + RTlnP_{Cl_2} = \overset{\circ}{g}_{Cl_2dis} + RTln(m_{Cl_2})$. Here, $P_{Cl_2} = 1$ where: $RTln(m_{Cl_2}) = \overset{\circ}{g}_{Cl_2gas} - \overset{\circ}{g}_{Cl_2dis}$, where $ln(m_{Cl_2}) = \frac{-1{,}650}{RT} = -\frac{1{,}650}{1.987 \times 298} = -2.7866$, which gives: $m_{Cl_2} = 6.16 \times 10^{-2}$.

4) In reality, at equilibrium: $m_{Cl_2} = m(1-x)$ and K.m $(1-x) = m^3x^3$ therefore m^3x^3 = K.$m_{Cl_2} = 4.37 \times 10^{-4} \times 6.16 \times 10^{-2} = 26.93 \times 10^{-6}$ and so $m.x = \sqrt[3]{26.93 \times 10^{-2}} = 2.997 \times 10^{-2}$. So, $m_{Cl_2} = m(1-x) = m - mx$ and therefore $m = m_{Cl_2} + m.x = 0.0616 + 0.02997$ # 0.092.

PROBLEM 4.1.– Nitridation of iron by ammonia

1) Nitrogen dissolves in its atomic form: $N_{2gas} \rightleftarrows 2N_{dis}$. Therefore, at equilibrium: $K(T) = \frac{a_N^2}{P_{N_2}} = \frac{(\gamma_N x_N)^2}{P_{N_2}}$; the amount of nitrogen dissolved is minimal, and so being in the application domain of Henry's law, we obtain: $x_N = k\sqrt{P_{N_2}} \Rightarrow x_N = k$ since $P_{N_2} = 1$ atm. Therefore, with X_N being the mass fraction of nitrogen:

$$\frac{x_N M_N}{x_N M_N + (1-x_N)M_{Fe}} = \frac{14x_N}{14x_N + 56(1-x_N)} = \frac{x_N}{x_N + 4(1-x_N)} = \frac{x_N}{4-3x_N},$$ whereby:

$x_N = \dfrac{4X_N}{1+3X_N}$ with $X_N = \dfrac{0.003}{100} \Rightarrow x_N = \dfrac{4 \times 0.003.10^{-2}}{1+3\times 0.003.10^{-2}}$ and so k #1.2 × 10^{-4}.

Second, we have two simultaneous equilibria $2NH_3 \rightleftarrows N_2 + 3H_2$ and $N_2 \rightleftarrows 2N_{dis}$, which equates to: $2NH_3 \rightleftarrows 2N_{\text{dis}} + 3H_2$ for which $K'(T) = a_N^2 \dfrac{P_{H_2}^3}{P_{NH_3}^2}$. The mass fraction X'_N is equal to 0.0045, whereby: $x'_N = 0.0178$; since we are always in the application domain of Henry's law, then: $(\gamma_N x'_N)^2 = K'(T)\dfrac{P_{NH_3}^2}{P_{H_2}^3}$, where

$x'_N = k'\sqrt{\dfrac{P_{NH_3}^2}{P_{H_2}^3}}$ and $k' = 0.0178\sqrt{\dfrac{0.9^3}{0.01}} = 0.1519$. This gives a mixture of 20% NH_3, in which $x'_N = 0.1519\dfrac{0.2}{\sqrt{0.8^3}} = 0.0425$, where $X'_N = 0.01097$, i.e. 1.097% of nitrogen by weight.

2) At equilibrium: $N_{2,gas} + 3H_{2,gas} \rightleftarrows 2NH_{3gas}$, which gives the equality:

$$RT\ln K(T) = 2\Delta G^\circ_{fNH_3} = 2\times 13.9 = 27.8 \text{ kcal/mol}.$$

which gives: $\ln K(T) = \dfrac{27800}{1933.8145} = 14.376$. We obtain the value of K: $K(T) = 1.755.10^{-6}$. So, at equilibrium $K(T) = P_{N_2}\dfrac{P_{H_2}^3}{P_{NH_3}^2}$, therefore, if we maintain this gaseous system such that $P_{H_2} = 0.9$ and $P_{NH_3} = 0.1$, we obtain the value: $P_{N_2} = 1.755.10^6 \dfrac{0.1^2}{0.9^3} = 24073$ atm.!. Here, nitrogen dissolves according to the law $x_N = 1.2.10^{-4}\sqrt{P_{N_2}} = 1.2.10^{-2} \times 155 = 0.0186$, which corresponds to a mass value of $X_N = 4.72.10^{-3}$, i.e. 0.472% of nitrogen by weight.

CONCLUSION.– The nitrogen content is therefore 155 times greater than that obtained with pure nitrogen at 1 atm.

PROBLEM 4.2.– Molar volumes of acetone/chloroform mixtures

1) Considering the ternary mixture formed of three pure species, the total molar volume is written as: $v = x_1 v_1 + x_2 v_2 + x_{12} v_{12}$. Here, the mixture is ideal and so $v_1 = v_1^p$, $v_2 = v_2^p$, $v_{12} = v_{12}^p$ and a_i are equal to x_i. Hence, $K_p = \dfrac{x_{12}}{x_1 x_2}$.

With the mixture being considered as a chemical equilibrium, we obtain the values in the following table:

	Acetone	Chloroform	Complex 1/2	Total
t = 0	n_1	n_2	0	$n_1 + n_2$
At equilibrium	$n_1 - \xi$	$n_2 - \xi$	Ξ	$n_1 + n_2 - \xi$

where the respective titers: $x_1 = \dfrac{n_1 - \xi}{n_1 + n_2 - \xi}$, $x_2 = \dfrac{n_2 - \xi}{n_1 + n_2 - \xi}$ and $x_{12} = \dfrac{\xi}{n_1 + n_2 - \xi}$, which gives $K_p = \dfrac{\xi(n_1 + n_2 - \xi)}{(n_1 - \xi)(n_2 - \xi)}$. The second-degree equation is $(K_p + 1)\xi^2 - (K_p + 1)(n_1 + n_2)\xi + K_p n_1 n_2 = 0$, where ξ is the root that gives the following expression at V:

$$V = (n_1 - \xi)v_1^p + (n_2 - \xi)v_2^p + \xi v_{12}^p$$

So $v = \dfrac{V}{n_1 + n_2}$ and $v = \left(\dfrac{n_1 - \xi}{n_1 + n_2}\right)v_1^p + \left(\dfrac{n_2 - \xi}{n_1 + n_2}\right)v_2^p + \left(\dfrac{\xi}{n_1 + n_2}\right)v_{12}^p$

2) By definition, $v_i = \left(\dfrac{\partial V}{\partial n_i}\right)_{n_j} = v_1^p + (v_{12}^p - v_1^p - v_2^p)\left(\dfrac{\partial \xi}{\partial n_1}\right)_{n_2}$. We must derive the previous equation with respect to n_1 with $dn_2 = 0$, where $\left(\dfrac{\partial \xi}{\partial n_1}\right)_{n_2} = \dfrac{(1 + K_p)\xi - K_p n_2}{(1 + K_p)[2\xi - (n_1 + n_2)]}$, which gives:

$$v_1 = v_1^p + (v_{1/2}^p - v_1^p - v_2^p)\frac{(1+K_p)\xi - K_p n_2}{(1+K_p)[2\xi - (n_1 + n_2)]}$$

(ξ is the root of the second degree equation).

PROBLEM 4.3.– Properties of carbonic anhydride

The calculation uses the values of the reduced variables, which are calculated: $T_r = \dfrac{373.2}{304.2} = 1.2268$ and $P_r = \dfrac{100}{72.8} = 1.3736.$

1) The molar volume (v):

a) Using the Redlich–Kwong equation in the form: $Z^3 - Z^2 + (A - B^2 - B)Z - AB = 0$ where $A = \dfrac{0.4275\ P_r}{T_r^{2,5}} = 0.3522$ and $B = \dfrac{0.08664\ P_r}{T_r} = 0.09701$. We obtain: $Z^3 - Z^2 + 0.2458\ Z - 0.03417 = 0$, and hence Z = 0.7265 and $v = Z\dfrac{RT}{P} = 0.7265 \times 306.2 = 222.5$ cm^3/mol.

b) The HWR tables give, as a function of Z_C and reduced variables T_r and P_r, the following table, from which we obtain: Z = 0.7348 and hence a volume $v = 0.7348 \times 306.2 = 225.0$ cm^3.mol^{-1}.

		Tr		
	Pr	1.20	1.2268	1.30
$Z_C = 0.27$	1.20	0.751	/	0.827
$Z_C = 0.274$	1.20	0.7516	0.7719	0.8274
	1.3736	/	0.7348	/
	1.40	0.705	0.7291	0.795

c) The L-K tables give:

	$Z^{(0)}$			$Z^{(1)}$		
	Tr			*Tr*		
Pr	1.20	1.2268	1.30	1.20	1.2268	1.30
1.20	0.7363	0.7564	0.8111	0.0991	0.1006	0.1048
1.3736	/	0.7167	/	/	0.1270	/
1.50	0.6605	0.6867	0.7624	0.1477	0.1462	0.1420

We obtain $Z = 0.7176 + (0.225 \times 0.1270) = 0.7453$ where:

$v = 0.7453 \times 306.2 = 228.2$ cm^3/mol

d) The generalized diagram of compressibility factors give $T_r = 1.23$ and $P_r = 1.37$, $Z = 0.73$ therefore $v = 0.73 \times 306.2$ # 223 cm^3.mol^{-1}, with the comparative table being:

	Exp	R-K	HWR	L-K	D.G	G.P.
v $(cm^3.mol^{-1})$	229.5	222.5	225.0	228.2	223.0	306.2
% difference	/	3	2	0.6	3	33

2) The molar enthalpy (h): difference in h from the ideal gas, is given by: $h_{TP} - h^*_{TP} = \int_0^P \left[\left(\frac{\partial h}{\partial P}\right)_T - \left(\frac{\partial h^*}{\partial P}\right)_T\right] dP = \int_0^P \left[v - T\left(\frac{\partial v}{\partial T}\right)_P\right] dP$, but $v\ T\left(\frac{\partial v}{\partial T}\right)_P$ cannot be easily deduced from a relation explicit in P, but it will be easier to calculate difference according to:

$$u_{Tv} - u^*_{Tv} = \int_\infty^v \left[\left(\frac{\partial u}{\partial v}\right)_T - \left(\frac{\partial u^*}{\partial v}\right)_T\right] dv = \int_\infty^v \left[T\left(\frac{\partial P}{\partial T}\right)_v - P\right] dv$$

because the function u depends only on T. Therefore: $u(T,P) = u(T,v)$ and, so: $h_{TP} - h^*_{TP} = u_{TP} - u^*_{TP} + Pv - (Pv)^* = u_{TP} - u^*_{TP} + Pv - RT$

a) We use the classical form of the Redlich–Kwong equation: $P = \frac{RT}{v - \mathrm{b}} - \frac{\mathrm{a}}{\sqrt{T}\ v(v+\mathrm{b})}$ from where we obtain the derivative: $\left(\frac{\partial P}{\partial T}\right)_v = \frac{R}{v-\mathrm{b}} + \frac{\mathrm{a}}{2T^{3/2}v(v+\mathrm{b})}$ and then the term $\left[T\left(\frac{\partial P}{\partial T}\right)_v - P\right] = \frac{3\mathrm{a}}{2\sqrt{T}v(v+\mathrm{b})}$, which gives the expression $u - u^* = \int_\infty^v \frac{3\mathrm{a}}{2\sqrt{T}v(v+\mathrm{b})} dv = \frac{3\mathrm{a}}{2\mathrm{b}\sqrt{T}} ln \frac{v}{v+\mathrm{b}}$. For the enthalpy, the difference from the ideal gas is given by the expression:

$$h - h^* = \frac{3\mathrm{a}}{2\mathrm{b}\sqrt{T}} \ln \frac{v}{v+\mathrm{b}} + RT\left(\frac{v}{v-\mathrm{b}} - 1\right) - \frac{\mathrm{a}}{\sqrt{T}(v+\mathrm{b})} = \frac{\mathrm{a}}{\sqrt{T}}\left(\frac{3}{2\mathrm{b}} ln \frac{v}{v+\mathrm{b}}\right) + RT\frac{\mathrm{b}}{v-\mathrm{b}}$$

with $\mathrm{a} = 0.4275\ \frac{R^2 T_c^{2.5}}{P_c}$ and $\mathrm{b} = 0.08664\ R\ \frac{T_c}{P_c}$. Therefore, $\frac{\mathrm{a}}{\sqrt{T}} = 79{,}992$ cal.cm^3.mol^{-1} and b = 29.7 cm^3.mol^{-1}, giving: $h_{TP} - h^*_{TP} = -\ 709$ cal/mol^{-1}.

b) The HWR tables give the values of $\frac{h^* - h}{T_c}$ / cal.mol^{-1}K^{-1} and as a result the following table:

			Tr	
	Pr	1.20	1.2268	1.30
$Z_C = 0.27$	/	2.25	/	1.74
$Z_C = 0.274$	1.20	2.206	2.073	1.708
	1.3736	/	2.379	/
	1.40	2.68	2.519	2.08

where: $h_{TP} - h^*_{TP} = -2.379 \times 304.2 = -724$ cal/mol.

c) The L-K tables give the values in the following table:

	$\left(\frac{h^* - h}{RT_c}\right)^0$			$\left(\frac{h^* - h}{RT_c}\right)^1$		
		Tr			*Tr*	
Pr	1.20	1.2268	1.30	1.20	1.2268	1.30
1.20	1.076	1.0181	0.860	0.381	/	0.218
1.3736	/	1.2132	/	/	0.3373	/
1.50	1.443	1.3554	1.116	0.381	/	0.218

Which gives:

$$h_{TP} - h^*_{TP} = -1.9872 \times 304.2\left[1.2132 + (0.225 \times 0.3373)\right] = -779 \text{ cal/mol}$$

This gives the following table:

	Exp	R-K	HWR	L-KC
h* – h	760	709	724	779
%difference	/	6.7	4.7	2.5

3) The fugacity (f): the difference from the ideal gas for the function $g(T,P)$ is written: $g_{TP} - g^*_{TP} = RT\,ln\frac{f}{P} = \int_0^P (v - v^*)dP$, and so:

a) $g_{TP} - g^*_{TP} = \int_0^P \left(v - \frac{RT}{P}\right)dP$ but, as for enthalpy, we cannot easily calculate $\left(v - \frac{RT}{P}\right)$ from the R-K equation. We therefore use the free energy function F since:

$$g_{TP} = g_{Tv}$$

$$g_{Tv} - g_{Tv}^{*} = \frac{1}{n}\left(F_{Tv} - F_{Tv}^{*}\right) + Pv - RT = \int_{\infty}^{v} -(P - P^{*})dv + Pv - RT$$

$$g_{Tv}^{*} = g_{TP^{*}}^{*} = g_{TP}^{*} + RT\,ln\frac{RT}{Pv} = g_{TP}^{*} - RT\,ln\,z$$

which finally allows us to write:

$$g_{TP} - g_{TP}^{*} = \int_{\infty}^{v}\left(\frac{RT}{v} - P\right)dv + Pv - RT - RT\,ln\,z = RT\,ln\frac{f}{P}$$

Therefore, using the Redlich–Kwong equation:

$$\int_{\infty}^{v}\left(\frac{RT}{v} - P\right)dv = \int_{\infty}^{v}\left[RT\left(\frac{1}{v} - \frac{1}{v - \mathrm{b}}\right) + \frac{\mathrm{a}}{\sqrt{T}\,(v + \mathrm{b})\,v}\right]dv$$

$$= \int_{\infty}^{v}\left(\frac{\mathrm{a}}{v(v + \mathrm{b})\sqrt{T}} - \frac{\mathrm{b}\,RT}{v(v - \mathrm{b})}\right)dv$$

which gives the following expression:

$$\int_{\infty}^{v}\left(\frac{RT}{v} - P\right)dv = \frac{\mathrm{a}}{\mathrm{b}\sqrt{T}}ln\frac{v}{v + \mathrm{b}} + RT\,ln\frac{v}{v - \mathrm{b}}$$

and finally: $Pv - RT = \frac{\mathrm{b\,RT}}{v - \mathrm{b}} - \frac{\mathrm{a}}{(v + \mathrm{b})\sqrt{T}}$, with the final relation being:

$$ln\frac{f}{P} = \frac{\mathrm{a}}{\mathrm{b}RT\sqrt{T}}ln\frac{v}{v + \mathrm{b}} + ln\frac{v}{v - \mathrm{b}} + \frac{\mathrm{b}}{v - \mathrm{b}} - \frac{\mathrm{a}}{(v + \mathrm{b})RT\sqrt{T}} - ln\,z$$

which gives $\ln\frac{f}{P} = -0.2652$ and $\frac{f}{P} = 0.767$, with $f = 76.7$ atm;

b) The HWR tables give the following values:

	Pr	*Tr*		
		1.20	1.2268	1.30
$Z_C = 0.27$	1.20	0.805	/	0.854
$Z_C = 0.274$		0.806	0.819	0.855
	1.3736	/	0.792	/
	1.40	0.773	0.788	0.830

with: $f = 79.2$ atm;

c) L-K tables provide the values for the following expressions:

	$\left(\log \frac{f}{P}\right)^0$			$\left(\log \frac{f}{P}\right)^1$		
	Tr			*Tr*		
Pr	1.20	1.2268	1.30	1.20	1.2268	1.30
1.20	–0.106	– 0.0990	–0.080	0.029	0.0314	0.038
1.3736	/	–0.1083	/	/	0.0355	/
1.50	–0.135	–0.1256	–0.100	0.041	0.0431	0.049

Here, $\log\frac{f}{P}$ = – 0.1083 + (0.225×0.0355 = – 0.1003, and so $\frac{f}{P}$ = 0.794 and $f = 79.4$ atm;

d) from the generalized diagram for $T_r = 1.23$ and $P_r = 1.37$, we obtain the following values: $\frac{f}{P} = 0.77$ and so $f \approx 77\,\text{atm}$.

Below is the resulting comparative table:

	R-K	HWR	L-K	D.G
f (atm)	76.7	79.2	79.4	77

which shows that the HWR and L-K tables provide the most acceptable values.

PROBLEM 4.4.– Methane/propane mixture

1) a) To calculate the total molar volume of the mixture, we solve the state equation by writing: $\frac{P}{RT}v^2 - v - \text{B} = 0$. If $\alpha = \frac{1}{v}$, then $\text{B}\alpha^2 + \alpha - \frac{P}{RT} = 0$ and

with $\begin{cases} P = 15\text{ atm} \\ T = 344\text{ K} \end{cases}$; we obtain $\frac{P}{RT}$ = 531.27 m^{-3}. Solving the second-degree equation $\text{B}\alpha^2 + \alpha$ – $531.27 = 0$ allows us to establish the values in the following table by calculating the corresponding values of B:

x_1	0	0.1	0.5	1
B	–290	–257	–126	–27
v (cm^3)	1,524.8	1,583.2	1,747	1,855.5

b) Lewis's rule is written as: $v = x_1 v_1^{pure} + x_2 v_2^{pure}$ with $v_1^{pure} = 1{,}855.5$ and $v_2^{pure} = 1{,}524.8$, which gives two values in the table below; similarly if $B_{12} = \frac{B_{11} + B_{22}}{2}$, we obtain two other values. The table below shows the different values of the three methods used:

	B_{ij}		$B_{12} = \frac{B_{11} + B_{22}}{2}$		Lewis's rule	
x_1	0.5	0.1	0.5	0.1	0.5	0.1
B	–126	–257	–158.5	–263.7	/	/
v (cm^3)	1,747	1,583.2	1,708	1,565.8	1,690.1	1,557.9

NOTE.– In the absence of a state equation, Lewis's rule gives valid results.

2) Helmholtz free energy (F) is given by: $dF = -SdT - PdV$ with $\left(\frac{\partial F}{\partial V}\right)_T = -P$, with the molar variable: $f = \frac{F}{n} = -sdT - Pdv$ and $\left(\frac{\partial f}{\partial v}\right)_T = -P$; for the corresponding ideal gas at the same T occupying the same volume v, we obtain $P^* = \frac{RT}{v}$ and $f^* = -sdT - P^* dv$ with: $\left(\frac{\partial f^*}{\partial v}\right)_T = -P^*$. Hence $f - f^* = \int_{\infty}^{v}\left[\left(\frac{\partial f}{\partial v}\right)_T - \left(\frac{\partial f^*}{\partial v}\right)_T\right]dv = -\int_{\infty}^{v}\left(P - \frac{RT}{v}\right)dv = -\int_{\infty}^{v} RT\frac{B}{v^2}$. This gives: $f - f^* = \frac{RTB}{v}$. So a mixture composed of n_1 moles of (1) and n_2 moles of (2) is represented by the relation:

$$F - F^* = (n_1 + n_2)(f - f^*) = (n_1 + n_2)\frac{RT}{v}\left(B_{11}x_1^2 + 2B_{12}x_1x_2 + B_{22}x_2^2\right)$$

So $v = \frac{V}{n_1 + n_2}$ therefore $F - F^* = \frac{RT}{V}\left(B_{11}n_1^2 + 2B_{12}n_1n_2 + B_{22}n_2^2\right)$.

3) The fugacity coefficient is defined by the following expression: $RT\, ln\frac{f_i(T,P,x_i)}{P_i} = g_i(T,P) - g_i^*(T,P)$, with by definition $g_i = \left(\frac{\partial U}{\partial n_i}\right)_{S,V,n_j}$

$= \left(\dfrac{\partial H}{\partial n_i}\right)_{S,P,n_j} = \left(\dfrac{\partial F}{\partial n_i}\right)_{T,V,n_j} = \left(\dfrac{\partial G}{\partial n_i}\right)_{T,P,n_j}$ so by adding and subtracting the term $g_i^*(T,P^*)$ from the previous relation:

$$g_i(T,P)-g_i^*(T,P)=g_i(T,P)-g_i^*(T,P^*)+g_i^*(T,P^*)-g_i^*(T,P)$$

which gives, at constant T, the following expressions:

$$g_i-g_i^*(P^*)=g_i(v)-g_i^*(v)=\left(\frac{\partial(F-F^*)}{\partial n_i}\right)_{T,V,n_j}=\frac{2RT}{v}\left(B_{ii}n_i+B_{ij}n_j\right)$$

and $g_i^*(P^*)-g_i^*(P)=RT\,ln\dfrac{P^*}{P}=-RT\,ln\dfrac{Pv}{RT}=-RT\,ln\,z$.

Therefore, $P_i=x_iP$ and so $RT\,ln\dfrac{f_i}{P_i}=RT\,ln\dfrac{f_i}{x_iP}=\dfrac{2RT}{v}\left(B_{ii}x_i+B_{ij}x_j\right)-RT\,ln\dfrac{Pv}{RT}$

giving the expressions:

$$f_1=x_1\frac{RT}{v}\exp\left[\frac{2}{v}\left(B_{11}x_1+B_{12}x_2\right)\right]\text{ and }f_2=x_2\frac{RT}{v}\exp\left[\frac{2}{v}\left(B_{22}x_2+B_{12}x_1\right)\right]$$

4) The term RT equals 2,861 J. We obtain the values giving in the following table:

	B_{ij}				$B_{12}=\dfrac{B_{11}+B_{22}}{2}$		Lewis's rule	
x_1	0	1	0.1	0.5	0.1	0.5	0.1	0.5
$f_1 10^3$ Pa	0	1.4981	0.1624	0.7644	0.1518	0.7515	0.150	0.749
f_2 10^3 Pa	1.2829	0	1.156	0.6575	1.1548	0.644	1.155	0.641

NOTE.– Assuming that the mixture behaves as an ideal gas, we obtain:

$$x_1=0.1:\ f_1=x_1P=0.759.10^3\text{Pa and } f_2=(1-x_1)P=1.367.10^3\text{Pa}$$

$$x_1=0.5:\ f_1=f_2=0.759.10^3\text{Pa}$$

5) a) Beginning with equation: $\frac{Pv}{RT} = 1 + B'P \Leftrightarrow v = \frac{RT}{P} + B'RT = v^* + RTB'$, where v^* is the molar volume of an ideal gas mixture at T and P; so, $dg = -sdT + vdP$, whereby $\left(\frac{\partial g}{\partial P}\right)_T = v$ and $\left(\frac{\partial g^*}{\partial P}\right)_T = v^*$, which gives the expression: $g(T,P) - g^*(T,P) = \int_0^P (v - v^*)dP = \int_0^P RTB'dP = RTB'P$ and therefore:

$$G_{TP} - G^*_{TP} = (n_1 + n_2)RTP\left(B'_{11}x_1^2 + B'_{12}x_1x_2 + B'_{22}x_2^2\right) \text{ and}$$

$$g_1 - g_i^* = \left(\frac{\partial(G - G^*)}{\partial n_1}\right)_{T,P,n_2}$$

Here, by definition, $RT\,ln\frac{f_1}{x_1P} = g_1 - g_i^*$ and so: $RT\,ln\frac{f_1}{x_1P} = RTP\left(2B'_{11}x_1 + 2B'_{12}x_2 - B'\right)$. This gives the fugacity: $f_1 = x_1P\exp\left[P(2B'_{11}x_1 + 2B'_{12}x_2 - B')\right]$.

b) Lewis's rule is expressed as: $v = x_1v_1^{pure} + x_2v_2^{pure}$ with: $v_1^{pure} = \frac{RT}{P} + RTB'_{11}$ and $v_2^{pure} = \frac{RT}{P} + RTB'_{22}$, where $v = \frac{RT}{P} + RTB'_{11}x_1 + B'_{22}x_2)$. So $v = \frac{RT}{P} + B'RT$, according to the state equation. To validate Lewis's rule, we must ensure that $B' = B'_{11}x_1 + B'_{22}x_2$. Therefore, for an equimolar mixture with the conditions of the first question, we should obtain: $\begin{cases} B' = \frac{1}{4}B'_{11} + \frac{1}{2}B'_{12} + \frac{1}{4}B'_{22} \\ B' = 0.5B'_{11} + 0.5B'_{22} \end{cases}$ and so $B'_{12} = \frac{1}{2}(B'_{11} + B'_{22})$.

PROBLEM 4.5.– Partial and mixing total molar volumes

1) a) The values of v_1^{pur} and v_2^{pur} are obtained directly from the intersection of the curve (Figure 4.10) with the axes of the pure bodies: v_1^{pur} = 2,047 cm^3/mol and v_2^{pur} = 1,748 cm^3/mol.

b) The partial molar *state function* is obtained from the graph using the Roozeboom method: the tangent at point ($x_1 = 0.8$) intersects the axes of the pure bodies in two points corresponding to the values: $v_1 = 1840$ cm^3/mol and $v_2 = 2060$ cm^3/mol.

2) By definition, $v^m = v - v^{pur} = v - \sum x_i v_i^{pur}$. We need to calculate, for each composition, the term $(2{,}047x_1+1{,}748x_2)$ and subtract it from the value of v, giving the following table of values:

x_1	0.1	0.2	0.3	0.4	0.5	0.6
v (cm^3/mol)	1,791.1	1,835.4	1,872.8	1,910.3	1,941.0	1,966.5
v^m (cm^3/mol)	13.2	27.6	35.1	42.7	43.5	39.1
x_1	0.7	0.8	0.9	/	/	/
v (cm^3/mol)	1,991.5	2,016.4	2,035.2	/	/	/
v^m (cm^3/mol)	34.2	29.2	18.1	/	/	/

3) By definition, $v_i^m = v_i - v_i^{pure}$, and we obtain the values of v_1 and v_2 for $x_1 = 0.8$. Here: $v_1^m = 2060 - 2047 = 13$cm^3/mol and $v_2^m = 1840 - 1748 = 92$ cm^3/mol.

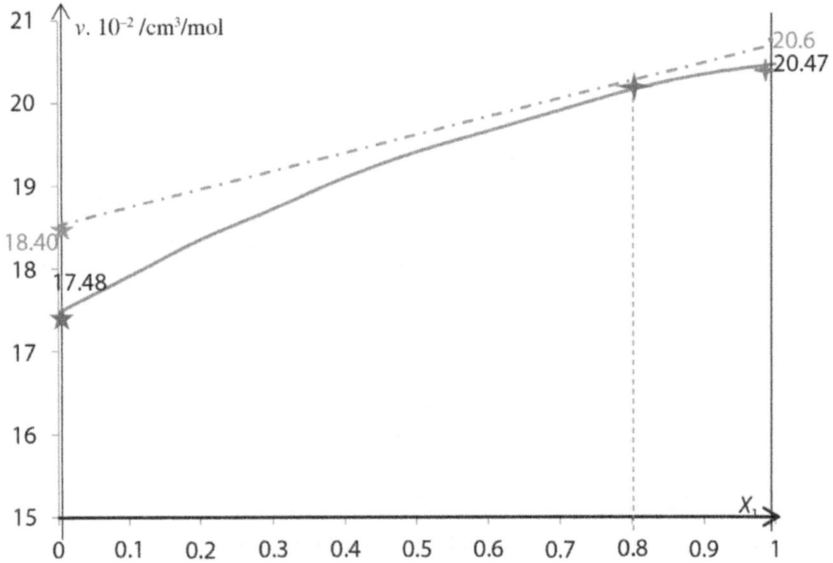

Figure 4.10. *v = f(x$_1$) curve*

PROBLEM 4.6.– Methane/propane mixture under pressure

We begin by testing the assessment methods of ω and m^2 from the values T_b and T_c. We obtain the following table of values:

CH_4	C_3H_8
$\theta = \frac{T_b}{T_c} = 0.586$	$\theta = \frac{T_b}{T_c} = 0.625$
$\omega = \frac{3}{7}$ and $\frac{\theta}{1-\theta}\log P_c - 1 = 0.005$	$\omega = \frac{3}{7}$ and $\frac{\theta}{1-\theta}\log P_c - 1 = 0.158$
$z_c = 0.2901 - 0.0879\ \omega = 0.289$	$z_c = 0.2901 - 0.0879\ \omega = 0.277$

1) Shortened virial equation: This allows us to write $z = \frac{Pv}{RT} = 1 + B\frac{P}{RT}$ with $B = B_{11}x_1^2 + 2B_{12}x_1x_2 + B_{22}x_2^2$. We must therefore calculate these coefficients.

a) Evaluating B_{11}, knowing that we have the following relation: $B_{11}\frac{P_c}{RT_c} = f^{(0)} + \omega.f^{(1)}$. We determine $f^{(0)}$ and $f^{(1)}$ with $T_r = 1.805$

$$f^{(0)} = 0.1445 - \frac{0.330}{T_r} - \frac{0.1385}{T_r^2} - \frac{0.0121}{T_r^3} - \frac{0.000607}{T_r^8} = -0.08292$$

$$\text{and } f^{(1)} = 0.0637 + \frac{0.331}{T_r^2} - \frac{0.423}{T_r^3} - \frac{0.008}{T_r^8} = 0.09329$$

$$\text{hence: } B_{11} = \frac{82.05 \times 190.6}{45.4}(0.09329 \times 0.008 - 0.08292) = -28.3\ \text{cm}^3\text{mol}^{-1}.$$

b) Evaluation of B_{22} with $T_r = 0.930$. We obtain: $f^{(0)} = 0.3866$ and $f^{(1)} = -0.09378$ and so $B_{22} = -290.3\ \text{cm}^3.\text{mol}^{-1}$.

c) Evaluation of B_{12}; for which we must determine the parameters relative to the binary (1)/(2), hence:

$$T_{c12} = \sqrt{190.6 \times 369.8} = 265.5\ \text{K};\ v_{c1} = \frac{0.228 \times 190.6 \times 82.05}{45.4} = -99.2\ \text{cm}^3\text{mol}^{-1}$$

$$v_{c2} = \frac{0.281 \times 369.8 \times 82.05}{41.8} = 203.9 \text{ cm}^3\text{mol}^{-1} \Rightarrow v_{c12} = 145.3 \text{ cm}^3\text{mol}^{-1}$$

$$T_{r12} = \frac{344}{265.5} = 1.296\ ;\ z_{c12} = \frac{0.288 + 0.281}{2} = 0.2845$$

$$P_{c12} = \frac{0,2845 \times 82.05 \times 265.5}{145.3} = 42.65 \text{ atm. and } \omega_{12} = \frac{0.008 + 0.152}{2} = 0.080$$

from where by: $f^{(0)} = -\ 0.19822$ and $f^{(1)} =\ 0.06584$; $B_{12} = -\ 98.4 \text{ cm}^3\text{mol}^{-1}$. And B = 0.25(– 28.3 – 2×98.4 – 290.3) (– 28,3 – 2 × 98,4 – 290,3) = – 129 cm³ /mol. Since $v = \frac{RT}{P} + B$, we obtain: $v = 2075–129 = 1946 \text{ cm}^3 .\text{mol}^{1}$.

2) Lewis's rule: It is expressed as $v = x_1 \cdot v_1^{pure} + x_2 \cdot v_2^{pure}$ with $v_i^{pure} = v_i^* + B_{ij} = \frac{RT}{P} + B_{ij} \Rightarrow v_1^{pure} = 2075 - 31 = 2044$ cm³.mol⁻¹. and $v_2^{pure} = 2075–330 = 1745$ cm³/mol, hence $v =$ 0.5 (2044 + 1745) = 1894.5 cm³/mol.

3) Corresponding states:

a) Kay's rule:

$$0.5 < \frac{T_{c1}}{T_{c2}} < 2 \quad \text{and} \quad P_{c1} \approx P_{c2} \quad \Rightarrow \quad \begin{cases} T_{c12} = \frac{1}{2}(T_{c1} + T_{c2}) = 280.2\ K \\ P_{c12} = \frac{1}{2}(P_{c1} + P_{c2}) = 43.65 \text{ atm. and} \\ \omega_{12} = \frac{1}{2}(\omega_1 + \omega_2) = 0.080 \end{cases}$$

$$\begin{cases} T_r = \frac{144}{280.2} = 1.228 \\ P_r = \frac{13.6}{43.65} = 0.312 \end{cases}$$ giving the following table of values:

	$Z^{(0)}$			$Z^{(1)}$		
	T_r			T_r		
P_r	1.20	1.228	1.30	1.20	1.228	1.30
0.2	0.9611	0.9637	0.9702	0.0084	0.0096	0.0125
0.312	/	0.9425	/	/	0.0160	/
0.4	0.9205	0.9259	0.9396	0.190	0.0212	0.0267

Therefore: $z = 0.9425 + (0.080 \times 0.0160) = 0.9438$ and so $v = 0.9438 \times 2075 =$ 1959 cm^3/mol.

b) The Gunn–Prausnitz rule: Here, we have the values for the parameters in the table below:

$$T_{c12} = 280,2K \Rightarrow P_{c12} = \frac{82.05 \times 0.2845 \times 280.2}{(99.2 + 203.9)/2} = 43.2 \text{ atm.}$$

so: $P_r = \frac{13.6}{43.2} = 0.315$ where $z^{(0)} = 0.9419$ and $z^{(1)} = 0.01622$, which gives $v = 1957\ cm^3mol^{-1}$.

c) Redlich–Kwong equation: We use the equation in the form $Z^3 - Z^2 + (A - B^2 - B)Z - AB = 0$ and we determine the coefficients A and B for pure bodies (1) and (2) and the binary (1)/(2). So we obtain:

$$\begin{cases} A_1 = 0.4275 \dfrac{P_r}{T_r^{2,5}} = 0.4275 \times \dfrac{0.3}{(1.805)^{2,5}} = 0.0293 \\ B_1 = 0.08664 \dfrac{P_r}{T_r} = 0.0144 \end{cases}$$

and $\begin{cases} A_2 = 0.167 \\ B_2 = 0.0144 \end{cases}$ so: $A_{12} = \left[\frac{1}{2}\left(\sqrt{A_1} + \sqrt{A_2}\right)\right]^2 = 0.0841$ and $B_{12} = \frac{1}{2}(B_1 + B_2) =$ 0.0224.

This allows us to write: for (1): $Z^3 - Z^2 + 0.0147.Z - 4.29192.10^{-4} = 0 \Rightarrow z = 0.9855$ hence $v = 2045\ cm^3mol^{-1}$; for (2): $Z^3 - Z^2 + 0.1358.Z - 5.060.10^{-3} = 0 \Rightarrow z = 0.8467$, hence: $v = 1757\ cm^3mol^{-1}$; for(1)/(2): $Z^3 - Z^2 + 0.0611.Z - 1.882.10^{-3} = 0 \Rightarrow z = 0.9369$, where: $v = 1944\ cm^3mol^{-1}$. The table below gives a comparison of results and shows that the R-K equation is the most satisfactory.

	Exp	Virial	Lewis	Kay	G-P	R-K
v_{1pure}	2,040	2,047	/	/	/	2,045
v_{2pure}	1,745	1,785	/	/	/	1,757
$v_{(1)/(2)}$	1,938	1,945	1,895	1,959	1,957	1,944

PROBLEM 4.7.– Water/methanol mixture

1) By definition, the partial variable: $v_i = \left(\frac{\partial V}{\partial n_i}\right)_{T,P,n_j}$ so for a binary:

$$v_1 = \left(\frac{\partial V}{\partial n_1}\right)_{T,P,n_2} = \frac{\partial\left(10^3 + 35n_1 + \frac{n_1^2}{2}\right)}{\partial n_1} = n_1 + 35.$$

To calculate v_2, we use Euler's theorem: $V = n_1v_1 + n_2v_2$, i.e. $v_2 = \frac{V - n_1v_1}{n_2}$

$$= \frac{10^3 + 35n_1 + 0.5n_1^2 - 35n_1 - n_1^2}{n_2} = \frac{10^3 - 0.5n_1}{n_2} \text{ with } n_2 = \frac{10^3}{18}.$$

This gives the expression: $v_2 = \frac{18\left(10^3 - 0.5n_1^2\right)}{10^3} = 18 - 9.10^{-3}n_1^2$.

2) By definition: $V^m = V - V^{pure} = V - \sum_i n_i v_i^{pure} = V - \left(n_1 v_1^{pure} + n_2 v_2^{pure}\right)$

$= 10^3 + 35n_1 + 0.5n_1^2 - 35.5n_1 - 10^3$.

3) $10^3 + 35n_1 + 0.5n_1^2 - 35.5n_1 - 10^3 = 0.5n_1^2 - 0.5n_1 = 0.5n_1(n_1 - 1)$. Therefore:

$v^m = \frac{V^m}{n} = \frac{V^m}{n_1 + n_2} = \frac{9n_1(n_1 - 1)}{18n_1 + 10^3}$ and $v_1^m = \left(\frac{\partial V^m}{\partial n_1}\right)_{T,P,n_2} = \frac{\partial\left(0.5n_1^2 - 0.5n_1\right)}{\partial n_1} =$

$n_1 - 0.5$ with $V^m = \sum_i n_i v_i^m$. We obtain the expression of v_2: $v_2^m = \frac{V^m - n_1 v_1^m}{n_2} =$

$$\frac{9n_1(n_1 - 1)}{n_2\left(18n_1 + 10^3\right)} - \frac{n_1(n_1 - 0.5)}{n_2} = -9.10^{-3}n_1^2.$$

4) Here, we simply follow the process in the opposite direction:

$$v_1 = v_1^m + v_1^{pure} = n_1 - 0.5 + 35.5 = n_1 + 35$$

$$v_2 = v_2^m + v_2^{pure} = -9.10^{-3}n_1^2 + 18$$

PROBLEM 4.8.– Activity of copper and zinc in alloys

1) Adding 0.1 mol of Cu to 10 mol of a mixture of the titer $x_{Zn} = 0.7$ does not change its composition, and the thermal effect observed corresponds to the transition of *Cu* in its pure state to a dissolved state. The amount of heat released per mole of Cu added is therefore $Q = -\left(h_{Cu}(0.7) - h^{*}_{Cu}\right) = -h^{m}_{Cu}(0.7)$. As a result, knowing h^{m} allows us to obtain, using Roozeboom's graphical method, the value of h^{m}_{Cu} for the titer 0.7 in *Zn*, as shown in Figure 4.11: the tangent to the curve $h^{m}(Zn)$ at point $x_{Zn} = 0.7$ intersects the axis of pure Cu at –3,520, therefore $h^{m}_{Cu} = -3520$ cal/mol. Hence, the amount of heat released from adding 0.1 mol will be: $q = 0.1 h^{m}_{Cu} = -$ 352 cal.

2) As the curve is symmetrical, we can use the analytical representation: $h^{m} = \mathrm{A} x_{Cu}(1 - x_{Cu})$, where $\mathrm{A} = \dfrac{h^{m}(x_{Cu})}{x_{Cu}(1 - x_{Cu})}$. We calculate the value of A for each point and calculate an average, which gives:

A = – 7161 cal./mol.

hence: $h^{m} = 7161(1 - x_{Cu})x_{Cu}$.

Here, $h^{m}_{Cu} = h^{m} + (1 - x_{Cu})\dfrac{dh^{m}}{dx_{Cu}}$.

Therefore, $h^{m}_{Cu} = \mathrm{A}(1 - x_{Cu})^2 = -7161 \times 0.49 =$ –3,509 cal/mol. The heat released is $q = 350.9$ cal.

3) The proposed analytical relation gives $\mathrm{B} = \dfrac{\log \gamma_{Zn}}{(1 - x_{Zn})^2}$.

We proceed as done previously and calculate an average value of B. We obtain $\mathrm{B} = -0.863$; followed by integration of the Gibbs–Duhem relation.

$x_{Cu} d \log \gamma_{Cu} + x_{Zn} d \log \gamma_{Zn} = 0$. We obtain: $\dfrac{\log \gamma_{Cu}}{(1 - x_{Cu})^2} = \dfrac{\log \gamma_{Zn}}{(1 - x_{Zn})^2}$, where $\log \gamma_{Cu} = \mathrm{B}(1 - x_{Cu})^2$ and for $x_{Cu} = 0.7$: $\log \gamma_{Cu} = -0.863 \times 0.09 = -0.07766$ and so $\gamma_{Cu} = 0.8363$. This gives: $a_{Cu} = \gamma_{Cu} . x_{Cu} = 0.5854$.

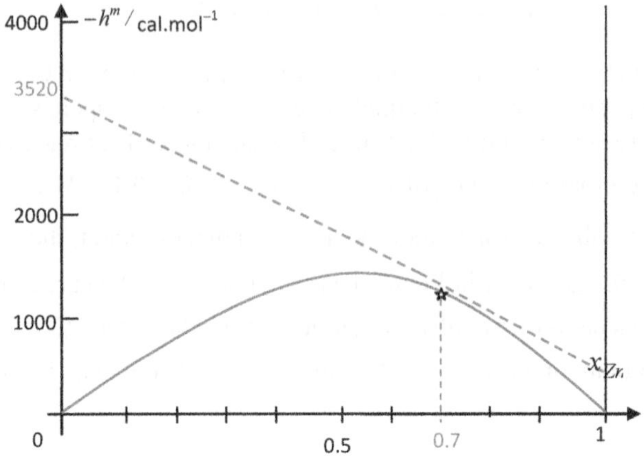

Figure 4.11. *Roozeboom's graphical method*

By definition: $g_{Cu}^{ex} = RT\, ln\, \gamma_{Cu} = h_{Cu}^{m} - Ts_{Cu}^{ex}$. A well-verified hypothesis considers that h_{Cu}^{m} and s_{Cu}^{ex} are independent of T, which means we can write: $s_{Cu}^{ex} = \dfrac{h_{Cu}^{m}}{T} - R\, ln\, \gamma_{Cu}$; at $T_0 = 1{,}360$ K, we obtain $s_{Cu}^{ex} = \dfrac{A(1-x_{Cu})^2}{T_0}$ - $4.576B(1-x_{Cu})^2 = -\ 1.3163(1-x_{Cu})^2$.

From this, we deduce an expression for the calculation of γ_i at any T: $\log\gamma_{Cu}(T) = (1-x_{Cu})^2\left(-\dfrac{A}{4.576\,T} + \dfrac{1.3163}{4.576}\right) = (1-x_{Cu})^2\left(-\dfrac{1565}{T} + 0.288\right)$ where for $x_{Cu} = 0.5$ and $T = 1{,}600$ K: $\log\gamma_{Cu} = -0.17255 \Rightarrow \gamma_{Cu} = 0.672$ and $a_{Cu} = 0.336$.

PROBLEM 4.9.– Dichloromethane/acetone solution degreasing

1) The partial molar variables can be calculated from the total molar variable using the following relations: $\begin{cases} h_1^m = h^m + x_2\left(\dfrac{\partial h^m}{\partial x_1} - \dfrac{\partial h^m}{\partial x_2}\right) \\ h_2^m = h^m + x_1\left(\dfrac{\partial h^m}{\partial x_2} - \dfrac{\partial h^m}{\partial x_1}\right) \end{cases}$. The partial derivatives are calculated as a function of each titer:

$$\begin{cases} \dfrac{\partial h^m}{\partial x_1} = x_2\left[A_0 + A_1(2x_1-1) + A_2(2x_1-1)^2\right] + 2x_1x_2\left[A_1 + 2A_2(2x_1-1)\right] \\ \dfrac{\partial h^m}{\partial x_2} = x_1\left[A_0 + A_1(2x_1-1) + A_2(2x_1-1)^2\right] \end{cases}$$

and after development, we obtain:

$$\begin{cases} h_1^m = x_2^2\left[A_0 - A_1 + A_2 + (4A_1 - 8A_2)x_1 + 12A_2x_1^2\right] \\ h_2^m = x_1^2\left[A_0 - 3A_1 + 5A_2 + (4A_1 - 16A_2)x_1 + 12A_2x_1^2\right] \end{cases}$$

We know that the heat released following an isobaric transformation Q_p is equal to $(-\Delta H)$ of the system and when we add a small amount, Δn_2 mol, of constituent(2) to the system, its enthalpy varies by:

$$\Delta H \;\#\; \Delta n_2\left(\frac{\partial H}{\partial n_2}\right)_{T,P,n_1} = h_2\Delta n_2 = \left(h_2^m + h_2^{pur}\right)\Delta n_2$$

so with respect to the initial state of the system, where the enthalpy is equal to $H + \Delta_2 h_2^{pur}$, the enthalpy of the final state after dissolution of Δn_2 mol of (2) will be equal to $H + \Delta H = H + \left(h_2^m + h_2^{pur}\right)\Delta n_2$. The variation in enthalpy that accompanies the dissolution corresponds to:

$$\left[H + \left(h_2^m + h_2^{pur}\right)\Delta n_2\right] - \left[H + h_2^{pur}.\Delta n_2\right] = h_2^m \Delta n_2$$

which gives: $Q = -\,h_2^m \dfrac{m_2}{M_2}$. Since $x_1 = 0.9956$, we can consider that h_2^m is the limiting value when $x_2 \to 0$ or when $x_1 \to 1$, and in this case: $h_2^m = A_0 + A_1 + A_2 = -3{,}736.7 \;\; -387.76 \;+\; 264.11 = -3{,}860.35$ J, where $Q = \dfrac{-\;3860.35\times 3}{58}$ $\# -200$ J.

To determine T_f, we must calculate C_p of the solution. As we know that they do not vary with T, we can write: $C_P = C_p^m + x_1 C_{p1}^{\circ} + x_2 C_{p2}^{\circ}$ with $C_p = \left(\dfrac{\partial h}{\partial T}\right)_P$ and $C_p^m = \left(\dfrac{\partial h^m}{\partial T}\right)_P$ and $h^m(T_2) - h^m(T_1) = \int_{T_1}^{T_2} C_p^m dT = C_p^m(T_2 - T_1)$.

Knowing the values of $h^m(x_1 = 0.5)$, we can calculate the value of $C^m_{p\,(x_1=0.5)} = \dfrac{0.25\times A_0 + 962}{(25-0)} = \dfrac{-\ 934.175 + 962}{25} = 1.113$ J/mol.K, and that of $C_{p(x_1=0.5)} = 1.113 + 0.5\times 126 + 0.5\times 101.4 = 114.8$ J./mol.K.

We can write the overall enthalpy of the equimolar solution of 10 mol of each pure constituent at $T_0 = 25°C$ using the following relation: $n_1 h_1^{pur}(T_0) + n_2 h_2^{pur}(T_0) = (n_1 + n_2)\,h(T_f)$ with $h_{Tf} = h_{T_0} + C_{p(x=0.5)}(T_f - T_0)$ and $h_{T_0} = h^m_{T_0} + x_1 h^m_{1T_0} + x_2 h^m_{2T_0}$. Since $x_1 = \dfrac{n_1}{n_1+n_2}$ and $x_2 = \dfrac{n_2}{n_1+n_2}$, we obtain: $n_1 h^{pur}_{1T_0}(T_0) + n_2 h^{pur}_{2T_0} = (n_1+n_2)\left[h^m_{T_0} + C_p(T_1 - T_0)\right] + n_1 h^{pur}_{1T_0} + n_2 h^{pur}_{2T_0}$, where $C_p(T_f - T_0) = -\ h^m_{T_0}$ and therefore $T_f = \dfrac{934.17}{114.8} + 25 = 33.14$ °C

NOTE.– We find that C^m_p represents 1% of C_p and normally we can disregard it in the enthalpy equations when the variation in temperature is minor. The calculation in this case is as follows: $T_f - T_0 = \dfrac{934.17}{113.7} = 8.21°C \Rightarrow T_f = 33.21°C$.

4) An adiabatic mixture is formed of identical masses, and therefore: $n_1 = \dfrac{m}{M_1}$ and $n_2 = \dfrac{m}{M_2}$, with the expression giving x_1, then x_2:

$x_1 = \dfrac{\frac{m}{M_1}}{\frac{m}{M_1} + \frac{m}{M_2}} = \dfrac{M_2}{M_1 + M_2} = \dfrac{58}{143} = 0.405$ and so $x_2 = 0.595$. The enthalpy equation is written as: $n_1 h^{pur}_{1T_1} + n_2 h^{pur}_{2T_2} = (n_1+n_2)\,h_{T_f}$ and $n_1 h^{pur}_{1T_1} + n_2 h^{pur}_{2T_2} = h_{T_f}$. Here, $h_{T_f} = h^m_{T_f} + x_1 h^{pur}_{1T_f} + x_2 h^{pur}_{2T_f}$ and $h^m_{T_f} = h^m_{T_0} + C^m_p(T_f - T_0)$, whereby disregarding $C^m_p(T_f - T_0)$: $h^m_{T_f} \# h^m_{T_0}$ and $h_{T_f} = h^m_{T0} + x_1 h^{pur}_{1T_f} + x_2 h^{pur}_{2T_f}$,and by replacing $x_i h^{pur}_{iT_f}$ with their expressions: $h_{T_f} = H^m_{T_0} + x_1 h^{pur}_{1T_1} + x_1 C^{pur}_{p1}(T_f - T_1) + x_2 h^{pur}_{2T_2} + x_2 C^{pur}_{p2}(T_f - T_2)$ and hence: $h^m_{T_0} + x_1 C^{pur}_{p1}(T_f - T_1) + x_2 C^{pur}_{p2}(T_f - T_2) = 0$. We obtain

$$T_f = \frac{-h^m_{T_0} + x_1 C^{pur}_{p1} T_1 + x_2 C^{pur}_{p2} T_2}{x_1 C^{pur}_{p1} + x_2 C^{pur}_{p2}}$$ and for $x_1 = 0.405$, we obtain $h^m_{T_f} = -880.4$ °C.

Therefore: $T_f = \dfrac{880.4 + 0.405 \times 101.4 \times 15 + 0.595 \times 126 \times 5}{0.405 \times 101.4 + 0.595 \times 126} = 16.1$ °C.

PROBLEM 4.10.– Adiabatic mixer: cyclohexane/methyl acetate

1) The expression of enthalpy h^m is written as: $h^m = h - x_1 h_1^{pur} - x_2 h_2^{pur}$ and so $h = h^m + x_1 h_1^{pur} + x_2 h_2^{pur}$ and C_p the mixture is defined by: $C_p = \left(\dfrac{\partial h}{\partial T}\right)_P = \left(\dfrac{\partial h^m}{\partial T}\right)_P + x_1\left(\dfrac{\partial h_1^{pur}}{\partial T}\right)_P + x_2\left(\dfrac{\partial h_2^{pur}}{\partial T}\right)_P$, which gives $C_p = C_p^m + x_1 C_{p1}^{pur} + x_2 C_{p2}^{pur}$. As coefficients A_i are independent of T, then $C_p^m = 0$.

2) We deduce the following expression of C_P: $C_p = x_1 C_{p1}^{pur} + (1 - x_1) C_{p2}^{pur}$ $= 66.944 x_1 + 50.626 - 506.2 x_1$, which gives: $C_P = 16.317 x_1 + 50.626$.

3) The overall relation for matter in the mixture is shown in Figure 4.12.

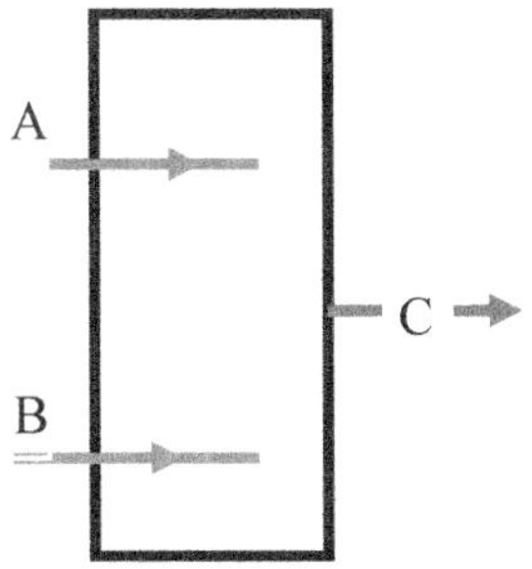

(A) component (2) pure: $\dot{n}_{1A} = 0$ and $\dot{n}_{2A} = \dfrac{10^4}{84} =$ 119.05 $\Rightarrow$ $\dot{n}_A = \dot{n}_{2A} = 119{,}05$ mol/min,

(B) mixture where $x_2 =$ $0.1 \Rightarrow \dot{n}_{1B} = 9\dot{n}_{2B}$ so $\dot{n}_B = 10\dot{n}_{2B}$,

(C) upon exit: $\dot{n}_C = \dot{n}_A + \dot{n}_B$.

Figure 4.12. *Schematic of the mixer*

a) The mass balance is written as: $\dot{m} = \dot{n}_{1B} M_1 + \dot{n}_{2B} M_2 =$ $\dot{n}_{2B}(9M_1 + M_2) = 750\dot{n}_{2B} = 5 \times 10^3$ g/min, thus $\dot{n}_{2B} = \dfrac{5000}{750} = 6.667$ mol/min; thus $\dfrac{dB}{dT}$ and $\dfrac{0.479 - 0.439}{308.15 - 318.15}$, hence $\dot{n}_C = 119.05 + 66.67 =$ 185.72 mol/min;

this results in the composition mixture: $x_{1C} = \frac{\dot{n}_{1C}}{\dot{n}_C} = \frac{(\dot{n}_{1A} + \dot{n}_{1B})}{\dot{n}_C} = \frac{60}{185.72} = 0.323 =$ 0.323 and $x_{2C} = 1 - x_{1C} = 0.677$.

b) The enthalpy equation is written as: $\dot{n}_A h_A + \dot{n}_B h_B = \dot{n}_C h_C$, knowing that:

$$h_A = h_2^{pur}(T_A)$$

$$h_B = h_B^m(T_B) + x_{1B} h_1^{pur}(T_B) + x_{2B} h_2^{pur}(T_B)$$

$$h_C = h_C^m(T_C) + x_{1C} h_1^{pur}(T_C) + x_{2C} h_2^{pur}(T_C)$$

$\dot{n}_{1B} = \dot{n}_B x_{1B}$; $\dot{n}_{2B} = \dot{n}_B x_{2B}$; $\dot{n}_{1C} = \dot{n}_{1B}$ and $\dot{n}_{2C} = \dot{n}_{2A} + \dot{n}_{2B}$

We obtain the expression:

$$\dot{n}_{2A}\left(h_{2(T_A)}^{pur} - h_{2(T_C)}^{pur}\right) + \dot{n}_{1B}\left(h_{1(T_B)}^{pur} - h_{1(T_C)}^{pur}\right) + \dot{n}_{2B}\left(h_{2(T_B)}^{pur} - h_{2(T_C)}^{pur}\right) = \dot{n}_C h_{C(T_C)}^m - \dot{n}_B h_{B(T_B)}^m$$

here h^m does not depend on T but only on composition. The relation above allows us to calculate T_C:

$$\dot{n}_{2A} C_{p2}(T_A - T_C) + \dot{n}_{1B} C_{p1}(T_B - T_C) + \dot{n}_{2B} C_{p2}(T_B - T_C) = \dot{n}_C h_C^m - \dot{n}_B h_B^m$$

where solving this gives the expression:

$$T = T_A + 20 \frac{\dot{n}_{1B} C_{p1} + \dot{n}_{2B} C_{p2}}{\dot{n}_{1C} C_{p1} + \dot{n}_{2C} C_{p2}} - \frac{\dot{n}_C h_C^m - \dot{n}_B h_B^m}{\dot{n}_{1C} C_{p1} + \dot{n}_{2C} C_{p2}}$$

with:

$$\begin{cases} \dot{n}_B h_B^m = \dot{n}_{1B} x_{2B}\left[A_0 + A_1(x_{1B} - x_{2B}) + A_2(x_{1B} - x_{2B})^2\right] \\ \dot{n}_C h_C^m = \dot{n}_{1C} x_{2C}\left[A_0 + A_1(x_{1C} - x_{2C}) + A_2(x_{1C} - x_{2C})^2\right] \end{cases}$$

With all calculations complete, we obtain: $T_C = 30 + 8.389 - 24.547 = 13.84$ °C.

The table below summarizes the results:

	x_1	x_2	$\dot{m}(g/min)$	$\dot{n}_1$	$\dot{n}_2$	$\dot{n}$	T (°C)	C_p	h^m
A	0	1	10^4	0	119.5	119.5	30	50.63	0
B	0.9	0.1	5×10^3	60	6.67	66.67	50	65.32	680.1
C	0.323	0.677	15×10^3	60	125.72	185.72	13.84	55.39	1616.1

PROBLEM 4.11.– Benzene/cyclohexane binary solution

1) a) By definition: $\frac{g^{ex}}{RT} = \frac{h^m}{RT} - \frac{s^{ex}}{R}$ and $s^{ex} = -\left(\frac{\partial g^{ex}}{\partial T}\right)_P$, where $\frac{h^m}{RT} = \frac{g^{ex}}{RT} - \frac{1}{R}\left(\frac{\partial g^{ex}}{\partial T}\right)_P$. Here, by replacing each term:

$$\frac{h^m}{RT} = A(T)x_1x_2 - \frac{1}{R}\left[R.A(T)x_1x_2 + RTx_1x_2\left(\frac{\partial A(T)}{\partial T}\right)_P\right] = -Tx_1x_2\frac{dA}{dT}$$

We get: $h^m = -x_1x_2RT^2\frac{dA}{dT} = Rx_1x_2\frac{dA}{d(1/T)} \Leftrightarrow \frac{h^m}{\text{RT}} = x_1x_2\frac{1}{T}\frac{dA}{d(1/T)}$. Therefore:

$$\frac{h^m}{\text{RT}} = x_1x_2\frac{dA}{d(\ln T)}$$

b) From $\frac{s^{ex}}{R} = \frac{h^m}{RT} - \frac{g^{ex}}{RT}$, we obtain $\frac{s^{ex}}{R} = x_1\text{x}_2\frac{1}{T}\frac{d\text{A}}{d(1/T)} - \text{A}x_1x_2$

$$= -x_1x_2\left(\text{A} - \frac{1}{T}\frac{d\text{A}}{d(1/T)}\right).$$

2) To determine h^m as a function of $x_1 x_2$ and T, we must establish the equation for A(T), as per the following table:

T (°C)	35	40	45	50
T (K)	308.15	313.15	318.15	323.15
$\frac{10^3}{T}$	3.2467	3.1949	3.1446	3.0960
A(T)	0.479	0.458	0.439	0.420

A(T) is a linear function in $\frac{1}{T}$ with the form: $A(T)=\frac{a}{T}+b$. We find that: a = 391.776 and b = −0.793, and the expression of h^m is: $\frac{h^m}{RT}=391.776\frac{x_1 x_2}{T}$ $\Rightarrow$ $h^m=391.776\,Rx_1x_2=3259.576\,x_1x_2$, which shows that h^m and is independent of T.

For the mixture of $n_1=0.8$ mol and $n_2=1.2$ mol, we obtain $n=2$ mol and so $x_1=0.4$ and $x_2=0.6$ and hence: h^m = 782.298 J.

We verify that: $\begin{cases}\text{at } 40\,°\text{C}: \dfrac{h^m}{RT}=1.2512x_1x_2\\ \text{at } 50\,°\text{C}: \dfrac{h^m}{RT}=1.2129x_1x_2\end{cases}$ and for $\begin{cases}x_1=0.4\\ x_2=0.6\end{cases}$, we obtain

$\begin{cases}h_{40}^m=782.661\text{ J}\\ h_{50}^m=782.643\text{ J}\end{cases}$, with the heat released by the mixture being:

$$Q=-n.h^m=-2\times 782.3=-1{,}564.6\text{ J}$$

3) The γ_i are given by: $RT\,ln\gamma_i=\bar{g}_i^{ex}=g^{ex}+x_j\left(\frac{\partial g^{ex}}{\partial x_i}-\frac{\partial g^{ex}}{\partial x_j}\right)$, which gives us: $ln\gamma_1=Ax_2^2$ and $ln\gamma_2=Ax_1^2$; since at T = 40 °C, we find that $A(T)=0.4581$ for $x_1=0.4$ and $x_2=0.6$ we obtain: $ln\gamma_1$= 0.1649 and so γ_1 = 1.1793 and $ln\gamma_2$ = 0.0733 and γ_2 = 1.0760.

The previous expressions are written as: $\begin{cases} ln\gamma_1 = \left(\frac{a}{T}+b\right)(1-x_1)^2 \\ ln\gamma_2 = \left(\frac{a}{T}+b\right)(1-x_2)^2 \end{cases}$, where

$$\begin{cases} \text{when } x_1 \to 0: \lim(ln\gamma_1) = ln\gamma_1^{\infty} = \dfrac{a}{T}+b \\ \text{when } x_1 \to 0: \lim(ln\gamma_2) = ln\gamma_2^{\infty} = \dfrac{a}{T}+b \end{cases}.$$

Finally, we obtain

$$\gamma_1^{\infty} = \gamma_2^{\infty} = e^{\left(\frac{391.776}{T} - 0.793\right)}.$$

PROBLEM 4.12.– N-heptane/aniline two-phase mixtures

1) The equilibrium between both phases can be represented by the equality of chemical potentials of constituents (1) and (2) in each phase, that is to say: $g_{1a} = g_{1b}$ and $g_{2a} = g_{2b}$; and the chemical potentials are expressed as:

$$g_1 = g_1^P + RT\ ln\ a_1 = g_1^P + RT\ lnx_1 + RT\ ln\ \gamma_1$$

$$g_2 = g_2^P + RT\ ln\ a_2 = g_2^P + RT\ lnx_2 + RT\ ln\ \gamma_2$$

where the terms $ln\,\gamma_1$ and $ln\,\gamma_2$ are partial *state functions* $\frac{g_i^{ex}}{RT}$ relative to the total *state function* $\frac{g^{ex}}{RT} = x_1 x_2$ (A'+B'x_1), and are obtained from the general relation:

$$g_i^{ex} = g^{ex} + (1-x_i)\left[\left(\frac{\partial g^{ex}}{\partial x_i}\right)_{TPx_j} - \left(\frac{\partial g^{ex}}{\partial x_j}\right)_{TPx_i}\right].$$

Therefore, if $A' = \frac{A}{RT}$ and $B' = \frac{B}{RT}$, then their expressions are: $ln\ \gamma_1 = x_2^2$ (A'+ 2B'x_1) and $ln\ \gamma_2 = x_1^2$ [(A'– B') + 2B'x_1)]. So both equations expressing the equality of g_i become:

$$lnx_{1a} + x_{2a}^2\,(A' + 2B'x_{1a}) = lnx_{1b} + x_{2b}^2\,(A' + 2B'x_{1b})$$

$$lnx_{2a} + x_{1a}^2\,[(A' - B') + 2B'x_{1a})] = lnx_{2b} + x_{1b}^2\,[(A' - B') + 2B'x_{1b})]$$

This is a system with two equations with two unknowns A' and B', which is written as:

$$\begin{cases} A'(x_{2a}^2 - x_{2b}^2) + 2B'(x_{1a}x_{2a}^2 - x_{1b}x_{2b}^2) = ln\dfrac{x_{1b}}{x_{1a}} \\ A'(x_{1a}^2 - x_{1b}^2) + B'\left[x_{1b}^2 - x_{1a}^2 + 2(x_{1a}^3 - x_{1b}^3)\right] = ln\dfrac{x_{2b}}{x_{2a}} \end{cases}$$

So by replacing the titers with their values, the system becomes:

$$\begin{cases} -0.829A' - 0.114B' = -2.377 \\ 0.773A' + 0.603B' = 2.060 \end{cases}$$

therefore $A' = \dfrac{2.060 - 0.603B'}{0.773}$ and so $-0,829\left(\dfrac{2.060 - 0.603B'}{0.773}\right)$ $-0.114B' = -2.377$, and finally: $-0.829\dfrac{2.060}{0.773} + \dfrac{0.829\times0.603B'}{0.773}$ $-0.114B' = -2.377$, where B' = – 0.3149 and $A' = \dfrac{2.060 + 0.603\times0.3149}{0.773} = 2.9093$. Therefore at $T = 312.15$ K, $RT = 620.242$, we obtain: A = 1,804.470 and B = – 194.736, giving the relation: $g^{ex} = x_1x_2(1,804.47 - 194.736x_1)$.

2) a) $ln\,\gamma_{1a} = x_{2a}^2$ $(A' + 2B'x_{1a}) = 0.032235$ and so $\gamma_{1a} = 1.0328$. $ln\,\gamma_{1b} = x_{2b}^2$ $(A' + 2B'x_{1b}) = 2.40634$, therefore $\gamma_{1b} = 11.095$.

b) $a_{1a} = \gamma_{1a}x_{1a} = 1.0328 \times 0.883 = \# 0.912$ and $a_{1b} = 11.095 \times 0.082 \# 0.910$.

CONCLUSION.– We confirm that the activity of each body is the same in each phase in equilibrium, particularly for n-heptane: $a_{1a} = a_{1b} = 0.91$.

3) The vapor tension of the mixture is given by $P = P_1 + P_2 = a_{1a}P_1^\circ + a_{2a}P_2^\circ$. Since $P_2^\circ \approx 0$, then $P = a_{1a}P_1^\circ$ where $P = 0.91\times89 = 80.99 \# 81$ mm Hg.

PROBLEM 4.13.– Activities of zinc/cadmium liquid alloys

1) If T_1 is the deposition temperature of pure Zn from a solution of titer x_{Zn} and T_0 the solidification temperature of pure component Zn, then the Zn deposition curve can be represented by Figure 4.13.

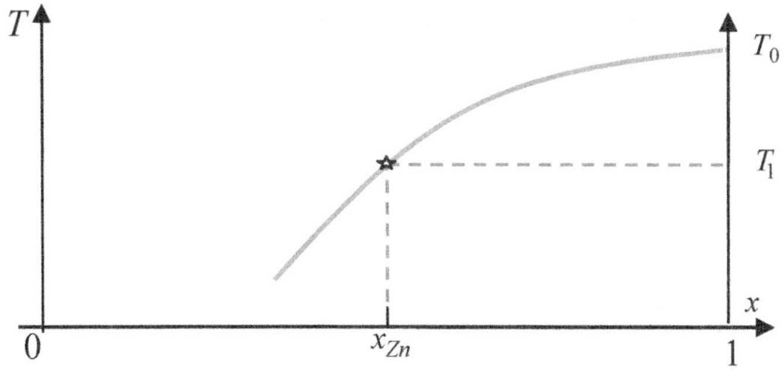

Figure 4.13. *Deposition curve of Zn*

T_1 is the deposition temperature of pure Zn form a solution of titer x_{Zn} and at T_0, the solidification temperature of pure *Zn*. The parameters T_0, T_1, L and x_{Zn} are linked by the relation: $ln\, a_{Zn} = ln\, \gamma_{Zn}.x_{Zn} = -\frac{L}{R}\left(\frac{1}{T_1} - \frac{1}{T_0}\right)$; so for $T = 546$ K, we obtain: $a_{Zn} = 0.707$ and $x_{Zn} = 0.3$, which gives: $\gamma_{Zn} = 2.36$.

2) By definition: $RT_1\, ln\, \gamma_1 = g_{T_1}^{ex} = h_{T_1}^{ex} - Ts_{T_1}^{ex}$; we must find an average calculation for $h_{T_1}^{ex}$ and $s_{T_1}^{ex}$ at a given *T*. We can then establish a development of $ln\, \gamma_1$, valid in this temperature domain. For this, we use the following:

i) $h_{T_1}^{ex} = h_{T_1}^{m}$: it is given by the calorimetry results by an expression independent of *T*. Therefore $h_{T_1}^{ex} = h^{ex}$;

ii) g^{ex} : it is given by the deposition curve and we find that s^{ex} is independent of *T*. We can write: $s_{T_1}^{ex} = \frac{h_{T_1}^{ex} - g_{T_1}^{ex}}{T_1} = \frac{h^{ex} - g_{T_1}^{ex}}{T_1}$, where at *T*: $g_T^{ex} = h^{ex} - \frac{T}{T_1}\left(h^{ex} - g_{T_1}^{ex}\right) = \frac{T}{T_1} g_{T_1}^{ex} - \left(\frac{T}{T_1} - 1\right) h^{ex}$ and $RT\, ln\, \gamma = \frac{T}{T_1} RT_1\, ln\, \gamma_1 - \left(\frac{T}{T_1} - 1\right) h^{ex}$, therefore $ln\, \gamma = ln\, \gamma_1 - \frac{h^{ex}}{R}\left(\frac{1}{T_1} - \frac{1}{T}\right)$. Adding $RT\, ln\, x$ to both members, we obtain the activity at *T* by the following development: $ln\, a = -\frac{L}{R}\left(\frac{1}{T_1} - \frac{1}{T_0}\right) - \frac{h^{ex}}{R}\left(\frac{1}{T_1} - \frac{1}{T}\right)$;

where for $x_{Zn} = 0.3$ and $T = 800$ K: $h^{ex} = 0.49(1200+1900\times 0.3) = 867$ cal/mol and $\log a_{zn} = -0.260$ with $a_{Zn} = 0.549$.

3) A regular solution can be characterized by the absence of excess entropy and so by: $RT\, ln\,\gamma = RT_1\, ln\,\gamma_1 = h^{ex}$ therefore $ln\,\gamma = \frac{T}{T_1} ln\,\gamma_1$; where for $x_{Zn} = 0.3$ ($\gamma_1 = 2.36$) and $T = 800$ K, we obtain: $\log\gamma = \frac{546}{800} ln 2.36 = 0.2543$ and $\gamma = 1.979$.

where $a_{Zn} = 1.979\times 0.3 = 0.593$.

PROBLEM 4.14.– Properties of a ternary mixture

Solving this requires knowing g^{ex} for the three binaries:

– (1)/(2): $g_{12}^{ex} = 0$ since the solution is ideal;

– (2)/(3): $g_{23}^{ex} = x_2 x_3 \left[A + B(x_2 - x_3)\right]$.

A and B can be calculated from the equality g_i of each constituent in both phases (a) and (b) in equilibrium, i.e. $g_{2a} = g_{2b}$ and $g_{3a} = g_{3b}$. All that is required is to develop these equations by introducing g_i^{ex} deduced from the relation g^{ex} by: $g_2^{ex} = x_3^2\left[A + B(4x_2 - 1)\right]$ and $g_3^{ex} = x_2^2\left[A + B(1 - 4x_3)\right]$ to ultimately obtain the two following equations of A and B:

$$\begin{cases} RT\, ln\, x_{3a} + x_{2a}^2\left[A + B(1 - 4x_{3a})\right] = RT\, ln\, x_{3b} + x_{2b}^2\left[A + B(1 - 4x_{3b})\right] \\ RT\, ln\, x_{2a} + x_{3a}^2\left[A + B(4x_{2a} - 1)\right] = RT\, ln\, x_{2b}^2 + x_{3b}^2\left[A + B(4x_{2b} - 1)\right] \end{cases}$$

which are rewritten for a system with two equations and two unknowns A and B:

$$\begin{cases} A(x_{3a}^2 - x_{3b}^2) + B\left[x_{3a}^2(4x_{2a} - 1) - x_{3b}^2(4x_{2b} - 1)\right] = RT\, ln\frac{x_{2b}}{x_{2a}} \\ A(x_{2a}^2 - x_{2b}^2) + B\left[x_{2a}^2(1 - 4x_{3a}) - x_{2b}^2(1 - 4x_{3b})\right] = RT\, ln\frac{x_{3b}}{x_{3a}} \end{cases}$$

Solving this gives: $A = 1707.64$ cal/mol and $B = -97.343$ cal/mol;

– (3)/(1): $g_{31}^{ex} = Cx_3x_1$. This choice is justified by the symmetry of the two-phase domain. C is determined in the same way as above and we obtain the equality: $RT\,ln\,x_{3a} + Cx_{1a}^2 = RT\,ln\,x_{3b} + Cx_{1b}^2$, where $C(x_{1a}^2 - x_{1b}^2) = RT\,ln\dfrac{x_{3b}}{x_{3a}}$. Therefore,

$x_{1a} + x_{1b} = 1$ and so $C(x_{1a} - x_{1b}) = RT\,ln\dfrac{x_{3b}}{x_{3a}} = = 1282.21$ cal/mol.

1) R-K proposes $g^{ex} = x_2x_3\left[A + B(x_2 - x_3)\right] + Cx_1x_3 + Dx_1x_2x_3$. We can determine g_1^{ex} by applying the relation that links them:

$$g_1^{ex} = g^{ex} + \frac{\partial\, g^{ex}}{\partial\, x_1} - x_1\frac{\partial\, g^{ex}}{\partial\, x_2} - x_2\frac{\partial\, g^{ex}}{\partial\, x_2} - x_3\frac{\partial\, g^{ex}}{\partial\, x_3}$$

which gives: $g_1^{ex} = D\,x_2x_3(1 - 2x_1) + C\,x_3(x_2 + x_3) - x_2x_3\left[A + 2B(x_2 - x_3)\right]$ and replacing A, B and C with their values, we obtain:

$$g_1^{ex} = Dx_2x_3(1 - 2x_1) + 1282.21.x_3(x_2 + x_3) - x_2x_3(1707.10 - 194.686(x_2 - x_3)$$

2) At point M, we obtain: $g_1^{ex}(M) = 0.0877D - 10.249$. Here, $g_1^{ex} = RT\,ln\,\gamma_1$ and so: $\ln\gamma_1 = \dfrac{0.0877}{RT}D - \dfrac{10.249}{RT}$.

3) Knowing P_1 and P_1^{pure}, we can calculate the activity of component (1) at point M: $a_1 = \gamma_1.x_1 = \dfrac{P_1}{P_1^{pure}} = \dfrac{4.91}{89}$; therefore, $\gamma_1 = \dfrac{4.91}{89\times 0.0517} = 1.0671$, where $ln\gamma_1 = 0.0649$ and $RTln\gamma_1 = 40.278 = 0.0877D - 10.249$, which gives $D = 576.62$ cal/mol.

4) By definition: $k_d = \dfrac{x_{1a}}{x_{1b}}$. Both phases are in equilibrium, hence $g_{1a} = g_{1b}$:

$RT\,ln\,x_{1a} + g_{1a}^{ex} = RT\,ln\,x_{1b} + g_{1b}^{ex}$ where $ln\dfrac{x_{1b}}{x_{1a}} = \dfrac{g_{1a}^{ex} - g_{1b}^{ex}}{RT}$; by placing it into the low concentrations hypothesis (1), that is to say, low x_1 in this case, g_{1a}^{ex} and g_{1b}^{ex} would have the limiting value corresponding to x_2 and x_3 of phases (a) and (b) in equilibrium in the binary (1)/(2). We can write:

$$\begin{cases} g_{1a}^{ex} \approx Dx_{2a}x_{3a} + Cx_{3a}(x_{2a} + x_{3a}) - x_{2a}x_{3a}\left[A + B(x_{2a} - x_{3a})\right] \\ g_{1b}^{ex} \approx Dx_{2b}x_{3b} + Cx_{3b}(x_{2b} + x_{3b}) - x_{2b}x_{3b}\left[A + B(x_{2b} - x_{3b})\right] \end{cases}$$. This

calculation gives: $\ln\frac{x_{1b}}{x_{1a}} = -1.67215$ and finally: $k_d = 0.187$.

PROBLEM 4.15.– Studying the tin/lead binary by the addition of mercury

1) a) Given that P_1° and P_2° are negligible, this implies that the pressure in top of the mixture is that of Hg. Therefore: $P = \gamma_3 x_3 P_3^\circ$ and $\gamma_3 = \frac{P}{x_3 P_3^\circ} = \frac{P}{430\,x_3}$.

b) Immediately from the experimental results, we establish the new table (below) of values using the relation proposed.

c) Using this table gives: binary (1/3): an average value of $\frac{\log \gamma_3}{(1-x_3)^2} =$ 0.586 hence parameter "a"; binary(2/3): an average value of $\log\frac{\gamma_3}{(1-x_3)^2} = 0.680$ hence parameter "c"; ternary(1/2/3): an average value of $\log\frac{\gamma_3}{(1-x_3)^2} = 0.539$.

Here, $x_1 = x_2 = \frac{1-x_3}{2}$, where the relation between the three parameters: a + b + c = 4 × 0.539 = 2.156, therefore b = 0.89. We obtain the final expression: $\ln\gamma_3 = 0.586\,x_1^2 + 0.89\,x_1x_2 + 0.68\,x_2^2$.

x_3	0.1	0.2	0.4	$\frac{\log\gamma_3}{(1-x_3)^2}$
	P (mm Hg)			
				0.5849
P_b^{pur}	128.0	204.0	279.5	0.5861
				0.5857
γ_3	2.9767	2.3721	1.625	0.5856

$\frac{x_2}{x_1} = 1$	117.5	190.0	269.0	0.5390
				0.5379
				0.5395
γ_3	2.7326	2.2093	1.5640	0.5388
S_b^{pure}	152.5	234.0	302.0	0.6788
				0.6792
				0.6777
γ_3	3.5465	2.7209	1.7538	0.6786

2) To determine the total molar variable g^{ex}, we use the Darken relation, which is written as: $\frac{g^{ex}}{RT} = (1-x_3)\int_1^{x_3}\left(\frac{g_3^{ex}/RT}{(1-x_3)^2}dx_3\right)_{\frac{x_1}{x_2}} + x_1\int_0^1\left(\frac{g_3^{ex}/RT}{(1-x_3)^2}dx_1\right)_{x_2=0} +$ $x_2\int_0^1\left(\frac{g_3^{ex}/RT}{(1-x_3)^2}dx_3\right)_{x_1=0}$. Now we calculate the expression between the parentheses: $\frac{g_3^{ex}/RT}{(1-x_3)^2} = \frac{\mathrm{a}x_1^2 + \mathrm{b}x_1x_2 + \mathrm{c}x_2^2}{(1-x_3)^2} = \frac{\mathrm{a}\left(\frac{x_1}{x_2}\right)^2 + \mathrm{b}\left(\frac{x_1}{x_2}\right) + \mathrm{c}}{\left(1+\frac{x_1}{x_2}\right)^2}$,

which shows that if $\frac{x_1}{x_2}$ is constant, then $\frac{g_3^{ex}/RT}{(1-x_3)^2}$ is also constant. If C is the constant, we can calculate the three integrals as follows:

– the first gives: $\mathrm{C}\int_1^{x_3}(dx_3)_{x_1/x_2} = \mathrm{C}(x_3 - 1)$;

– the second with $x_2 = 0$ gives: $\int_0^1\left(\frac{\mathrm{a}x_1^2}{x_1^2}\right)_{x_2=0} dx_3 = \mathrm{a}$;

– the third with $x_1 = 0$ gives: $\int_0^1\left(\frac{\mathrm{a}x_2^2}{x_2^2}\right)_{x_1=0} dx_3 = \mathrm{c}$.

Finally: $\frac{g^{ex}}{RT} = -(1-x_3)^2 \frac{\left(ax_1^2 + bx_1x_2 + cx_2^2\right)}{(1-x_3)^2} + ax_1 + cx_2$ and so:

$\frac{g^{ex}}{RT} = -\left(ax_1^2 + bx_1x_2 + cx_2^2\right) + ax_1 + cx_2$, which is written in the symmetrical form:

$$\frac{g^{ex}}{RT} = ax_1x_3 + (a-b+c)x_1x_2 + cx_2x_3.$$

In the binary mixture *Sn/Pb*, there is no Hg, therefore $x_3 = 0$, and hence the expression: $\frac{g^{ex}}{RT} = (a-b+c)x_1x_2$. This allows us to deduce the activity coefficients of both metals in solution: $ln\,\gamma_{Sn} = (a-b+c)x_{Sn}^2$ and $ln\,\gamma_{Pb} = (a-b+c)x_{Pb}^2$.

PROBLEM 4.16.– Properties of magnesium/bismuth/lead alloys

1) n_2 and n_3 prior to mixing are at 900 K and the solution forms at this temperature. Therefore, the heat released is given by: $-Q = (n_2 + n_3)h^m = (n_2 + n_3)Bx_2x_3 = B\left(\frac{n_2 n_3}{n_2 + n_3}\right) = 4{,}310 \times 0.48 = 2{,}068.8$ J.

2) The enthalpy of dissolution of solid Mg at each addition is: $Q' = \frac{-Q}{\Delta n_1}$ and it corresponds to the difference in enthalpy: $h_1(\text{mixture}) - h_1(\text{pure solid}) = h_1 - h_{1l}^{pur} - h_{1s}^{pur}$, which gives the equality: $\frac{-Q}{\Delta n_1} = h^m + h_{1l}^{pur} - h_{1s}^{pur}$ and so $\frac{-Q}{\Delta n_1} = h_1^m + L_{f1}$, where $h_1^m = -\left(\frac{Q}{\Delta n_1} + L_{f1}\right)$, which gives the following table of values:

x_1	0.0	0.1	0.2	0.3	0.4	0.45	0.5
$-h_1^m$ (J/mol)	24,686	19,522	15,049	11,234	8,044	6,670	5,439

3) a) The number of moles n_1 represents the total Δn_1 (including those for which the corresponding thermal effect was not measured). We must therefore determine

this form $x_1 = \dfrac{n_1}{n_1 + n_2 + n_3}$ where the only variable is n_1. As a result: $n_1 = (n_2 + n_3)\dfrac{x_1}{1 - x_1} = \dfrac{2x_1}{1 - x_1}$, which gives the following table:

x_1	0.0	0.1	0.2	0.3	0.4	0.45	0.5
n_1	0.0	0.222	0.50	0.857	1.333	1.6364	2.00
x_1 (J/mol)	24,686	19,522	15,049	11,234	8,044	6,670	5,439
$\dfrac{-h_1^m}{(1-x_1)^2}$	24,686	24,101	23,514	22,927	22,344	22,050	21,756

The graph representative of these results is shown in Figure 4.14.

b) From the graph, we find that $H^m = H^m_{n_1=0} + \int_{n_1=0}^{n_1} h_1^m dn_1$ and for the binary (2)/(3), we know that: $H^m_{n_1=0} = (n_2 + n_3)h^m_{n_1=0} = (n_2 + n_3)\mathrm{B}x_2x_3 = \mathrm{B}\dfrac{n_2n_3}{n_2+n_3}$, therefore $H^m = \mathrm{B}\dfrac{n_2n_3}{n_2+n_3} + \int_0^{n_1} h_1^m dn_1$. Since $h^m = \dfrac{H^m}{\sum_i n_i}$, we obtain

$$h^m = \frac{\mathrm{B}n_2n_3}{(n_2+n_3)(n_1+n_2+n_3)} + \frac{1}{n_1+n_2+n_3}\int_0^{n_1} h_1^m dn_1 .$$

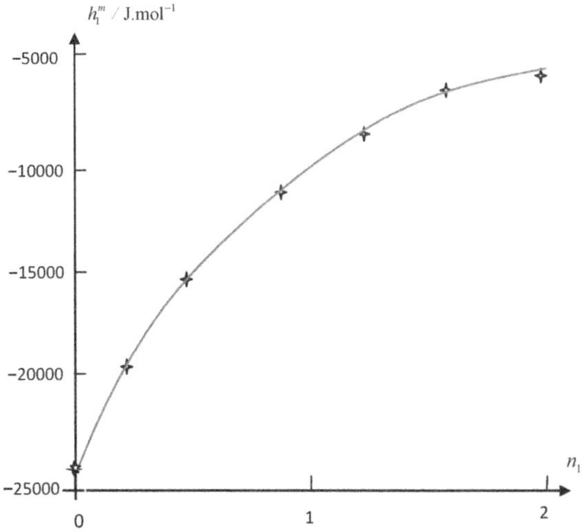

Figure 4.14. *Curve showing variation in h_1^m as a function of n_1*

c) For $n_1 = 1.4$, it gives the result: $\int h_1^m dn_1 = -$ 19,400 J and $h^m = -$ 6,314 J.

4) Darken's integration is calculated according to the function $\frac{h_1^m}{(1-x_1)^2}$ as a function of x_1, the graph (from the previous table) is given by Figure 4.15 and shows that we have a straight line.

The base relation is as follows: $\left(\frac{h^m}{1-x_1}\right)_{x_1} = \left(\frac{h^m}{1-x_1}\right)_{x_1=0} + \int_0^{n_1} \frac{h_1^m}{(1-x_1)^2} dx_1 = \left(\frac{h^m}{1-x_1}\right)_{x_1=0} + I(x_1)$, where the expression of enthalpy of a mixture: $h^m = (1-x_1)\left[\frac{Bn_2n_3}{(n_2+n_3)^2} + I(x_1)\right]$ with $n_1 = 1.4$; $n_2 = 0.8$ and $n_3 = 1.2$, we obtain: $h^m = \frac{(-4310 \times 0.24) - 9668}{1.7} =$ –6,296 J, a value fairly close to the least accurate obtained in (3c).

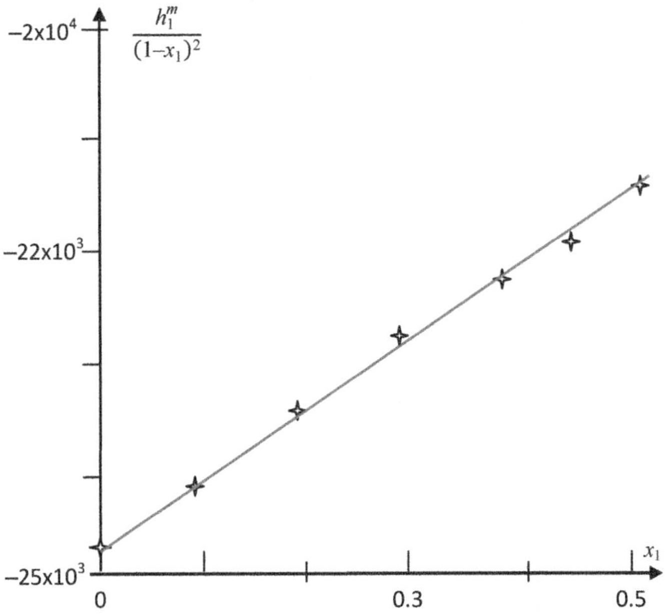

Figure 4.15. *Darken curve* $\frac{h_1^m}{(1-x_1)^2}$ *as a function of* x_1

Appendix

Generalized diagram

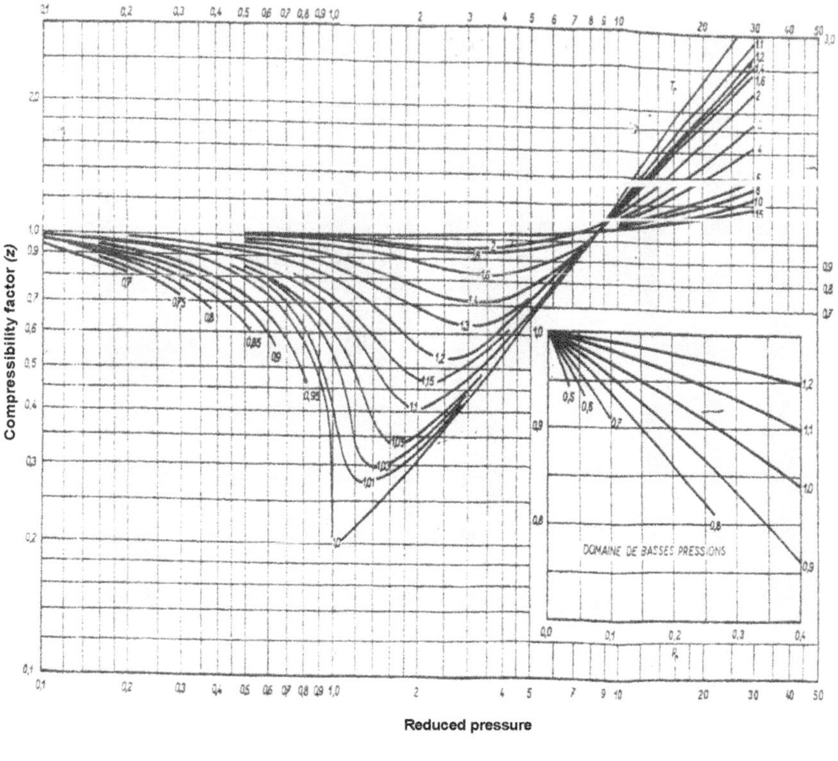

Nomenclature

Letters

(A), (B), (C), (D): Part of a system (compartment, balloon, etc.).

a, b, c, d, A, B, C, D: Parameters; coefficients.

a: Activity.

A: Affinity.

STP: Normal conditions of temperature and pressure.

C_p, C_v: Calorific capacity at constant P, constant V.

Cte: Constant.

d: Density; distance.

d: Exact total differential.

E: Electromotive force of a cell; eutectic.

E: Energy.

F: Free energy (of Helmholtz).

f: Molar free energy, fugacity.

f: Mathematical function.

F: Propulsion force.

G: Gibbs free energy (free enthalpy); *g*: molar free enthalpy.

H: Enthalpy; *h*: molar enthalpy.

h: Height; calorimetric coefficient.

I: Electrical intensity.

K: Equilibrium constant; K_P: equilibrium constant at constant *P*.

k: Kilo; proportionality coefficient; Sievert's constant.

k_i: Equilibrium coefficient; molale constant.

k_d: Distribution coefficient.

l: Calorimetric coefficient.

L: Latent heat of state change.

l: Length, distance.

m: Mass.

M: Molar mass, M_{spc}: critical specific mass.

m: Molality.

$\dot{m}$: Mass flow rate.

n: Number; normal.

n: Number of moles; $\dot{n}$: molar flow rate.

N: Rotation velocity.

P: Pressure; P_c: critical pressure; P_a: atmospheric pressure.

$\mathcal{P}$: Power.

P_{tc}: Critical point.

Q: Heat (calorific energy); q: molar heat.

r: Thermal yield; number of reactions; stopcock.

R: Electrical resistance.

S: Entropy; s: molar entropy.

S: Straight section; valve.

T: Temperature; T_c: critical temperature.

t: Time.

U: Internal energy; u: molar internal energy.

$\mathcal{U}$: Electrical voltage.

V: Standard potential.

v: Variance; velocity.

V: Volume; v: molar volume; $\bar{v}$: mass volume.

X: Weight composition.

x: Distance between two points.

x: Molar fraction (or titer); mathematical variable.

y: Molar fraction (or titer).

$\dot{w}$: Electricity quantity; energy flow.

W: Work (mechanical energy).

Z: Any state function.

Indices

aq: Aqueous.

av: Average.

c: Kinetic; critical.

comb: Combustion.

comp: Compression.

cr: Crystal.

cre: Created.

eb: Boiling.

enc.: Enclosure.

eq: Equilibrium.

exp: Experimental.

ext: External.

f: Formation; final.

f: Fusion.

i: Initial.

i, j: Numeration indices.

int: Internal.

irrev: Irreversible.

l or liq: Liquid.

lim: Limit.

max: Maximal; maximum.

min: Minimal; minimum.

p: Potential.

pist: Piston.

react: Reaction.

rg: Real gas.

rev: Reversible.

s: Solid; solidification.

sol: Solution; solvent.

sub: Sublimation.

sys: System.

tot: Total.

tr: Transition; triple.

univ: Universe.

z: Compressibility factor.

Exponents

*: Ideal gas state.

∞: Infinite.

°: Standard.

cal: Calculated.

dis: Dissolved.

ex: Excess.

m: Mixing.

pur: Pure body.

Units

A: Ampere.

Ah: ampere hour = 3,600 C.

at: Atom.

atm: Atmosphere = 101,325 N/m^2 = 760 mm Hg = 760 tor

= 1.01325 bar = 1.033 kg/cm^3 = 101,325 Pa.

C: Coulomb.

°C: degrees Celsius or centigrade.

cal: Calorie = 4.184 J.

cL: Centiliter.

cm: Centimeter.

g: Gram.

h: Hour.

J: Joule.

K: Kelvin.

kg: Kilogram.

L: Liter.

L.atm: Liter-atmosphere = 101,325 J.

m: Meter.

mA: Milliampere.

mg: Milligram.

min: Minute.

mmHg: Millimeter of mercury = 133.322 N/m^2.

mol: Mole.

N: Newton.

p.p.m.: Parts per million = 10^{-6}.

Pa: Pascal.

s: Second.

torr = 1 mm Hg.

u.e.: Unit of entropy.

u.t.: Unit of time.

V: Volt.

W: Watt.

Ω: Ohm.

Symbols

γ: Activity coefficient; ratio C_P/C_V.

α: Linear or volumetric expansion coefficient; degree of dissociation.

β: Thermoelastic coefficient.

υ: Stoichiometric coefficient.

ξ: Degree of chemical progress.

∂: Partial derivative.

ω: Acentric factor.

χ: Compressibility factor.

$\int$: Integral.

ρ: Density.

Λ: Matrix parameter.

ϕ: Phase; number of phases.

ρ: Yield.

Ω: Section.

∑: Sum.

σ: System; heat source.

ε: Decomposition rate.

θ: Debye characteristic temperature.

δ: Infinitesimal variation.

Δ: Finite variation.

Universal constants

$\mathcal{F}$: Faraday number = 96,485 C/mol.

g: Universal gravitation = 9.80665 m/s^2.

$\mathcal{N}$: Avogadro's number = 6.02214×10^{23}.

R: Ideal gas constant = 1.98717 cal/mol.K

= 8.31447 J/mol.K = 0.0820574 L.atm/mol.K.

z_c: Critical compressibility coefficient # 0.27.

References

[BEL 82] BELAADI S., HOURIEZ J., BALESDENT D., *Thermodynamic Properties of Solid Solutions between n-Paraffins from 18 to 24 Atoms of Carbon*, IUPAC 8, London, 1982.

[BEL 88] BELAADI S., BASTIDE J.-P., PERRACHON G., *Prediction of the Excess Properties Using Redlich-Kister's Equation*, ISFPE, Prague, 1988.

[BJO 75] BJORNBOM P.-H., "The independent reactions in calculations of complex chemical equilibria", *Industrial & Engineering Chemistry Research*, vol. 14, no. 2, p. 102, 1975.

[BOR 14] BOREL L., FAVRAT D., *Thermodynamique et énergétique 1 et 2*, PPUR, Lausanne, 2014.

[CAL 10] CALECKI D., DIU B., *Exercices et problèmes de thermodynamique*, Hermann, Paris, 2010.

[CLE 14] CLEYNEN O., *Thermodynamique de l'ingénieur*, Framabook, Lyon, 2014.

[DUB 12] DUBARRY-BARBE J.-P., *Thermodynamique: exercices et problèmes corrigés*, Ellipses, Paris, 2012.

[DUC 96] DUCAN J., JONES J.-C., *Engineering Thermodynamics*, Prentice Hall International, Upper Saddle River, NJ, 1996.

[GEO 76] GEORGE B., BROWN L.-P., FARMER C.-H., "Computation of multicomponent, multiphase equilibrium", *Industrial & Engineering Chemistry Process Design and Development*, vol. 15, no. 3, pp. 372–377, 1976.

[KAL 78] KALITVENTZEFF B., HEYEN G., Exercice d'application, Document CPIC-Ensic, Nancy, 1978.

[MAR 63] MARQUARDT W.-D., "An algorithm for least-squares estimation of nonlinear parameters", *Journal of Society for Industrial Applied Mathematics*, vol. 11, no. 2, pp. 431–441, 1963.

[PER 01] PEREZ J.-P., *Thermodynamique: fondamentaux et applications*, Dunod, Malakoff, 2001.

[SAN 73] SANDERSON R.-V., CHIEN H.-H.-Y., "Simultaneous chemical and phase equilibrium calculation", *Industrial & Engineering Chemistry Research*, vol. 12, no. 1, pp. 81–85, 1973.

[SMI 49] SMITH T.-E., BONNER R.-F., "Vapor-liquid equilibrium still for partially miscible liquids", *Industrial & Engineering Chemistry Research*, vol. 41, p. 2867, 1949.

[STU 71] STULL D.-R., PROPHET H., JANAF Thermochemical Tables, U.S. Department of Commerce/National Bureau of Standards, Washington, DC, 1971.

[TAB 94] TABET D., SABBAH R., BELAADI S., "Étude thermodynamique de l'acridine", *Thermochimica Acta*, no. 247, pp. 201–207, 1994.

[TOU 05] TOUIL D., BELAADI S., "Heat Exchange Modeling of grate clinker cooler and entropy production analysis", *International Journal of Heat and Technology*, vol. 25, no. 1, pp. 61–66, 2005.

[WHI 58] WHITE W.-B., JOHNSON S.-M., DENTZIG J.-B., "Chemical equilibrium in complex mixtures", *Journal of Chemical Physics*, vol. 28, no. 5, 1958, pp. 213–241, 1958.

[ZEM 51] ZEMANSKY M.-W., *Heat and Thermodynamics*, McGraw-Hill, New York, 1951.

Index

D

F

G

H

I

N, O

P

W

Y, Z

Other titles from

ISTE

in

Chemical Engineering

2019

HARMAND Jérôme, LOBRY Claude, RAPAPORT Alain, SARI Tewfik
Chemostat and Bioprocesses Set
Volume 3 – Optimal Control in Bioprocesses: Pontryagin's Maximum Principle in Practice

2018

LOBRY Claude
Chemostat and Bioprocesses Set
Volume 2 – The Consumer–Resource Relationship: Mathematical Modeling

2017

HARMAND Jérôme, LOBRY Claude, RAPAPORT Alain, SARI Tewfik
Chemostat and Bioprocesses Set
Volume 1 – The Chemostat: Mathematical Theory of Microorganims Cultures

2016

SOUSTELLE Michel
Chemical Thermodynamics Set
Volume 5 – Phase Transformations
Volume 6 – Ionic and Electrochemical Equilibria
Volume 7 – Thermodynamics of Surfaces and Capillary Systems

2015

SOUSTELLE Michel
Chemical Thermodynamics Set
Volume 1 – Phase Modeling Tools
Volume 2 – Modeling of Liquid Phases
Volume 3 – Thermodynamic Modeling of Solid Phases
Volume 4 – Chemical Equilibria

2014

DAL PONT Jean-Pierre, AZZARO-PANTEL Catherine
New Approaches to the Process Industries: The Manufacturing Plant of the Future

HAMAIDE Thierry, DETERRE Rémi, FELLER Jean-François
Environmental Impact of Polymers

2012

DAL PONT Jean-Pierre
Process Engineering and Industrial Management

2011

SOUSTELLE Michel
An Introduction to Chemical Kinetics

2010

SOUSTELLE Michel
Handbook of Heterogenous Kinetics

Printed and bound by CPI Group (UK) Ltd, Croydon, CR0 4YY